Applied Mathematics and Computational Physics

Applied Mathematics and Computational Physics

Editor

Aihua Wood

MDPI • Basel • Beijing • Wuhan • Barcelona • Belgrade • Manchester • Tokyo • Cluj • Tianjin

Editor
Aihua Wood
Department of Mathematics and
Statistics, Air Force Institute of
Technology
USA

Editorial Office
MDPI
St. Alban-Anlage 66
4052 Basel, Switzerland

This is a reprint of articles from the Special Issue published online in the open access journal *Mathematics* (ISSN 2227-7390) (available at: https://www.mdpi.com/si/mathematics/Appl_Math_Comput_Phys).

For citation purposes, cite each article independently as indicated on the article page online and as indicated below:

LastName, A.A.; LastName, B.B.; LastName, C.C. Article Title. *Journal Name* **Year**, *Volume Number*, Page Range.

ISBN 978-3-0365-2305-7 (Hbk)
ISBN 978-3-0365-2306-4 (PDF)

© 2021 by the authors. Articles in this book are Open Access and distributed under the Creative Commons Attribution (CC BY) license, which allows users to download, copy and build upon published articles, as long as the author and publisher are properly credited, which ensures maximum dissemination and a wider impact of our publications.

The book as a whole is distributed by MDPI under the terms and conditions of the Creative Commons license CC BY-NC-ND.

Contents

Preface to "Applied Mathematics and Computational Physics" vii

Benjamin Akers, Tony Liu and Jonah Reeger
A Radial Basis Function Finite Difference Scheme for the Benjamin–Ono Equation
Reprinted from: *Mathematics* 2021, 9, 65, doi:10.3390/math9010065 1

Philip Cho, Aihua Wood, Krishnamurthy Mahalingam, Kurt Eyink
Defect Detection in Atomic Resolution Transmission Electron Microscopy Images Using Machine Learning
Reprinted from: *Mathematics* 2021, 9, 1209, doi:10.3390/math9111209 13

Cui Guo, Yinglin Wang and Yuesheng Luo
A Conservative and Implicit Second-Order Nonlinear Numerical Scheme for the Rosenau-KdV Equation
Reprinted from: *Mathematics* 2021, 9, 1183, doi:10.3390/math9111183 29

Ian Holloway, Aihua Wood, and Alexander Alekseenko
Acceleration of Boltzmann Collision Integral Calculation Using Machine Learning
Reprinted from: *Mathematics* 2021, 9, 1384, doi:10.3390/math9121384 45

Iskandar Waini, Anuar Ishak and Ioan Pop
Nanofluid Flow on a Shrinking Cylinder with Al_2O_3 Nanoparticles
Reprinted from: *Mathematics* 2021, 9, 1612, doi:10.3390/math9141612 61

Jeremiah Bill, Lance Champagne, Bruce Cox and Trevor Bihl
Meta-Heuristic Optimization Methods for Quaternion-Valued Neural Networks
Reprinted from: *Mathematics* 2021, 9, 938, doi:10.3390/math9090938 75

Abdulaziz S. Alkabaa, Ehsan Nazemi, Osman Taylan and El Mostafa Kalmoun
Application of Artificial Intelligence and Gamma Attenuation Techniques for Predicting Gas–Oil–Water Volume Fraction in Annular Regime of Three-Phase Flow Independent of Oil Pipeline's Scale Layer
Reprinted from: *Mathematics* 2021, 9, 1460, doi:10.3390/math9131460 99

Abdulrahman Basahel, Mohammad Amir Sattari, Osman Taylan and Ehsan Nazemi
Application of Feature Extraction and Artificial Intelligence Techniques for Increasing the Accuracy of X-ray Radiation Based Two Phase Flow Meter
Reprinted from: *Mathematics* 2021, 9, 1227, doi:10.3390/math9111227 113

Derek G. Spear, Anthony N. Palazotto, and Ryan A. Kemnitz
Modeling and Simulation Techniques Used in High Strain Rate Projectile Impact
Reprinted from: *Mathematics* 2021, 9, 274, doi:10.3390/math9030274 129

Amjad Ali, Muhammad Umar, Zaheer Abbas, Gullnaz Shahzadi, Zainab Bukhari and Arshad Saleem
Numerical Investigation of MHD Pulsatile Flow of Micropolar Fluid in a Channel with Symmetrically Constricted Walls
Reprinted from: *Mathematics* 2021, 9, 1000, doi:10.3390/math9091000 159

Omar Guillén-Fernández, María Fernanda Moreno-López, Esteban Tlelo-Cuautle
Issues on Applying One- and Multi-Step Numerical Methods to Chaotic Oscillators for FPGA Implementation
Reprinted from: *Mathematics* **2021**, *9*, 151, doi:10.3390/math9020151 . . . . **179**

Iskandar Waini, Anuar Ishak and Ioan Pop
Hybrid Nanofluid Flow over a Permeable Non-Isothermal Shrinking Surface
Reprinted from: *Mathematics* **2021**, *9*, 538, doi:10.3390/math9050538 . . . . **193**

Tony Liu and Rodrigo B. Platte
Node Generation for RBF-FD Methods by QR Factorization
Reprinted from: *Mathematics* **2021**, *9*, 1845, doi:10.3390/math9161845 . . . . **211**

Pablo Pereira Álvarez, Pierre Kerfriden, David Ryckelynck and Vincent Robin
Real-Time Data Assimilation in Welding Operations Using Thermal Imaging and Accelerated High-Fidelity Digital Twinning
Reprinted from: *Mathematics* **2021**, *9*, 2263, doi:10.3390/math9182263 . . . . **237**

Preface to "Applied Mathematics and Computational Physics"

In an age of ever-increasing computing power, there has been a rapid development of powerful computational methods in all areas of engineering and physics. In some cases, increased processing power allows us to address complex problems that not long ago had been considered out of reach for practical purposes. In other cases, novel computing techniques have been developed to approximate more accurately solutions for problems that otherwise would remain intractable for practical applications. Of particular note is the success of machine learning techniques, which provide a new avenue to approach many computational challenges and have been widely utilized in a variety of engineering and physics disciplines.

As faster and more efficient numerical algorithms become available, the understanding of the physics and the mathematical foundation behind the new methods will play an increasingly important role. In this Special Issue, we provide a platform for researchers from both academia and industry to present their new and novel computational methods that have engineering and physics applications.

Aihua Wood
Editor

Article

A Radial Basis Function Finite Difference Scheme for the Benjamin–Ono Equation

Benjamin Akers [1], Tony Liu [1,*] and Jonah Reeger [2]

[1] Department of Mathematics and Statistics, Air Force Institute of Technology, Dayton, OH 45433, USA; Benjamin.akers@afit.edu

[2] Independent Researcher, Bellbrook, OH 45305, USA; jonah.reeger@gmail.com

* Correspondence: tony.liu@afit.edu; Tel.: +1-937-255-3636 (ext. 4722)

Abstract: A radial basis function-finite differencing (RBF-FD) scheme was applied to the initial value problem of the Benjamin–Ono equation. The Benjamin–Ono equation has traveling wave solutions with algebraic decay and a nonlocal pseudo-differential operator, the Hilbert transform. When posed on $\mathbb{R}$, the former makes Fourier collocation a poor discretization choice; the latter is challenging for any local method. We develop an RBF-FD approximation of the Hilbert transform, and discuss the challenges of implementing this and other pseudo-differential operators on unstructured grids. Numerical examples, simulation costs, convergence rates, and generalizations of this method are all discussed.

Keywords: radial basis functions; finite difference methods; traveling waves; non-uniform grids

Citation: Akers, B.; Liu, T.; Reeger, F. A Radial Basis Function Finite Difference Scheme for the Benjamin–Ono Equation. *Mathematics* **2021**, *9*, 65. https://doi.org/10.3390/math9010065

Received: 24 November 2020
Accepted: 18 December 2020
Published: 30 December 2020

Publisher's Note: MDPI stays neutral with regard to jurisdictional claims in published maps and institutional affiliations.

Copyright: © 2020 by the authors. Licensee MDPI, Basel, Switzerland. This article is an open access article distributed under the terms and conditions of the Creative Commons Attribution (CC BY) license (https://creativecommons.org/licenses/by/4.0/).

1. Introduction

In this paper we use the Benjamin–Ono equation as a test-bed for new radial basis function-finite differencing (RBF-FD) simulations of nonlocal wave equations on non-uniform grids. The Benjamin–Ono equation presents the numerical challenges of numerical stiffness, a nonlocal pseudo-differential operator, and localized traveling solutions with slow decay. The equation

$$u_t - \mathcal{H} u_{xx} + u u_x = 0, \qquad (1)$$

in which $\mathcal{H}$ is the Hilbert transform, has many known exact solutions. For example, on $\mathbb{R}$, Equation (1) supports traveling solitary waves solutions:

$$u(x,t) = \frac{4}{(x-t)^2 + 1}.$$

The Benjamin–Ono equation is known to be well-posed [1] and integrable. It can be solved with inverse scattering, and many exact solution profiles are known [2,3]. It has been numerically simulated many times, both in the periodic setting [4] and on $\mathbb{R}$ [5,6].

In this work we develop an RBF-FD scheme for the Benjamin–Ono equation. Common practice for the simulation of (1) on $\mathbb{R}$ is to use periodic boundary conditions, allowing for Fourier collocation, on a large spatial domain [7]. Global radial basis functions (RBFs) have been used as a basis set for simulation of Benjamin–Ono [8]; instead, this work is the first example of the application of RBF-FD to this model. In many cases, RBFs allow for high orders of accuracy while taking advantage of non-uniform spacing in the node set when approximating linear operators. There are an increasing number of texts discussing the use of RBFs in the approximation of differential operators (see, e.g., [9,10]) while presenting much of their history and theory in detail. Recently, the concept of RBF-FD has been further extended to the approximation of definite integrals—first over smooth surfaces [11–13] and then over the volume of the ball [14]. In this paper we look at an extension of RBF-FD now to pseudo-differential operators. The method presented in this paper utilizes as a basis for approximation the so-called polyharmonic spline RBF augmented by shifted monomials. In this case, reference [15] explains that if the shifted monomials up to degree m are included in the basis for approximation, then all of the terms in the Taylor series up to degree m will be handled exactly for the function being interpolated. Therefore, for functions with rapidly decaying terms in the Taylor series, the linear operator will be applied to an approximation with $O(\Delta x^m)$ accuracy on a node set with step size Δx.

RBF-FD simulation of the Benjamin–Ono equation presents the challenge of creating a local approximation of a non-local pseudo-differential operator—the Hilbert transform. The process used herein is generalizable to other pseudo-differential operators, but is the primary cost of the method as it requires diagonalization of an RBF-FD differentiation matrix. Further, the spectrum of the RBF-FD differentiation matrix, particularly when constructed on non-uniform spaced node sets, can often include spurious eigenvalues (for example, with a positive, real part when approximating an operator with pure imaginary spectrum), similar to those observed in [16,17]. We observe that these spurious eigenvalues are the result of floating point errors due to the conditioning of the RBF-FD discretization of the linear operators.

Another complication is the slow decay of the solution as $|x|$ increases. To deal with this complication and with a localized steep gradient in the solutions, we increase the point density where the wave amplitude is large and decrease point density in the far field. This allows for increased accuracy over uniform grids with the same number of nodes. Consideration of non-uniform node sets is a key advantage of RBF based approximations. Even in the context of approximating a non-local operator with local approximations and slowly decaying solutions, we demonstrate $O(\Delta x^m)$ accuracy where Δx is the smallest spacing between adjacent points in a node set. We report errors based on this smallest step size, rather than the largest or the number of nodes as in [18,19], because the mapping we use to refine the node set both decreases the step size near important features of the solution and increases the large step sizes elsewhere while keeping the total number of nodes fixed. To further illustrate this method, we present a brief example in another model equation [20].

In the process of simulating the Benjamin–Ono equation, we present a simple framework for using RBF-FD to approximate pseudo-differential operators. The procedure extends the applicability of RBF methods beyond the purely differential equations previously simulated (see [16,21]) to a host of other pseudo-differential model equations—e.g., Whitham [22], Akers-Milewksi [20], and many more [23–25]. Many of these pseudo-differential equations exhibit coherent structures which are computed with quasi-Newton iteration [26,27]. The simulation of the dynamics near these coherent structures is the application where we believe this method will be most useful. The diagonalization cost required to approximate the pseudo-differential operators in our simulations (a pre-processing step) is comparable to the cost of the quasi-Newton iteration already being done to compute these waves [28].

The paper is organized as follows. Section 2 describes the RBF-FD based numerical method for simulating the Benjamin–Ono equation. This includes a discussion of RBF-FD, a node placement strategy, the approximation of a pseudo-differential operator, and the time-stepping scheme. Then, Section 3 presents numerical results when applying the method to both the Benjamin–Ono equation and the Akers–Milewski equation. Finally, Section 4 draws some conclusions about the use of these methods for approximation pseudo-differential operators.

2. Numerical Method

The numerical method begins in the familiar way of partitioning an interval using N subintervals. The endpoints of these subintervals are the sets of points $\mathcal{S}_N = \{x_k\}_{k=1}^N$. In this paper periodic boundary conditions on a large domain are imposed as a proxy for the slow decay of the solution as $|x| \to \infty$. To implement these boundary conditions, the method creates two periodic images of the set $\mathcal{S}_N$. These are defined by $\mathcal{S}_N^\pm = \{x_k^\pm\}_{k=1}^N :=$

$\{x_k \pm L\}_{k=1}^{N}$, where the signs each define a separate set and L is the period. Considering a point $x_k \in \mathcal{S}_N$, define $\mathcal{N}_k^n = \{x_{k,j}\}_{j=1}^{n}$ to be the set of n points in $\mathcal{S}_N \cup \mathcal{S}_N^+ \cup \mathcal{S}_N^-$ nearest to x_k. The proposed method approximates the application of a linear operator $\mathcal{L}$ to $u: \mathbb{R} \mapsto \mathbb{R}$ by interpolating u over the points in $\mathcal{N}_k^n$ and then applying the linear operator to the interpolant. Traditional finite difference methods are defined in this way, where a polynomial interpolant of degree $n-1$ is constructed over the set x_k and $n-1$ prescribed neighbors, and the linear operator is applied to the interpolant. The method proposed here, however, utilizes local RBF interpolation. RBF interpolation has been used successfully in the approximation of differential operators over subsets of scattered data through the concept of RBF-FD in, for instance, [10,29–31].

2.1. RBF-FD Weight Calculations for Linear Operators

For the simplicity of discussion, consider approximating $\mathcal{L}$ applied to u at a point $x_k \in \mathcal{S}_N$. That is, consider

$$\mathcal{L}_k(u) := \mathcal{L}u(x)|_{x=x_k}.$$

Following common RBF/RBF-FD procedures, the interpolant is constructed via

$$s_k(x) := \sum_{j=1}^{n} c_{k,j} \phi\left(|x - x_{k,j}|\right) + \sum_{l=0}^{m} d_{k,l} (x - x_k)^l$$

with ϕ being function dependent only on the distance from the point $x_{k,j}$. Note that the shift in the monomial terms is included for numerical stability when inverting the matrix A_k in what follows. The interpolation coefficients $c_{k,j}$, $j=1,2,\ldots,n$, and $d_{k,l}$, $l=0,1,\ldots,m$, are chosen to satisfy the interpolation conditions $s(x_{k,j}) = u(x_{k,j})$, $j=1,2,\ldots,n$, along with constraints $\sum_{j=1}^{n} c_{k,j} (x_{k,j} - x_k)^l = 0$, for $l=0,1,\ldots,m$. By applying $\mathcal{L}$ to the interpolant and then evaluating at x_k, we wish to reduce the desired approximation to

$$\mathcal{L}_k(u) \approx \sum_{i=1}^{N} w_{k,i} u(x_{k,i}). \tag{2}$$

A simple derivation can be carried out to show that the weights can be found by solving the system of linear equations $A_k \mathbf{v}_k = \mathbf{b}_k$. This system of equations includes the $(n+m+1) \times (n+m+1)$ matrix

$$A_k = \begin{bmatrix} \Phi_k^T & P_k \\ P_k^T & 0 \end{bmatrix},$$

where $\Phi_{k,ij} = \phi\left(|x_{k,i} - x_{k,j}|\right)$ and $P_{k,il} = (x_{k,i} - x_k)^l$, for $i,j=1,2,\ldots,n$ and $l=0,1,\ldots,m$ ([10], Section 5.1.4). The right-hand side is the length $n+m+1$ column vector

$$\mathbf{b}_k = \begin{bmatrix} \mathcal{L}_k(\phi(|x - x_{k,1}|)) & \mathcal{L}_k(\phi(|x - x_{k,2}|)) & \cdots & \mathcal{L}_k(\phi(|x - x_{k,n}|)) & \mathcal{L}_k(\pi_0) & \mathcal{L}_k(\pi_1) & \cdots & \mathcal{L}_k(\pi_m) \end{bmatrix}^T.$$

The system of linear equations is uniquely solvable in our present context ([32], Theorem 8.21) and the weights $w_{k,i}$, $i=1,2,\ldots,n$ are the first n entries of the solution vector $\mathbf{v}_k$.

It is typical to approximate the action of $\mathcal{L}$ on u at each point in $\mathcal{S}_N$ simultaneously through the product $D\mathbf{u}$, where D is an $N \times N$ matrix and

$$\mathbf{u} = \begin{bmatrix} u(x_1) & u(x_2) & \cdots & u(x_N) \end{bmatrix}^T.$$

The entries of D are found row by row (something that is easily parallelized), so that

$$D_{ki} = \begin{cases} w_{k,j} & \text{if } x_{k,j} = x_i \text{ or } x_{k,j} = x_i \pm L \\ 0 & \text{otherwise} \end{cases};$$

that is, entry i of row k is nonzero only if x_i (one of the points in $\mathcal{S}_N$) or one of its periodic images, $x_i \pm L$, appears in the set $\mathcal{N}_k^n$ of the points nearest x_k.

2.2. Node Placement

To construct node sets With non-uniform spacing that take advantage of the features of the solution, we first create a spatial node set with equal spacing $\tilde{x}$ on the domain $[-L/2, L/2]$. We then apply a nonlinear transformation:

$$x = \tilde{x}\left(1 + a/L^2(\tilde{x} - L/2)(\tilde{x} + L/2)\right). \tag{3}$$

The parameter a dictates the variation in step size in this transformation. For $a = 0$, the node set has uniform spacing; for $a = 4$ the transformation degenerates by making the step size near the origin equal to zero. The transformation in Equation (3) is designed to preserve the overall domain length. It is by no means the only transformation which places a larger density of points near the origin and fewer far from the origin. We also ran experiments with the generalization

$$x = \tilde{x}\left(1 + a/L^2(\tilde{x} - L/2)(\tilde{x} + L/2)\right)^p. \tag{4}$$

for $p = 2, 3, 4$. Increasing either p or a causes increased node density near the origin; however, it also leads to issues with the spectrum of the approximation of the linear operators, as we discuss in next section. For the numerical results presented in this work, we used only (3); examples of step-sizes for a sampling of a values in Equation (3) are in Figure 1.

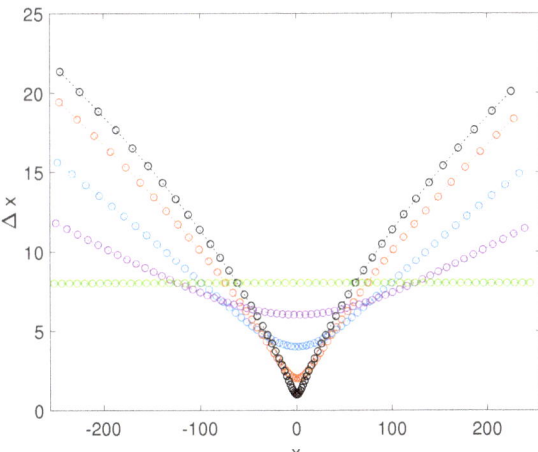

Figure 1. The step sizes from (3) as a function of x. The horizontal equi-spaced case is $a = 0$. The step size becomes increasing variable as a ascends through the samples $a = 1, 2, 3$, and 3.5 (the extreme graph). The step size near the origin vanishes as $a \to 4$. The domain has length $L = 512$, with $N = 64$ points.

2.3. Approximating the Hilbert Transform

In this work we use degree seven polyharmonic splines as the radial basis functions, complemented with polynomials as described above, so that

$$\phi(r) = r^7.$$

These are a common choice for RBF-FD [10], but make the computation of pseudo-differential operators such as the Hilbert transform less straightforward than some other RBFs [8]. The classic procedure to approximate a linear operator using RBFs includes a step where the linear operator is exactly applied to the basis function. This has been done in a previous RBF-based numerical study of the Benjamin–Ono equation using gaussian basis functions on which the Hilbert transform can be exactly calculated [8]. For polyharmonic splines, an exact formula for the Hilbert transform is unavailable. Instead, we use ideas motivated by the Fourier transform definition of the Hilbert transform:

$$\widehat{\mathcal{H}u} = -i\,\text{sign}(k)\hat{u}.$$

The Fourier transform definition reveals a relationship between the spectrum of the derivative operator, $\lambda_{\partial x} = ik$, and the spectrum of the Hilbert transform $\lambda_{\mathcal{H}} = -i\,\text{sign}(k)$, so that

$$\lambda_{\mathcal{H}} = -i\,\text{sign}(Im(\lambda_{\partial x})). \tag{5}$$

Using this relationship, if the spectrum of the derivative operator is known, then the spectrum of the Hilbert transform can be computed by an algebraic transformation of the spectrum of the derivative. We use this transformation to compute an approximation to the Hilbert transform by first computing a matrix which approximates the derivative, as described in Section 2.1. Next we diagonalize this matrix using QR iteration to get its eigenvalues $\lambda_{\partial x}$ and eigenvectors. The Hilbert transform can then be approximated using the same eigenvectors and Equation (5) applied to the computed λ_{dx} to get its eigenvalues. This computation of the Hilbert transform, a pre-processing step, costs $O(N^3)$ flops, where N is the number of spatial points. The time-stepping portion of the RBF-FD method is $O(N^2)$ flops per step, due to the full matrix that approximates the Hilbert transform. The resulting method however, allows for non-uniform point spacing, so that larger domains can be accurately simulated with a fixed number of points. We expect this method to be competitive only in cases where there are fine local features of interest and slow decay to a remote boundary, as in the Benjamin–Ono solitary wave on $\mathbb{R}$. We admit that the Benjamin–Ono equation has special properties, i.e., integrability, which make other simulation methods such as inverse scattering available [3]. The above algorithm makes no use of integrability, and thus can be trivially extended to other equations with pseudo-differential operators, such as the intermediate-long wave [25], Whitham [22], and Akers–Milewski equations [20].

2.4. Time Stepping

For the time evolution of the system of differential equations induced by the RBF-FD spatial discretization of (1), we use a second order IMEX method [33],

$$\frac{3u^{q+1} - 4u^q + u^{q-1}}{2\Delta t} = 2f(u^q) - f(u^{q-1}) + g(u^{q+1}), \tag{6}$$

in which $f(u)$ is the nonlinear term (explicitly treated) and $g(u)$ is the linear term (implicitly treated). This method is sometimes called SBDF [33], or extrapolated GEAR [34]. The linear stability region for this scheme is exterior to an egg-shaped region in the right half plane; it is unconditionally stable for wave equations—such as the Benjamin–Ono equation—that have pure imaginary linear spectra. We chose this method over the competing IMEX scheme CNAB (Crank–Nicholson Adams–Bashforth) so that small, real perturbations of the pure imaginary spectrum do not leave the stability region (as is the case for the CNAB).

As we will see later, floating point errors due to the condition number of the matrices involved in RBF-FD discretization of the linear operators can cause changes in the spectrum of the approximation of the linear operators involved. We chose (6) to be as robust as possible to such errors.

The method we used to approximate the linear operators begins with an eigenvalue calculation. As such, other numerical time steppers are available—for example, exponential time differencing (ETD) [35] and integrating factors (IF) [36]. Since the equation is nonlinear, after diagonalizing the linear part, both ETD and IF methods require full matrix multiplications to evaluate the nonlinearity at each time step. The IMEX method used here has the inversion of a full matrix (for the implicit linear term); however, this can be done as a pre-processing step, so it also has a cost which scales like a full matrix multiplication per time step (and thus is comparable with ETD or IF). We chose the IMEX scheme due to its unconditional stability, to ameliorate the stiffness of this equation in explicit time steppers and to be robust to perturbations of the spectrum off the imaginary axis. In future work we plan to explore other time steppers, including higher order IMEX methods with unbounded stability regions [33].

3. Results

In this section, the numerical method described herein is evaluated via a number of numerical tests. We show the effects of the numerical discretization and truncation error, and the effect of truncating $\mathbb{R}$ to finite length. Errors are measured as a function the minimal step size of our non-uniform parameterization. The effect of the non-uniform step-size on the spectrum of approximation to the linear operator is also discussed. Example simulations of algebraically decaying Benjamin–Ono solitary waves are shown, along with an oscillatory exponentially decaying wave in the Akers–Milewski equation.

In Figure 2, we display an example of a simulation of the Benjamin–Ono solitary wave:

$$u(x,t) = \frac{4}{1+(x)^2}. \tag{7}$$

The simulation was computed in a frame which travels with the wave so that the wave appears stationary. As is common practice, a large domain size with periodic boundary conditions was used as a proxy for $\mathbb{R}$ [7] since the solution decays as $|x| \to \infty$. The crest of the wave profile, in the left panel, is marked with a solid black line, to highlight the lack of oscillation in time. The initial profile is also marked with a solid black line for highlighting purposed. The right panel shows the step sizes used for this simulation, which range over an order of magnitude with $\Delta x \in [0.125, 1.25]$. The node spacing is concentrated near the origin, where the wave (and its derivatives) is largest; the largest spacing occurs for large x, where the wave is small. This allows for increased accuracy for the same number of points as compared to uniform step sizes.

In Figure 3, we study the accuracy of the spatial discretization of the linear operator

$$\mathcal{L}u = \mathcal{H}u_{xx} + u_x, \tag{8}$$

when applied to the solitary wave (7). There are two competing parameters which determine the spatial accuracy, the domain size L, and the minimal space step Δx. We present two experiments, one where the domain size is varied for a sampling of parameterizations (left panel of Figure 3) and one where the parameterization is varied for a fixed domain size (right panel of Figure 3). In the left panel, all discretizations show an initial decrease in error as the domain size increases, scaling like $O(L^{-4})$, up to a point where the truncation error of the discretization of the linear operator grows to be larger than the domain discretization error. For each discretization, including Fourier, there is a length L for which the method is most accurate (for a fixed number of points). All of the RBF-FD discretizations, based on (3), are able to give a more accurate discretization than a Fourier discretization (with the best observed improvement at $a = 3.5$). That these methods outperform a Fourier discretization on the Benjamin–Ono solitary wave is natural, since a periodic tiling of this wave has a discontinuous first derivative at the boundary, limiting the accuracy of a Fourier discretization. The variable space step RBF-FD is not limited by this discontinuity, even with an approximation that assumes more than one continuous derivative at the boundary. The accuracy of the RBF-FD discretization of the linear operator as a function of minimal step size (controlled by the parameter a) is depicted in the right panel of Figure 3, in which we observe $O(\Delta x^8)$ accuracy. Given that the approximation of the linear operator was formed by applying the linear operator to an interpolant with a basis set of polynomials terms up to degree eight and polyharmonic splines of r^7, it is not surprising to see $O(\Delta x^8)$ accuracy since the interpolation procedure provides for at least that order of accuracy [15].

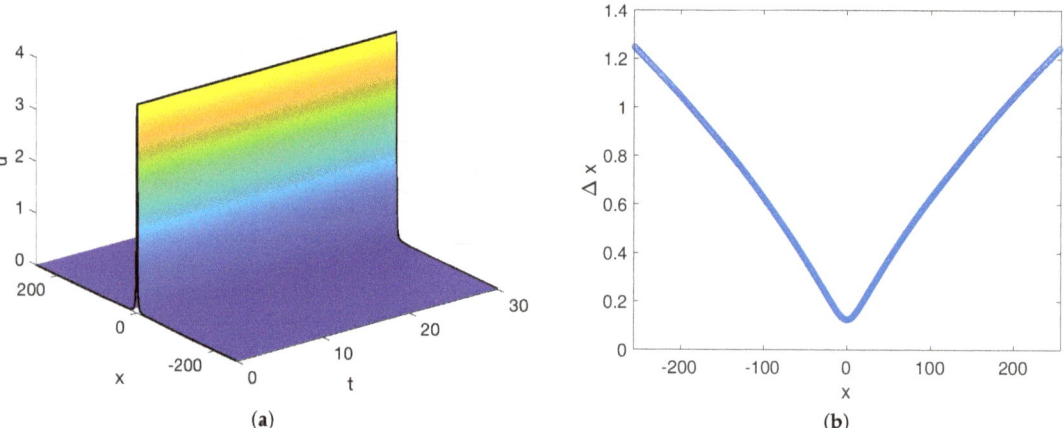

Figure 2. (a) The soliton solution propagated on a domain $x \in [-256, 256]$. (b) The step-sizes used for the RBF-FD discretization used in the left simulation.

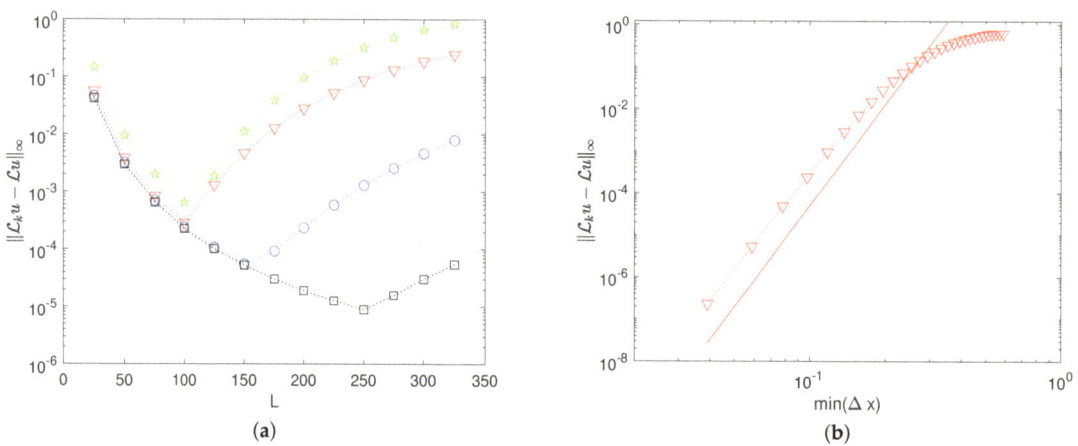

Figure 3. The infinity norm of the error of the numerical approximation of the linear operator, $\mathcal{L}_N u$, against the exact $\mathcal{L}u$ on the solitary wave $u(x) = 4/(1+x^2)$ is depicted. (a) The error's dependence on domain size, L, is compared for the RBF-FD discretization using the spatial points in (3) with $a=2$ (red triangles), $a=3$ (blue circles), and $a=3.5$ (black squares), and an equi-spaced Fourier discretization (green stars). All of these computations use $N=512$. (b) The error dependence on the minimal step size for $N=1024, L=800$, altering the step size with the parameter "a". The numerical results are marked with triangles; the continuous line marks $O(\Delta x^8)$.

The method used here to discretize the linear operator in Equation (8) is based on a diagonalization of a RBF-FD differentiation matrix. First we compute an RBF-FD approximation of a derivative matrix. Next we compute the eigenvalues and eigenvectors of this matrix. As we knew the exact spectrum of the derivative and the Hilbert transform in the infinite dimensional problem from Fourier analysis,

$$\hat{\partial}_x = ik \quad \text{and} \quad \hat{\mathcal{H}} = -i\text{sign}(k),$$

we constructed eigenvalues of the discrete Hilbert transform by applying the same relationship to the discrete problem. If λ_{Dx} is an eigenvalue of the differentiation matrix, then the eigenvalues of the discretized Hilbert transform $\lambda_{\mathcal{H}}$ are defined to satisfy

$$\lambda_{\mathcal{H}} = i\text{sign}(\text{imag}(\lambda_{Dx})).$$

Eigenvalue manipulation based on the above definition is used to find all the eigenvalues of the approximation to (8); the eigenvector matrix (and its inverse) of the RBF-FD differentiation matrix is then multiplied by diagonal matrices with this approximate spectrum to get an approximation for (8). Although the eigenvalues were constructed to be pure imaginary, as is the spectrum of the exact problem, the result of the matrix multiplication (by the matrix of eigenvectors and its inverse) can perturb the spectrum due to machine precision errors. These errors scale with the condition number of the matrix of the eigenvectors of the RBF-FD differentiation matrix. As is the case for classic polynomial interpolation, the condition number of this matrix grows as the points get closer together. This increase in condition number results in a corresponding increase in the size of spurious real eigenvalues of the discretized linear operator (which poses a stability problem for numerical time stepping algorithms). To observe this phenomenon, after creating the approximate linear operator $\mathcal{L}_k$, we apply a QR iteration to compute the eigenvalues of this matrix. That these computed eigenvalues differ at all from the desired spectrum is a direct result of the conditioning of the eigenspace of RBF-FD differentiation matrix. An example of the computed spectrum of the matrix $\mathcal{L}_k$ used to evolve the Benjamin–Ono solitary wave in Figure 2 is in the left panel of Figure 4. This spectrum includes spurious eigenvalues with a real part as large as 10^{-9}. In the right panel of Figure 4, we observe the dependence of the eigenvalue with the largest real part as a function of the parameter a from Equation (3). The match between the size of the eigenvalues and the condition number of the matrix of eigenvectors, F, times the machine precision, is displayed in the right panel of Figure 4.

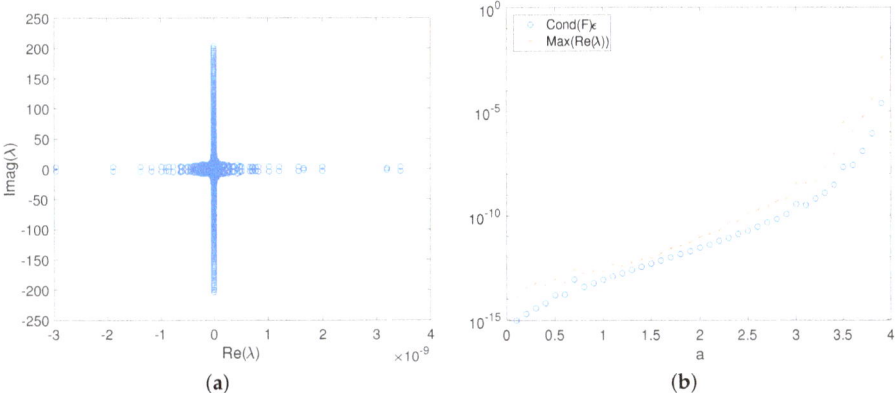

Figure 4. (a) The spectrum of the linear operator used for the evolution depicted in Figure 2. (b) The condition number and the maximally real eigenvalue of the approximation of the linear operator as a function of the step-size parameter "a" in Equation (3), marked with xs, is compared to the condition number of eigenvector matrix, F, of the RBF-FD differentiation matrix times the machine precision, ϵ, marked with circles.

Examples of the discretized eigenfunctions are plotted in Figure 5. These eigenfunctions were all computed on a domain of length $L = 800$; the panels each show different subsets of the computational domain. As the discretization becomes coarser near the boundary ($a \to 4$), globally supported eigenfunctions are more poorly resolved. This is natural, as a trade-off for better resolution near the origin. The exact relationship between the resolution of the eigenfunctions and the errors in the argument of the corresponding eigenvalues is unknown. This does, however, point to a possible explanation for the ill-conditioning of the eigenspace of the differentiation matrix. Poor resolution of globally supported eigenfunctions could be the cause; the ill-conditioned eigenspace could be due to some points being too far apart, rather than too close together (as is typical in polynomial interpolation matrices).

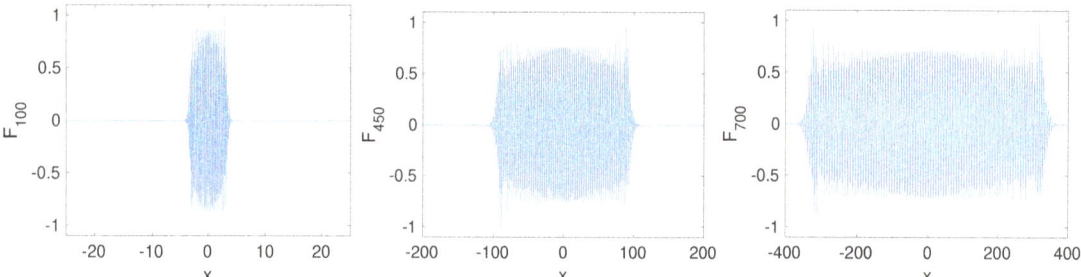

Figure 5. The real part of three sample eigenfunctions is depicted. These eigenfunctions have eigenvalues with "spurious" real parts $Re(\lambda_{100}) \approx 10^{-9}$, $Re(\lambda_{450}) \approx 10^{-8}$, and $Re(\lambda_{700}) \approx 10^{-6}$ respectively. These values were computed with $a = 3.9$, $N = 1024$, and $L = 800$.

A few references have studied the behavior of eigenvalue stability with regard to node placement. This includes [16] which investigates special node distributions that improve eigenvalue stability for global RBF methods, and [37–42] which investigate the effects of using mapped nodal sets on the accuracy and eigenvalue stability of finite-difference and pseudo-spectral methods. These works may provide the framework for future research avenues which resolve the relationship between node placement the ill-conditioning of the eigenspace of differential operators.

The method presented here and error analysis are presented in the context of the Benjamin–Ono equation, where the combination of algebraic decay of the solitary wave and nonlocal nature of the Hilbert transform make a challenging testbed for a numerical scheme. The same ideas generalize easily to other nonlocal equations; it is trivial to apply this method to the Whitham equation [22] or the Akers–Milewski equation [20,43]. The Akers–Milewski equation,

$$u_t + \mathcal{H}u - \mathcal{H}u_{xx} + uu_x = 0, \qquad (9)$$

supports traveling, wavepacket-type solitary waves [44]. These waves decay exponentially in space, and thus do not present the same challenges for numerical simulation, (Fourier collocation is spectrally accurate) as the Benjamin–Ono solitary wave (7). As evidence of the ease of generalizing this approach, we include an example of the evolution of such a wave using RBF-FD discretization a non-uniform grid in Figure 6.

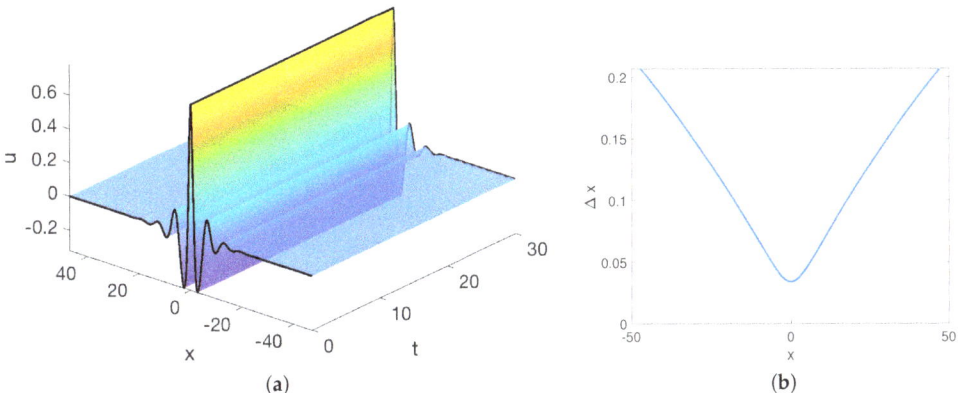

Figure 6. (a) An example of the evolution of a traveling wavepacket-type solitary wave in the Akers–Milewski Equation (9). (b) The step sizes used for this simulation.

Although the method is presented in one spatial dimension with periodic boundary conditions, its application is not exclusive to either. The exact methodology can be directly applied for higher dimensional problems (for example, the two-dimensional solutions of Akers–Milewski [20]). In order to consider different boundary conditions, one needs a relationship between the spectrum of the derivative operator and that of the pseudo-differential operator with the new boundary conditions. This relationship may be simple to determine (for example, the relationship for homogeneous Dirichelet boundary conditions agreeing with that of the periodic case), in which case the definition of the pseudo-differential operator is unchanged and the boundary conditions may be implemented with standard RBF-FD methods. For general boundary conditions the relationship between the derivative's spectra and the pseudo-differential operator's spectra should be investigated before applying the method.

4. Conclusions

In this paper an RBF-FD implementation of the Benjamin–Ono equation was presented. This required the development of an RBF-FD implementation of the Hilbert transform. An approximation of the Hilbert transform was built by manipulating the spectrum of a differentiation matrix. This approach generalizes simply to other pseudo-differential operators, but incurs an $O(N^3)$ pre-processing cost, meaning that this method is expensive compared to a Fourier implementation. The approach, however, allows for arbitrary boundary conditions and non-uniform grids. We expect it to be most useful in problems where an $O(N^3)$ pre-processing cost is already being paid, for example, in the evolution of a coherent structure which was computed via quasi-Newton iteration. A future avenue for research is in the relationship between the node placement and the conditioning of the eigenspace of the differentiation matrix; we observe that this plays a key role in the accuracy of the approximation of the spectrum of the linear operator.

Author Contributions: Conceptualization, B.A.; methodology, B.A. and J.R.; software, B.A., J.R., and T.L.; validation, B.A., J.R., and T.L.; formal analysis, B.A., J.R., and T.L.; writing—original draft preparation, B.A., J.R., and T.L.; writing—review and editing, B.A., J.R., and T.L.; visualization, B.A.; supervision, B.A.; project administration, B.A.; funding acquisition, B.A. All authors have read and agreed to the published version of the manuscript.

Funding: This research received no external funding.

Acknowledgments: All three authors acknowledge the support of Office of Naval Research via the project Atmospheric Propagation Sciences for High Energy Lasers. B.F. Akers acknowledges support from the Air Force Office of Sponsored Research's Computational Mathematics Program via the project Radial Basis Functions for Numerical Simulation.

Conflicts of Interest: This report was prepared as an account of work sponsored by an agency of the United States Government. Neither the United States Government nor any agency thereof, nor any of their employees, make any warranty, express or implied, or assume any legal liability or responsibility for the accuracy, completeness, or usefulness of any information, apparatus, product, or process disclosed, or represent that its use would not infringe privately owned rights. Reference herein to any specific commercial product, process, or service by trade name, trademark, manufacturer, or otherwise does not necessarily constitute or imply its endorsement, recommendation, or favoring by the United States Government or any agency thereof. The views and opinions of authors expressed herein do not necessarily state or reflect those of the United States Government or any agency thereof.

References

1. Tao, T. Global well-posedness of the Benjamin–Ono equation in H1 (R). *J. Hyperbolic Differ. Equat.* **2004**, *1*, 27–49. [CrossRef]
2. Matsuno, Y. Interaction of the Benjamin-Ono solitons. *J. Phys. A Math. Gen.* **1980**, *13*, 1519. [CrossRef]
3. Kaup, D.; Matsuno, Y. The inverse scattering transform for the Benjamin–Ono equation. *Stud. Appl. Math.* **1998**, *101*, 73–98. [CrossRef]
4. Ambrose, D.M.; Wilkening, J. Computation of time-periodic solutions of the Benjamin–Ono equation. *J. Nonlinear Sci.* **2010**, *20*, 277–308. [CrossRef]
5. Feng, B.F.; Kawahara, T. Temporal evolutions and stationary waves for dissipative Benjamin–Ono equation. *Phys. D Nonlinear Phenom.* **2000**, *139*, 301–318. [CrossRef]
6. Pelloni, B.; Dougalis, V.A. Numerical solution of some nonlocal, nonlinear dispersive wave equations. *J. Nonlinear Sci.* **2000**, *10*, 1–22. [CrossRef]
7. Kalisch, H. Error analysis of a spectral projection of the regularized Benjamin–Ono equation. *BIT Numer. Math.* **2005**, *45*, 69–89. [CrossRef]
8. Boyd, J.P.; Xu, Z. Comparison of three spectral methods for the Benjamin–Ono equation: Fourier pseudospectral, rational Christov functions and Gaussian radial basis functions. *Wave Motion* **2011**, *48*, 702–706. [CrossRef]
9. Fasshauer, G.E. *Meshfree Approximation Methods with MATLAB*; World Scientific: Singapore, 2007; Volume 6.
10. Fornberg, B.; Flyer, N. *A Primer On Radial Basis Functions With Applications To The Geosciences*; SIAM: Philadelphia, PA, USA, 2015.
11. Reeger, J.A.; Fornberg, B. Numerical quadrature over the surface of a sphere. *Stud. Appl. Math.* **2016**, *137*, 174–188. [CrossRef]
12. Reeger, J.A.; Fornberg, B.; Watts, M.L. Numerical quadrature over smooth, closed surfaces. *Proc. R. Soc. A Math. Phys. Eng. Sci.* **2016**, *472*. doi: 10.1098/rspa.2016.0401. [CrossRef]
13. Reeger, J.A.; Fornberg, B. Numerical quadrature over smooth surfaces with boundaries. *J. Comput. Phys.* **2018**, *355*, 176–190. [CrossRef]
14. Reeger, J.A. Approximate Integrals Over the Volume of the Ball. *J. Sci. Comput.* **2020**, *83*. [CrossRef]
15. Bayona, V. An insight into RBF-FD approximations augmented with polynomials. *Comput. Math. Appl.* **2019**, *77*, 2337–2353. doi:10.1016/j.camwa.2018.12.029. [CrossRef]
16. Platte, R.B.; Driscoll, T.A. Eigenvalue stability of radial basis function discretizations for time-dependent problems. *Comput. Math. Appl.* **2006**, *51*, 1251–1268. [CrossRef]
17. Sarra, S.A. A numerical study of the accuracy and stability of symmetric and asymmetric RBF collocation methods for hyperbolic PDEs. *Numer. Methods Part. Differ. Equ. Int. J.* **2008**, *24*, 670–686. [CrossRef]
18. Libre, N.A.; Emdadi, A.; Kansa, E.J.; Shekarchi, M.; Rahimian, M. A multiresolution prewavelet-based adaptive refinement scheme for RBF approximations of nearly singular problems. *Eng. Anal. Bound. Elem.* **2009**, *33*, 901–914. [CrossRef]
19. Davydov, O.; Oanh, D.T. Adaptive meshless centres and RBF stencils for Poisson equation. *J. Comput. Phys.* **2011**, *230*, 287–304. [CrossRef]
20. Akers, B.; Milewski, P.A. A model equation for wavepacket solitary waves arising from capillary-gravity flows. *Stud. Appl. Math.* **2009**, *122*, 249–274. [CrossRef]
21. Shen, Q. A meshless method of lines for the numerical solution of KdV equation using radial basis functions. *Eng. Anal. Bound. Elem.* **2009**, *33*, 1171–1180. [CrossRef]
22. Ehrnstr'om, M.; Kalisch, H. Traveling waves for the Whitham equation. *Differ. Integral Equat.* **2009**, *22*, 1193–1210.
23. Akers, B.; Milewski, P.A. Dynamics of three-dimensional gravity-capillary solitary waves in deep water. *SIAM J. Appl. Math.* **2010**, *70*, 2390–2408. [CrossRef]
24. Akers, B.; Milewski, P.A. Model equations for gravity-capillary waves in deep water. *Stud. Appl. Math.* **2008**, *121*, 49–69. [CrossRef]
25. Ablowitz, M.; Fokas, A.; Satsuma, J.; Segur, H. On the periodic intermediate long wave equation. *J. Phys. A Math. Gen.* **1982**, *15*, 781. [CrossRef]

26. Parau, E.; Vanden-Broeck, J.M.; Cooker, M. Nonlinear three-dimensional gravity-capillary solitary waves. *J. Fluid Mech.* **2005**, *536*, 99–105. [CrossRef]
27. Oliveras, K.; Curtis, C. Nonlinear travelling internal waves with piecewise-linear shear profiles. *J. Fluid Mech.* **2018**, *856*, 984–1013. [CrossRef]
28. Claassen, K.M.; Johnson, M.A. Numerical bifurcation and spectral stability of wavetrains in bidirectional Whitham models. *Stud. Appl. Math.* **2018**, *141*, 205–246. [CrossRef]
29. Flyer, N.; Lehto, E.; Blaise, S.; Wright, G.B.; St-Cyr, A. A Guide to RBF-generated finite-differences for nonlinear transport: Shallow water simulations on a sphere. *J. Comput. Math.* **2012**, *231*, 4078–4095. [CrossRef]
30. Flyer, N.; Wright, G.; Fornberg, B. Radial Basis function-generated finite differences: A mesh-free method for computational geosciences. In *Handbook of Geomathematics*; Freeden, W., Nashed, M.Z., Sonar, T., Eds.; Springer: Berlin/Heidelberg, Germany, 2014; doi: 10.1007/978-3-642-27793-1 61-1. [CrossRef]
31. Fornberg, B.; Flyer, N. Solving PDEs with radial basis functions. *Acta Numer.* **2015**, *24*, 215–258. [CrossRef]
32. Wendland, H. *Scattered Data Approximation*; Cambridge University Press: Cambridge, UK, 2005.
33. Ascher, U.; Ruuth, S.; Wetton, B. Implicit-explicit methods for time-dependent partial differential equations. *SIAM J. Numer. Anal.* **1995**, *32*, 797–823. [CrossRef]
34. Varah, J.M. Stability restrictions on second order, three level finite difference schemes for parabolic equations. *SIAM J. Numer. Anal.* **1980**, *17*, 300–309. [CrossRef]
35. Kassam, A.K.; Trefethen, L. Fourth-order time-stepping for stiff PDEs. *SIAM J. Sci. Comput.* **2005**, *26*, 1214–1233. [CrossRef]
36. Milewski, P.; Tabak, E. A pseudospectral procedure for the solution of nonlinear wave equations with examples from free-surface flows. *SIAM J. Sci. Comput.* **1999**, *21*, 1102–1114. [CrossRef]
37. Kosloff, D.; Tal-Ezer, H. A modified Chebyshev pseudospectral method with an O (N-1) time step restriction. *J. Comput. Phys.* **1993**, *104*, 457–469. [CrossRef]
38. Bayliss, A.; Turkel, E. Mappings and accuracy for Chebyshev pseudo-spectral approximations. *J. Comput. Phys.* **1992**, *101*, 349–359. [CrossRef]
39. Shen, J.; Wang, L.L. Error analysis for mapped Legendre spectral and pseudospectral methods. *SIAM J. Numer. Anal.* **2004**, *42*, 326–349. [CrossRef]
40. Zhong, X.; Tatineni, M. High-order non-uniform grid schemes for numerical simulation of hypersonic boundary-layer stability and transition. *J. Comput. Phys.* **2003**, *190*, 419–458. [CrossRef]
41. Shukla, R.K.; Zhong, X. Derivation of high-order compact finite difference schemes for non-uniform grid using polynomial interpolation. *J. Comput. Phys.* **2005**, *204*, 404–429. [CrossRef]
42. Orszag, S.A. Accurate solution of the Orr–Sommerfeld stability equation. *J. Fluid Mech.* **1971**, *50*, 689–703. [CrossRef]
43. Diorio, J.; Cho, Y.; Duncan, J.H.; Akylas, T. Gravity-capillary lumps generated by a moving pressure source. *Phys. Rev. Lett.* **2009**, *103*, 214502. [CrossRef]
44. Akers, B.F.; Seiders, M. Numerical Simulations of Overturned Traveling Waves. In *Nonlinear Water Waves*; Springer: Berlin/Heidelberg, Germany, 2019; pp. 109–122.

Article

Defect Detection in Atomic Resolution Transmission Electron Microscopy Images Using Machine Learning

Philip Cho [1,*], Aihua Wood [1,*], Krishnamurthy Mahalingam [2] and Kurt Eyink [2]

[1] Air Force Institute of Technology, Department of Mathematics & Statistics, 2950 Hobson Way, Wright-Patterson AFB, OH 45433, USA
[2] Air Force Research Lab, Material and Manufacturing Directorate, Wright-Patterson AFB, OH 45433, USA; krishnamurthy.mahalingam.ctr@us.af.mil (K.M.); kurt.eyink@us.af.mil (K.E.)
* Correspondence: philip.cho@afit.edu (P.C.); aihua.wood@afit.edu (A.W.)

Citation: Cho, P.; Wood, A.; Mahalingam, K.; Eyink, K. Defect Detection in Atomic Resolution Transmission Electron Microscopy Images Using Machine Learning. *Mathematics* **2021**, *9*, 1209. https://doi.org/10.3390/math9111209

Academic Editor: Amir Mosavi

Received: 09 April 2021
Accepted: 24 May 2021
Published: 27 May 2021

Publisher's Note: MDPI stays neutral with regard to jurisdictional claims in published maps and institutional affiliations.

Copyright: © 2021 by the authors. Licensee MDPI, Basel, Switzerland. This article is an open access article distributed under the terms and conditions of the Creative Commons Attribution (CC BY) license (https://creativecommons.org/licenses/by/4.0/).

Abstract: Point defects play a fundamental role in the discovery of new materials due to their strong influence on material properties and behavior. At present, imaging techniques based on transmission electron microscopy (TEM) are widely employed for characterizing point defects in materials. However, current methods for defect detection predominantly involve visual inspection of TEM images, which is laborious and poses difficulties in materials where defect related contrast is weak or ambiguous. Recent efforts to develop machine learning methods for the detection of point defects in TEM images have focused on supervised methods that require labeled training data that is generated via simulation. Motivated by a desire for machine learning methods that can be trained on experimental data, we propose two self-supervised machine learning algorithms that are trained solely on images that are defect-free. Our proposed methods use principal components analysis (PCA) and convolutional neural networks (CNN) to analyze a TEM image and predict the location of a defect. Using simulated TEM images, we show that PCA can be used to accurately locate point defects in the case where there is no imaging noise. In the case where there is imaging noise, we show that incorporating a CNN dramatically improves model performance. Our models rely on a novel approach that uses the residual between a TEM image and its PCA reconstruction.

Keywords: transmission electron microscopy (TEM); convolutional neural networks (CNN); anomaly detection; principal component analysis (PCA); machine learning; deep learning; neural networks; Gallium-Arsenide (GaAs)

1. Introduction

Point defects are zero-dimensional defects in crystalline materials that have a strong influence on their atomic structure and properties. The engineering of point defects in materials, by creation of specific defect types and by the control of spatial location and number density, is foundational in the development of novel materials for advanced electronic and photonic applications. Transmission electron microscopy (TEM) is a widely used technique for imaging defects, due to its versatility for many different modes of imaging and spectroscopy at high spatial resolution. However, detection of point defects in TEM images continues to remain a challenge in many material systems, since the contrast due to the defect is affected by various factors such as its local environment and imaging conditions.

Recent efforts to develop machine learning methods for the detection of point defects in TEM images focused on supervised methods that require labeled training data that is generated via simulation. These methods treat the defect detection problem as a pixel-level classification problem [1,2]. In contrast, we treat the defect detection problem as an anomaly detection problem and propose two self-supervised machine learning methods that can be trained solely using defect-free TEM images. Importantly, since our models only require defect-free images for training, it allows for our models to be trained directly on experimental images of samples that are manufactured to be defect-free. The first method we propose uses principal components analysis (PCA) and the second method uses both PCA and convolutional neural networks (CNN). We assess the performance of these methods by introducing hypothetical anomalies that mimic point defects of different types in simulated images of defect-free GaAs (Figure 1) [3] and examining the detection accuracy of our proposed algorithms. We use GaAs as the test material. We also note that atomic resolution TEM imaging is performed in two different modes, wherein the electron beam is in parallel illumination (conventional high-resolution transmission electron microscopy) or as a focused probe (scanning transmission electron microscopy). The present work is based on parallel beam mode, since images simulated for this mode exhibits widely varying (although distinct) patterns for different imaging conditions, providing a large dataset for training and testing purposes. However, the results are also applicable to focused probe mode images.

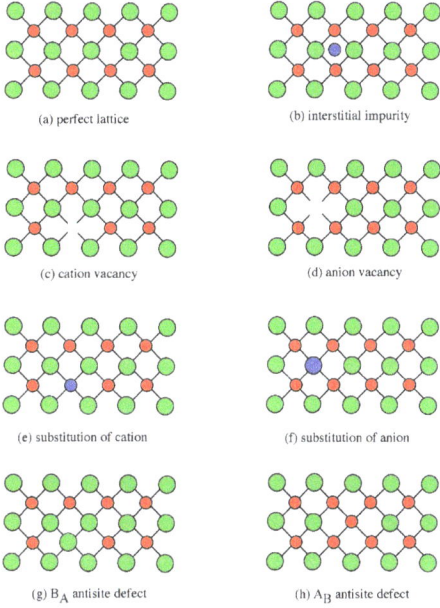

Figure 1. Examples of different types of defects that could occur. Green corresponds to A and red corresponds to B. Blue represents a dopant [3].

Related Works

In recent years, CNNs have proven to be a highly effective tool for image analysis. Applications include image classification, object detection, pose estimation, and text recognition [4,5]. Given the data-intensive nature of TEM imagery, there have been recent efforts to employ CNNs in the analysis of TEM images. Examples of using neural networks for analyzing TEM images include using CNNs for denoising TEM images [6,7], generating TEM images from partial scans [8], enhancing TEM images [9,10], classifying types of crystalline structures [11], locating defects in non-crystalline materials [12], mapping atomic structures and defects [1], and mapping general structures of interest [2].

The latter two studies [1,2] are of particular interest because they propose CNN models that can be used to identify point defects in TEM images of crystalline materials. In both studies, the framework is to train a CNN using simulated TEM images and then apply the trained models to experimental images. Additionally, both propose training a multi-class classification CNN that outputs pixel-wise classifications, i.e., every pixel in a TEM image is assigned a predicted class. The classes can be vacancies, dopants, and defect-free lattice [1] or general, non-overlapping structural characteristics such as the column height of the sample [2]. Both of these models require extensive simulated data where the true label for each pixel is known. After training the pixel-wise classification model with pixel-by-pixel truth data, the models are shown to produce strong results on experimental TEM images.

Similar to the aforementioned work, we seek to develop a model that can detect local structures of interest in TEM images, namely defects, in crystalline materials. However, we choose to take an approach that does not rely on a training data set with pixel-by-pixel truth labels. Rather, we rely on a training set of TEM images that are only known to be free of defects. In both prior works, the authors acknowledge the difficulty in acquiring experimental images where the true defect locations are known and, therefore, propose models that are solely trained on simulated data with known defect locations. Since we only use a training set consisting of TEM images without defects, the proposed methods can be trained on simulated images or experimental images. Using TEM images of samples that are known to be free of defects, the proposed methods seek to locate areas of a TEM image that are anomalous and, thus, are most likely to contain a defect.

Due to a wide range of applications, anomaly detection has been well studied and there are numerous proposed anomaly detection techniques. Anomaly detection methods, also known as novelty detection methods, are trained using only "normal" observations and the goal is to accurately determine whether a new observation is anomalous or normal. Anomaly detection problems commonly arise in practice when there is an abundance of normal training data and a limited number of anomalous observations. Examples of well-known applications include medical imaging [13–15] and fraud detection [16–18]. A thorough review and taxonomy of machine learnings methods for anomaly is provided by Pimentel, et al. [19].

In recent years, CNNs have become a leading tool for anomaly detection in image data. Many of these efforts have been focused on benchmark datasets such as CIFAR-10 and ImageNet where the general approach is to use a subset of classes as normal data and then testing whether an image from a new class is correctly identified as an anomaly [20–22]. In this approach, a new image is considered an anomaly if the object in the new image does not match the objects included in the training data. For example, an anomaly detection model would attempt to distinguish between an airplane and a bird. One common framework for anomaly detection uses autoencoders to generate anomaly scores based on reconstruction error [23,24]. A thorough survey of deep learning methods for anomaly detection are provided in several works [25,26].

Defect detection can be considered a more specific type of anomaly detection problem where the goal is recognize subtle abnormalities in an image where the background object is normal. For example, a defect detection model would attempt to distinguish between a piece of fabric with and without a tear. While much progress has been made on general anomaly detection methods, recent work has shown that these methods do not generalize well to defect detection problems [27]. Tailored methods for defect detection [27,28] have been shown to outperform general anomaly detection models when using the MVTec benchmark dataset, a dataset specifically designed for defect detection [29]. Given that anomaly detection methods may generalize poorly to defect detection problems, our contribution is a novel method that is specifically intended for point defect detection in TEM images. While our methods are specifically tailored for defect detection in TEM images, our proposed PCA-CNN model has several parallels to recently proposed state-of-the-art defect detection methods [27].

In addition to our PCA-CNN model, we also propose a baseline defect detection method that uses principal components analysis (PCA) and reconstruction error to locate point defects in TEM images. The PCA model serves primarily as a performance baseline for the PCA-CNN model. PCA is a commonly used method for anomaly detection and is preferred for its simplicity [30–32]. PCA-based anomaly detection methods generally involve measuring the reconstruction error between a data point and its reconstruction. The reconstruction is generated via a linear transformation that is fitted on a training data set of normal observations [33]. The PCA method is presented in more detail in a later section. The concept of using reconstruction error for anomaly detection is also applicable to deep learning models that generate reconstructions via an autoencoder instead of PCA [20,23,24].

2. Materials and Methods

In this section, we introduce the simulated data used for model development and propose two methods for detecting defects in TEM images of GaAs. We specifically consider defect detection using high-resolution transmission electron microscopy (HRTEM) images, hereafter referred to as TEM images. The first method involves using PCA and reconstruction error, measured by mean squared error (MSE), to detect defects. The second method involves using PCA in combination with a weakly supervised CNN classification model to detect defects. Both models are trained using simulated TEM images of GaAs samples that are free of defects and then used to determine the location of a point defect in a simulated image of a GaAs sample. For each of the models, we consider the case when imaging noise is present and when there is no imaging noise.

2.1. Data Processing

The first step in developing a model for predicting the location of point defects is to generate simulated TEM images. We note that atomic resolution TEM imaging is performed in two different modes, wherein the electron beam is in parallel illumination (conventional high-resolution transmission electron microscopy) or as a focused probe (scanning transmission electron microscopy). The present work is based on parallel beam mode, since images simulated for this mode exhibits widely varying patterns for different imaging conditions, providing a large dataset for training and testing purposes. However, the results are also applicable to focused probe mode images. TEM images for GaAs were simulated using the TempasTM software. The TempasTM software has been developed in collaboration with the Material and Manufacturing Directorate, Air Force Research Lab (AFRL) and AFRL has validated the simulation results against experimental images of GaAs. The output of the simulation for a crystal projected along the [110] zone axis for TEM accelerating voltage of 300 kV and up to specimen thickness of 15 nm. The imaging parameters for the objective lens were set such that the spherical aberration coefficient was $-15\,\mu m$ and defocus ranging from -20 nm to $+20$ nm.

Ideally, experimental data would be used for this study, but due to the difficulty in acquiring experimental data, we use simulated TEM images to train and test our defect detection models. The use of simulated data is a start towards developing a method that can be trained directly on experimental data. A key consideration, then, is an understanding of the extent to which we can control defects in experimental images. As discussed earlier, it is possible to produce experimental GaAs samples that are defect-free so we assume it is feasible to acquire experimental TEM images that are known to be defect-free. In contrast, when defects, such as dopants, are added to experimental GaAs samples during the production process, the true locations of the dopant atoms in the GaAs sample are unknown. Thus, it is infeasible to generate a set of TEM images for which we know the true location of the point defects. The lack of knowledge about the true location of the defects in an experimental image is crucial. In light of this lack of defect truth data, the goal is to develop a defect detection method that is trained solely on defect-free TEM images.

Our dataset consists of simulated TEM images of GaAs using 8 different thickness conditions and 21 different defocus conditions. The thickness is varied from 1 nm to 15 nm in 2 nm steps. The defocus condition ranges from -20 nm to 20 nm in 2 nm steps. Thus, there are a total of 168 unique imaging conditions. These 168 imaging conditions are split into a set of 112 train conditions (66%) and 56 test conditions (33%). The splitting of the train and test conditions is done in a nonrandom manner. A third of the defocus conditions, $\{-18\,nm, -12\,nm, -6\,nm, 0\,nm, +6\,nm, +12\,nm, +18\,nm\}$, are assigned to the test set and the remaining are assigned to the training set. The imaging conditions have a significant impact on the resulting TEM image so splitting on the imaging conditions ensures that model performance generalizes beyond conditions only in the set of training conditions. For the remainder of the paper, we refer to these sets as the train and test conditions.

We use the train and test conditions to further generate the training and tests data for our models. For each of the 112 train conditions, we simulate a single TEM image of dimension 1007×1024. The image is represented as a matrix of dimension 1007×1024 where each entry represents a grayscale pixel value. Since the TEM image consists of a repeating lattice structure, we choose to analyze the TEM images in smaller segments of dimension 84×118. Each of these image segments is large enough to include two sets of GaAs pairs in both the vertical and horizontal direction. At the same time, these image segments are small enough such that accurately identifying the presence of a defect in a particular image segment is nearly equivalent to determining the location of the defect. Thus, after generating the larger simulated TEM images, we generate 50 random crops from each training set image where each crop is an image segment of dimension 84×118. Please note that the crops are random so the location of the GaAs atoms differs within each image segment. These 5600 image segments constitute the training data for the PCA and form the basis for the training data for the CNN. During the training of the CNN, we apply data augmentation and randomized circular defects to the 5600 image segments to generate labeled training data. This process is described in more detail when we present the PCA-CNN model.

Next we use the test conditions to generate the test data. For each of the 56 test conditions, we generate 30 TEM images that are each 1007×1024. Specifically, each simulated image contains a single point defect that can be one of three defect types. For each of these three defect types, 10 replicates are generated where the defect location is randomized for each replicate. This results in a total of 1680 test images that are each 1007×1024. The three types of defects are (1) an antisite complex where the Gallium and Arsenic atoms are reversed, (2) substitutional defect where a dopant has an approximately 5% larger radius, (3) an arbitrary circular defect. Figure 2a shows an example of each of the three defect types. We choose to consider these three types of defects because it includes a very subtle defect in the substitutional defect, a more obvious defect in the antisite defect, and a general defect in the circular defect. The circular defect is located randomly in an image segment while the other two located appropriately. The circular defect is meant to capture any general point defect such as an interstitial defect or a vacancy. The circular defect is unique in that it is easily added to any TEM image, either simulated or experimental. This flexibility plays an important role in the CNN model that introduced in a later section. For each combination of imaging condition and defect type, we generate 10 simulated TEM images with a randomly located defect. This results in 1680 test images where the defect location is known. Unlike the smaller image segments used in the training set, the images in the test set are 1007×1024. The test set images are used to evaluate whether or not the defect detection methods can accurately predict the location of the defect in the test image. Specifically, a 84×118 sliding window is used to determine the likelihood that each image segment in the 1007×1024 image contains a defect. Using a stride length of 4, each 1007×1024 image results in over 50,000 image segments that must be individually analyzed. The process for generating the training and test data is summarized in Figure 3.

The simulated TEM images do not include imaging noise. However, experimental TEM images can have varying degrees of noise that make it difficult to identify defects in a TEM image. Therefore, it is desirable for our proposed defect detection methods to be robust to imaging noise. To account for the presence of imaging noise in experimental images, Gaussian noise is used in both the training and test sets. Specifically, Gaussian noise with $\varepsilon \sim \mathcal{N}(\mu = 0, \sigma^2 = 0.05)$ is added to each pixel value for images in the training set. For the test set, varying levels of Gaussian noise, where $\sigma^2 = 0.00, 0.05, 0.10$, are added to the TEM images and model performance is evaluated for each noise level. Figure 2b shows the effect of the Gaussian noise on a TEM image.

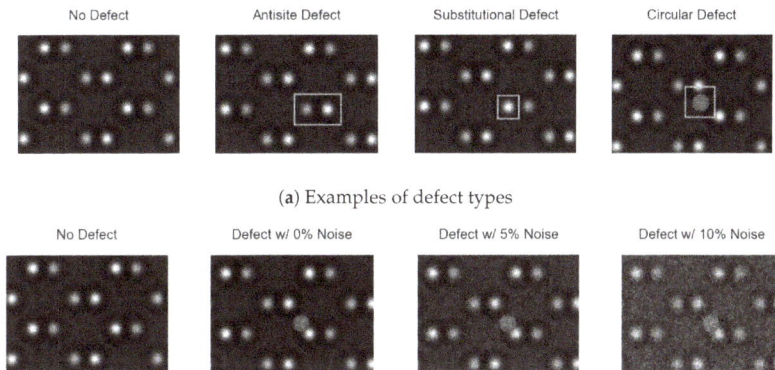

(a) Examples of defect types

(b) Examples of effect of increasing imaging noise

Figure 2. (a) Three different types of defects are considered. For each imaging condition in the test set, each of the three defect types is added to the test image. (b) Examples of increasing levels of Gaussian noise. The noise percentage level corresponds to the variance, σ^2, of the Gaussian noise that is added the image. A circular defect is shown for reference.

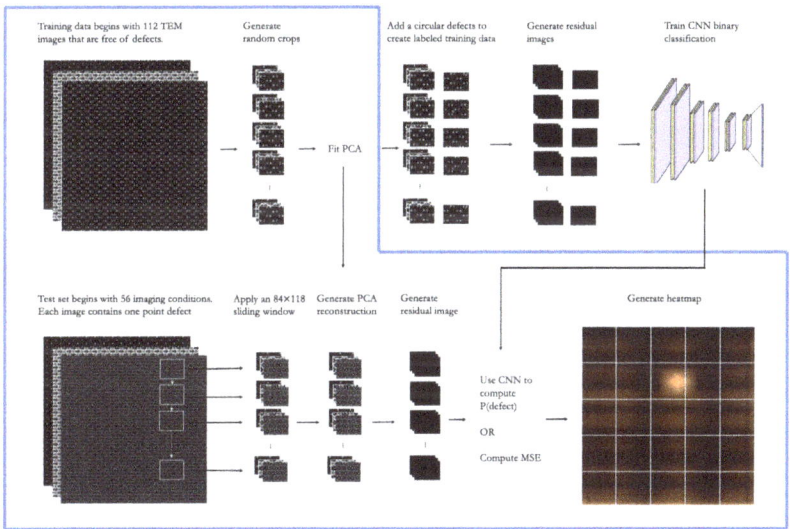

Figure 3. The methodology of the PCA-based defect detection method is summarized by the steps bordered in blue. The steps outside the blue border are the additional necessary to incorporate the CNN classification model into the defect detection methodology.

2.2. PCA Model

We present a method of detecting defects using PCA reconstructions. We fit a PCA transformation on the 5600 defect-free 84×118 image segments in the training set. Then we apply an 84×118 sliding window across each 1007×1024 test set image and, for each window, we generate a PCA reconstruction of the image segment in the window. Since the PCA transformation (and inverse transformation) is only fitted on defect-free TEM images, the assumption is that PCA will struggle to reconstruct an image with a defect. Thus, we expect that the reconstruction error of image segments with a defect to be greater than the reconstruction error of images without defects. We can predict the location of a defect by identifying the image segment with the highest MSE. With this framework in mind, we present the method in more detail below.

In general, PCA is a method for transforming a data matrix, $\mathbf{X}$, with dimensions $m \times n$ to a lower-dimensional representation, $\mathbf{X_k}$, with dimensions $m \times k$ where $k < n$. Specifically, PCA involves a linear transformation, $\mathbf{X_k} = \mathbf{X W_k}$, where the transformation matrix is defined as $\mathbf{W_k} = \arg\max_{W_k} \left\| \mathbf{X} - \mathbf{X_k W_k}^T \right\|$ with the constraint that $\mathbf{W_k}$ is orthogonal, $\mathbf{W_k}^T \mathbf{W_k} = \mathbf{I}$. Notice that $\mathbf{X_k W_k}^T$ is an $m \times n$ matrix that can be interpreted as a reconstruction of the original data using the lower-dimensional representation. Thus, $\mathbf{W_k}$ is a transformation matrix that minimizes reconstruction error for a given data matrix $\mathbf{X}$ and dimension k [33].

In our PCA-based model, the training data consists of 50 randomly cropped image segments from each of the 112 larger TEM images in the training set. These 5600 training image segments can be represented by the data matrix $\mathbf{Q} \in \mathbb{R}^{5600 \times 9912}$ where the rows represent individual image segments and the columns represent mean-centered values at each pixel location. The orthogonal linear transformation $\mathbf{Q_k} = \mathbf{Q W_k}$ projects the original data, $\mathbf{Q}$, to a lower k-dimensional representation, $\mathbf{Q_k}$. In PCA, the weight matrix $\mathbf{W_k} \in \mathbb{R}^{9912 \times k}$ is constructed such that the reconstruction MSE, $\left\| \mathbf{Q} - \mathbf{Q_k W_k}^T \right\|_F^2$, is minimized. Notice that $\hat{\mathbf{Q}} = \mathbf{Q_k W_k^T}$, a matrix of dimension 5600×9912 represents the reconstructed images. The projection to the lower-dimensional space and the reconstruction back to the original dimensional space are both determined by $\mathbf{W_k}$. Once $\mathbf{W_k}$ is fit using the training data, it can be used to generate the reconstruction of any 84×118 image segment.

We set the value of k using reconstruction mean-squared error (MSE) of a test set. Specifically, we fit the PCA using the 5600 image segments in the training set and then apply the fitted PCA to image segment from the test conditions to compute the average reconstruction MSE. For each of the 56 test conditions, 50 random crops are taken where each crop is known to be free of defects. Figure 4 shows the effect of increasing the number of components on MSE. To prevent overfitting to the noise in the training set, we set $k = 150$. Figure 4 shows several examples of image segments under various imaging conditions as well as the associated reconstruction with $k = 150$. Figure 5 also shows examples of circular defects and the effect of the PCA reconstruction on the defect. The circular defects in the raw image are not visible in the PCA reconstruction which indicates that PCA reconstruction struggles to accurately reconstruct anomalous point defects.

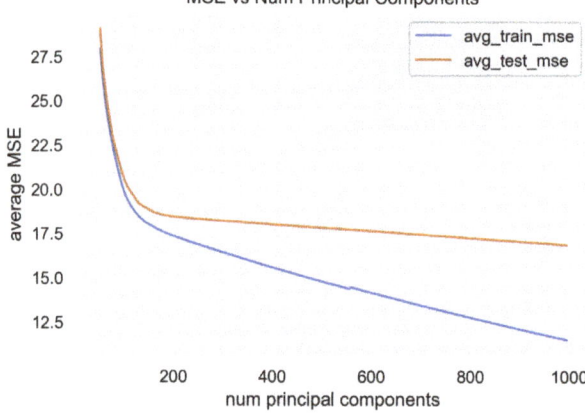

Figure 4. The number of components used to fit the PCA is determined using the average reconstruction MSE of the test set images. The average reconstruction MSE for test set images falls rapidly and levels off after the number of components exceeds 150.

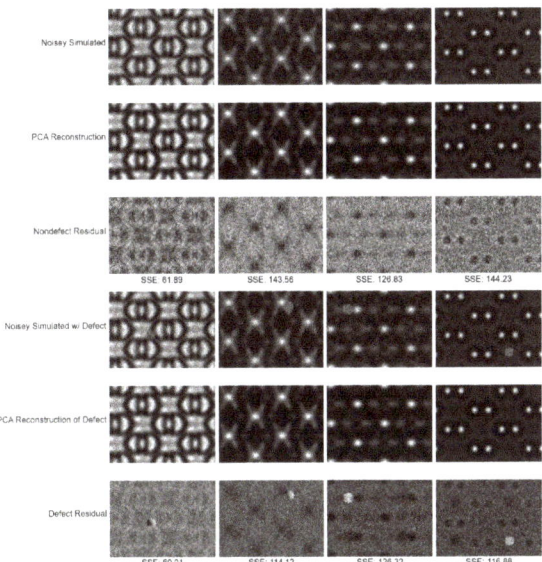

Figure 5. The first three rows show (1) the defect-free image segments with noise, (2) the PCA reconstruction, and (3) the residual between the raw and reconstructed image, respectively, for a range of imaging conditions. The bottom three rows show the same sequence images except the raw image contains a circular defect that has been randomly inserted. Notably, the PCA reconstructed image does not accurately reconstruct the defect since the PCA transformation was fitted only on images without defects.

The difference between an image segment and its reconstruction is referred to as the residual image. The residual image, intuitively, shows what is remaining when the general lattice structure is "subtracted" from the original image. Thus, the residual images consists of noise and any anomalies in the lattice structure. The reconstruction MSE can be regarded as a scalar that summarizes the residual image. For each of the 5600 images in the training set, we can compute the reconstruction MSE with and without a circular defect to understand the distribution of reconstruction MSE. Figure 6a shows how the presence of a defect changes the reconstruction MSE for each training example. In addition, Figure 6b shows how the addition of imaging noise affects the reconstruction MSE distribution with and without a defect. The concept of a residual image plays an important role in the CNN model that is presented in the next section.

After fitting the PCA transformation, we apply the resulting W_k to the test set images via a sliding window. Recall that each test set image is of dimension 1007×1024 and contains a single point defect with known location. We use a 84×118 sliding window across the 1007×1024 image and, for each window, we complete the following three steps: (1) generate the PCA reconstruction, (2) generate the residual between the original image segment and the reconstruction, (3) compute the pixel-wise mean squared error (MSE). We then generate a heatmap that shows the average reconstruction MSE for each pixel in the full-size TEM image. The predicted location of the defect corresponds to the area of the heatmap that has the largest reconstruction MSE. Figure 7 shows an example of a test image and the corresponding MSE heatmap. The defect in the test image is a substitutional defect where a single Gallium atom is replaced with a dopant atom that has a 5% larger radius. The defect is difficult to identify visually, but the heatmap accurately locates the defect. This method is applied to all imaging conditions in the test set and we evaluate the accuracy in predicting the location of each type of defect. Figure 3 summarizes the process for predicting defect location using PCA.

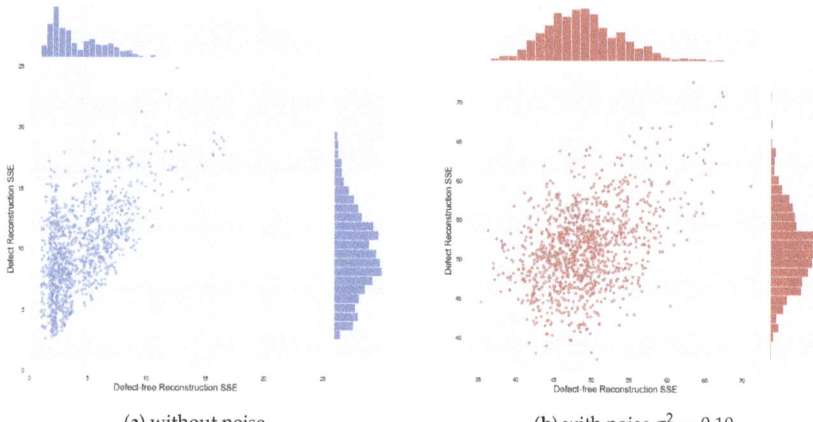

(a) without noise   (b) with noise $\sigma^2 = 0.10$

Figure 6. The scatterplot shows the reconstruction MSE of 5600 defect-free image segments (x-axis) in the PCA training set and the corresponding reconstruction MSE for the same image segment with a circular defect inserted. Points that are close to the $x = y$ represent image segments where the reconstruction MSE does not differ much with or without a defect. The marginal plots show the distribution of reconstruction MSEs with and without defects. On left, without imaging noise. On right, with imaging noise.

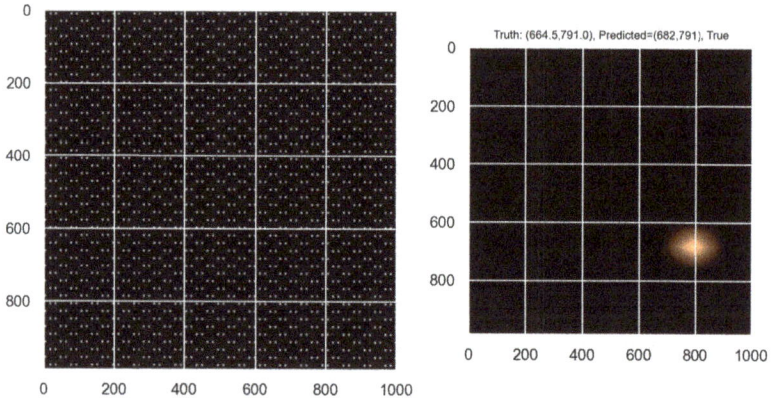

Figure 7. Heatmap that shows the pixels with the largest average MSE based on the PCA reconstruction. Bright spots correspond to areas that are mostly likely to have a defect.

2.3. PCA-CNN Model

In this section, we supplement the PCA-based detection method with a CNN classifier to improve the accuracy of the defect location predictions. This combined method significantly improves the prediction accuracy of the PCA model, especially in the case when there is imaging noise.

The PCA-based defect detection method has the benefit of being straightforward. However, in the presence of imaging noise, using PCA reconstruction error can lead to issues. Figure 5 shows the PCA residual images of segments with and without defects. In these particular examples, the reconstruction MSE for the defect images is actually lower than the reconstruction of the MSE for the defect-free images. Notably, if we visually inspect the residual images, the residual images clearly show the presence of a point defect. To address this shortcoming, we introduce a CNN classification model fitted on the PCA residual images. Intuitively, reconstruction MSE is equivalent to adding up the squared values in the residual image and it ignores any local patterns in the residual image. A

CNN, on the other hand, can be trained to look for the presence of local patterns in the residual image that may be evidence of a defect. To the best of our knowledge, the use of the residual image for defect detection is a novel approach.

A CNN is a type of neural network that is commonly used for analyzing image data ([4]). The key concept in a CNN involves the use of small filters or kernels to extract local information from an image. A filter, often of dimension 3×3, is a matrix consisting of weights. The filter is applied to an image by sliding the filter across the image and taking the sum of the element-wise product between the filter weights and the image pixel values. The sums of these element-wise products are then stored in a new matrix, commonly referred to as a feature map, which can once again be analyzed using another set of filters. The training process of a CNN involves optimizing the filter weights to minimize a given loss function. In our application, our goal is to train a CNN to identify the presence of a defect within a residual image.

The training data for the CNN model begins with the same set of defect-free training images used to fit the PCA. Recall that 50 random crops from each of the 112 training images were used to fit the PCA. These same 5600 images are used to build a set of labeled training data for the CNN classifier. Since the training data only includes image segments that are defect-free, a set of labeled training data with defects is generated by adding random, circular defects to each of the 5600 training images. These circular defects could be representative of an interstitial defect or a vacancy, but they are not necessarily meant to represent a realistic defect that would be observed in an experimental image. Instead, the hope is that the CNN will learn to classify any residual image with an abnormal local pattern as one containing a defect. Since the circular defects are arbitrary and are added post-hoc to the simulated image, this method can easily be applied to experimental TEM images as well. After generating the labeled training, a CNN classification model is trained such that for an input PCA residual image, the model outputs a scalar $\hat{y} = $ P(defect) where P(defect) $\in [0, 1]$ is the probability that the image segment contains a defect. A summary of the CNN model development process is visualized in Figure 3.

Our primary CNN architecture is adapted from the classic LeNet-5 architecture [34] and has 58,000 trainable parameters. Figure 8 shows the details of each layer of the CNN. It contains four convolutional layers with max-pooling following by two dense layers. We use a binary cross-entropy loss function and is optimized using nAdam. The model is trained for 200 epochs. Importantly, the training data are generated randomly for each batch so the location of the circular defects and noise patterns in the training set are randomized during training. The CNN is trained using Python 3.7 and Keras 2.3 with a TensorFlow 2.4.1 backend. The model achieves >99% training accuracy and test accuracy in less than 100 epochs. At the completion of 200 epochs, the test accuracy is 99.8% (Figure 9. Since the test set images are generated using a separate set of imaging conditions (focal length and thickness), the strong performance on the test set suggests that the trained CNN generalizes well to imaging conditions that were not included in the training set.

In addition to the LeNet based architecture, a VGG-16 architecture [35] was also implemented for comparison. The VGG-16 model was pretrained on ImageNet and the top dense layers were retrained using the TEM images. This resulted in 14.7 million fixed parameters and 3.2 million trainable parameters. After training for 100 epochs, the VGG-16 model achieved an accuracy of 98.2%. Given the much smaller size of the LeNet-based model and the better test set performance, the LeNet-based model was chosen as the preferred model.

```
Layer (type)                 Output Shape              Param #
=================================================================
conv2d_3 (Conv2D)            (None, 82, 116, 16)       160
_____
batch_normalization_3 (Batch (None, 82, 116, 16)       64
_____
activation_5 (Activation)    (None, 82, 116, 16)       0
_____
max_pooling2d_3 (MaxPooling2 (None, 41, 58, 16)        0
_____
conv2d_4 (Conv2D)            (None, 39, 56, 16)        2320
_____
batch_normalization_4 (Batch (None, 39, 56, 16)        64
_____
activation_6 (Activation)    (None, 39, 56, 16)        0
_____
max_pooling2d_4 (MaxPooling2 (None, 19, 28, 16)        0
_____
conv2d_5 (Conv2D)            (None, 17, 26, 16)        2320
_____
batch_normalization_5 (Batch (None, 17, 26, 16)        64
_____
activation_7 (Activation)    (None, 17, 26, 16)        0
_____
max_pooling2d_5 (MaxPooling2 (None, 8, 13, 16)         0
_____
flatten_1 (Flatten)          (None, 1664)              0
_____
dense_2 (Dense)              (None, 32)                53280
_____
activation_8 (Activation)    (None, 32)                0
_____
dense_3 (Dense)              (None, 1)                 33
_____
activation_9 (Activation)    (None, 1)                 0
=================================================================
Total params: 58,305
Trainable params: 58,209
Non-trainable params: 96
```

Figure 8. CNN architecture motivated by LeNet-5.

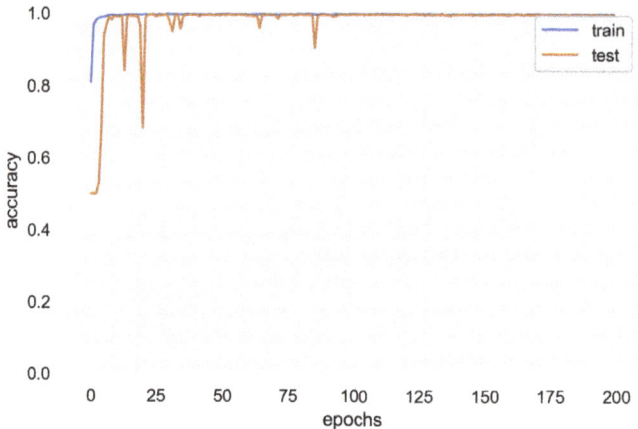

Figure 9. Train and test set accuracy during the CNN training process. The input to the CNN model is a residual image of dimension 84 × 118 and the output is the probability that the image contains a defect, P(defect).

After training the CNN, an 84 × 118 sliding window is applied to each of the 168 test images that are 1007 × 1024 with one hidden point defect. Using a stride of four pixels, this process results in 50,000 image segments that must be classified as having a defect or not. For each 84 × 118 window, we apply the following three steps: (1) generate a PCA reconstruction, (2) generate a residual image between the original image segment and the PCA reconstruction, and (3) pass the residual image into the trained CNN to generate

P(defect). For each pixel in the 1007 × 1024 test image, we compute the average P(defect) for all sliding windows that contain the pixel. This results in a smoothed heatmap for the entire test image. The location of the defect is then predicted to be the area of the heatmap that has the highest average P(defect). The heatmap shown earlier in Figure 3 is an example of a heatmap generated using the CNN classification model with a sliding window.

In many applications of CNNs for anomaly detection, the output of the CNN classifier, P(defect), is compared to a fixed threshold value to determine if a particular input contains an anomaly or not [19]. Please note that a threshold is not necessary here since the predicted defect location is simply the pixel value with the largest average P(defect). If we generalize to the case where there are n defects in a GaAs sample, then the locations corresponding to the n largest average P(defect) would be the predicted locations of the defects.

3. Results

In this section, we compare the performance of the two defect detection methods discussed above. Recall that there are 56 imaging conditions that were reserved for the test set and there are three defect types. For each combination of imaging condition and defect type, we generate 10 simulated TEM images, each of dimension 1007 × 1024, where the defect location is randomized. This results in 1680 test images where the defect location is known. For each of the 1680 test images (540 images for each of the three defect types), we apply the PCA and PCA-CNN defect detection methods to predict the location of the defect. We compare the predicted defect location to the true defect location to determine whether the model successfully located the defect.

Table 1 shows the accuracy of both methods in predicting the defect location for various levels of imaging noise. The PCA defect detection method performs particularly well in the case of no imaging. It accurately locates all three defects types at nearly >97% and generally outperforms the CNN model. However, as the imaging noise increases, we observe the superior performance of the CNN model. Specifically, when imaging noise rises to $\sigma^2 = 0.10$, the PCA model achieves an accuracy of 56% and 57% on antisite and circular defects, respectively, while the CNN model achieves 75% and 93% accuracy.

Table 1. Accuracy of the PCA and PCA-CNN model in locating point defects in the test set images. Table (a) shows the accuracy results when including all images in the test set. Table (b) shows the accuracy results when only the nominal defocus conditions are included. In both cases, the CNN model is more robust to imaging noise.

Method	Noise	Substitution $n = 560$	Antisite $n = 560$	Circular $n = 560$
(a) Location detection accuracy including all imaging conditions				
PCA	$\sigma^2 = 0.00$	0.97	1.00	1.00
	$\sigma^2 = 0.05$	0.16	0.80	0.94
	$\sigma^2 = 0.10$	0.04	0.56	0.57
PCA-CNN	$\sigma^2 = 0.00$	0.71	0.86	1.00
	$\sigma^2 = 0.05$	0.64	0.90	0.99
	$\sigma^2 = 0.10$	0.14	0.75	0.93

Method	Noise	Substitution $n = 240$	Antisite $n = 240$	Circular $n = 240$
(b) Location detection accuracy for central defocus conditions, {−6 nm, 0 nm, +6 nm}				
PCA	$\sigma^2 = 0.00$	1.00	1.00	1.00
	$\sigma^2 = 0.05$	0.24	0.91	0.92
	$\sigma^2 = 0.10$	0.04	0.70	0.61
PCA-CNN	$\sigma^2 = 0.00$	1.00	0.98	1.00
	$\sigma^2 = 0.05$	0.89	0.99	0.99
	$\sigma^2 = 0.10$	0.23	0.89	0.91

The results in Table 1a report the performance of the two methods under all test imaging conditions. Recall that the test set includes an equal number of TEM images for a range of defocus conditions. In practice, extreme defocus conditions are relatively uncommon and are actively avoided. Narrowing the focus on the central range of defocus conditions, $\{-6 \text{ nm}, 0 \text{ nm}, +6 \text{ nm}\}$, provides a better representation of expected performance on experimental images. Table 1b shows the defect location accuracy of both methods under nominal defocus conditions. Under the restricted set of defocus conditions, the CNN model remains more robust in the presence of imaging noise. Specifically, when $\sigma^2 = 0.10$, the CNN model achieves 89% and 91% accuracy for antisite and circular defects, respectively, while the PCA model achieves 70% and 61% accuracy.

Based on these preliminary results, it appears that the substitution defects are more challenging to identify compared to the antisite and circular defect. This is unsurprising given that the substitution defects are also the most challenging to identify from visual inspection. The substitution defects were purposely subtle so as to determine the effectiveness of the proposed methods for a wide range of defects. In practice, the substitution defects are unlikely to sit precisely in a gallium or arsenic site. If the substitution defect is slightly misaligned, then it is likely that the proposed methods would be more effective in locating the defect. The antisite and random circular defects are more readily identified visually which is reflected in the accuracy results. Although the circular defect is not representative of a particular defect, the circular defect could be representative of an interstitial defect or a vacancy.

4. Discussion

In this paper, we introduce two methods for determining the location of a point defect in a TEM image of GaAs. Compared to recent applications of using CNNs for defect detection ([1,2], and references therein), the proposed PCA and PCA-CNN methods of defect detection are unique in that they can be trained on TEM images that are defect-free. Unlike prior approaches to defect detection, this opens the door to training these models using experimental data. After training both models using a set of simulated images that are free of defects, we demonstrate the performance of both methods in locating a simulated defect in an HRTEM image. In the case of no imaging noise, we show the PCA method is sensitive to minor defects such as a subtle substitution defect (97% accuracy). However, as imaging noise is introduced, the performance of the PCA method declines rapidly. Supplementing the PCA method with a CNN classification model improves the performance of the model dramatically. The CNN classification model achieves >89% accuracy for both antisite and circular defects at the highest level of imaging noise ($\sigma^2 = 0.10$). These results suggest that the CNN approach has the potential to be highly effective in analyzing experimental images.

Our PCA-CNN classification model is unique in that it is trained on PCA residual images. Using the PCA reconstruction to generate a residual image is a novel approach that has notable benefits. One of the benefits is that it allows for a single pre-trained CNN to be used for a wide range of imaging conditions. This is in contrast to prior studies that rely on condition-specific models for defect detection. Imaging conditions, such as thickness and defocus condition, change the overall "pattern" that is visible in an TEM image. By taking the difference between an image segment and its reconstruction, we are, intuitively, "subtracting" the pattern that is associated with a set of imaging conditions. The residual images are then less correlated with the imaging conditions used to generate the TEM image and can be analyzed using a single pre-trained CNN. Another benefit is that using the residual images allows a CNN to more effectively classify defects. Specifically, when we trained a CNN classification model directly on image segments in the training set without using residual images, the trained model far underperformed our model that uses residual images. This suggests that the use of residual images is a key step in training an effective CNN classification model in the context of TEM images.

The results presented in this paper are based on simulated TEM images. However, the goal is to implement and adapt these methods for experimental images as they become available. We observe that experimental images pose unique challenges compared to simulated images. In the simulated TEM images, the imaging conditions and the imaging noise were assumed to be consistent across the entire image. In contrast, the thickness of a sample can vary in an experimental image and the imaging noise is unlikely to be consistent across an entire image. While additional steps will be necessary to account for these variations, we believe the key ideas of using PCA reconstructions and residual images will remain an integral part of analyzing defects in experimental TEM images.

5. Conclusions

In this paper, we propose an anomaly detection method for locating point defects in crystalline materials using TEM images. The proposed method involves using a PCA reconstruction to generate a residual image and then a self-supervised CNN classifier to detect the presence of an anomaly in the residual image. Unlike earlier works that rely on extensive pixel-by-pixel labeled training data via simulation ([1,2]), our proposed method is a self-supervised method that only requires defect-free TEM images in the training set. Since the method only requires defect-free TEM images, it allows for the possibility of training a defect detection model directly on experimental TEM images that are defect-free. Additionally, our novel use of a residual image allows for strong results using a simple, computationally efficient CNN architecture that generalizes well to imaging conditions that are not included in the training set. Using simulated TEM images with a single point defect, we show that our PCA-CNN method is able to accurately locate point defects and it outperforms reconstruction error-based methods, particularly in the case when there is significant imaging noise.

Author Contributions: Conceptualization, P.C., A.W., K.M. and K.E.; methodology, P.C. and A.W.; software, P.C. and K.M.; investigation, P.C., A.W. and K.M.; formal analysis, P.C. and A.W.; investigation, P.C. and A.W.; writing—original draft preparation, P.C.; writing—review and editing, P.C., A.W., K.M. and K.E.; visualization, P.C.; supervision, A.W. All authors have read and agreed to the published version of the manuscript.

Funding: This research received no external funding.

Institutional Review Board Statement: Not applicable.

Informed Consent Statement: Not applicable.

Data Availability Statement: The simulated TEM images used in this study are being reviewed for public release by AFRL and are not currently available for public release. Individual data requests can be submitted to the corresponding author.

Conflicts of Interest: The authors declare no conflict of interest.

References

1. Ziatdinov, M.; Dyck, O.; Maksov, A.; Li, X.; Sang, X.; Xiao, K.; Unocic, R.R.; Vasudevan, R.; Jesse, S.; Kalinin, S.V. Deep Learning of Atomically Resolved Scanning Transmission Electron Microscopy Images: Chemical Identification and Tracking Local Transformations. *ACS Nano* **2017**, *11*. [CrossRef] [PubMed]
2. Madsen, J.; Liu, P.; Kling, J.; Wagner, J.B.; Hansen, T.W.; Winther, O.; Schiøtz, J. A Deep Learning Approach to Identify Local Structures in Atomic-Resolution Transmission Electron Microscopy Images. *Adv. Theory Simul.* **2018**, *1*. [CrossRef]
3. Barron, A.; Raja, P.M.J. Crystal Structure. 2021. Available online: https://chem.libretexts.org/@go/page/55904 (accessed on 8 April 2021).
4. Zhang, Q.; Zhang, M.; Chen, T.; Sun, Z.; Ma, Y.; Yu, B. Recent advances in convolutional neural network acceleration. *Neurocomputing* **2019**, *323*, 37–51. [CrossRef]
5. Goodfellow, I.J.; Bengio, Y.; Courville, A. *Deep Learning*; MIT Press: Cambridge, MA, USA, 2016.
6. Mohan, S.; Manzorro, R.; Vincent, J.L.; Tang, B.; Sheth, D.Y.; Simoncelli, E.P.; Matteson, D.S.; Crozier, P.A.; Fernandez-Granda, C. Deep Denoising For Scientific Discovery: A Case Study In Electron Microscopy. *arXiv* **2020**, arXiv:2010.12970.
7. Wang, F.; Henninen, T.R.; Keller, D.; Erni, R. Noise2Atom: Unsupervised denoising for scanning transmission electron microscopy images. *Appl. Microsc.* **2020**. [CrossRef]

8. Ede, J.M.; Beanland, R. Partial Scanning Transmission Electron Microscopy with Deep Learning. *Sci. Rep.* **2020**, *10*. [CrossRef] [PubMed]
9. De Haan, K.; Ballard, Z.S.; Rivenson, Y.; Wu, Y.; Ozcan, A. Resolution enhancement in scanning electron microscopy using deep learning. *Sci. Rep.* **2019**, *9*, 1–7. [CrossRef] [PubMed]
10. Suveer, A.; Gupta, A.; Kylberg, G.; Sintorn, I.M. Super-resolution reconstruction of transmission electron microscopy images using deep learning. In Proceedings of the 2019 IEEE 16th International Symposium on Biomedical Imaging (ISBI 2019), Venice, Italy, 8–11 April 2019. [CrossRef]
11. Aguiar, J.A.; Gong, M.L.; Unocic, R.R.; Tasdizen, T.; Miller, B.D. Decoding crystallography from high-resolution electron imaging and diffraction datasets with deep learning. *Sci. Adv.* **2019**, *5*. [CrossRef] [PubMed]
12. Li, W.; Field, K.G.; Morgan, D. Automated defect analysis in electron microscopic images. *Npj Comput. Mater.* **2018**. [CrossRef]
13. Clifton, L.; Clifton, D.A.; Watkinson, P.J.; Tarassenko, L. Identification of patient deterioration in vital-sign data using one-class support vector machines. In Proceedings of the 2011 Federated Conference on Computer Science and Information Systems, FedCSIS 2011, Szczecin, Poland, 18–21 September 2011.
14. Quinn, J.A.; Williams, C.K. Known unknowns: Novelty detection in condition monitoring. In *Lecture Notes in Computer Science (including Subseries Lecture Notes in Artificial Intelligence and Lecture Notes in Bioinformatics)*; Springer: Berlin/Heidelberg, Germany, 2007; [CrossRef]
15. Tarassenko, L.; Hayton, P.; Cerneaz, N.; Brady, M. Novelty detection for the identification of masses in mammograms. In Proceedings of the 1995 Fourth International Conference on Artificial Neural Networks, Cambridge, UK, 26–28 June 1995. [CrossRef]
16. Jyothsna, V.; Rama Prasad, V.; Munivara Prasad, K. A Review of Anomaly based Intrusion Detection Systems. *Int. J. Comput. Appl.* **2011**. [CrossRef]
17. Samrin, R.; Vasumathi, D. Review on anomaly based network intrusion detection system. In Proceedings of the 2017 International Conference on Electrical, Electronics, Communication, Computer, and Optimization Techniques (ICEECCOT), Mysuru, India, 15–16 December 2017. [CrossRef]
18. Ganeshan, R.; Kolli, C.S.; Kumar, C.M.; Daniya, T. A Systematic Review on Anomaly Based Intrusion Detection System. In Proceedings of the IOP Conference Series: Materials Science and Engineering, Warangal, India, 9–10 October 2020. [CrossRef]
19. Pimentel, M.A.; Clifton, D.A.; Clifton, L.; Tarassenko, L. A review of novelty detection. *Signal Process.* **2014**, *99*, 215–249. [CrossRef]
20. Cho, P.; Farias, V.; Kessler, J.; Levi, R.; Magnanti, T.; Zarybnisky, E. Maintenance and flight scheduling of low observable aircraft. *Nav. Res. Logist.* **2015**. [CrossRef]
21. Ruff, L.; Vandermeulen, R.A.; Görnitz, N.; Deecke, L.; Siddiqui, S.A.; Binder, A.; Müller, E.; Kloft, M. Deep one-class classification. In Proceedings of the 35th International Conference on Machine Learning, ICML 2018, Stockholm, Sweden, 10–15 July 2018; Volume 10, pp. 6981–6996.
22. Chalapathy, R.; Menon, A.K.; Chawla, S. Anomaly detection using one-class neural networks a preprint. *arXiv* **2018**, arXiv:1802.06360.
23. Thompson, B.B.; Marks, R.J.; Choi, J.J.; El-Sharkawi, M.A.; Huang, M.Y.; Bunje, C. Implicit learning in autoencoder novelty assessment. In Proceedings of the International Joint Conference on Neural Networks, Honolulu, HI, USA, 12–17 May 2002. [CrossRef]
24. Hawkins, S.; He, H.; Williams, G.; Baxter, R. Outlier detection using replicator neural networks. In *Lecture Notes in Computer Science (including Subseries Lecture Notes in Artificial Intelligence and Lecture Notes in Bioinformatics)*; Springer: Berlin/Heidelberg, Germany, 2002. [CrossRef]
25. Pang, G.; Shen, C.; Cao, L.; van Den Hengel, A. Deep Learning for Anomaly Detection: A Review. *arXiv* **2020**, arXiv:2007.02500.
26. Chalapathy, R.; Chawla, S. Deep learning for anomaly detection: A survey. *arXiv* **2019**, arXiv:1901.03407.
27. Li, C.L.; Sohn, K.; Yoon, J.; Pfister, T. CutPaste: Self-Supervised Learning for Anomaly Detection and Localization. *arXiv* **2021**, arXiv:2104.04015.
28. Yi, J.; Yoon, S. Patch SVDD: Patch-Level SVDD for Anomaly Detection and Segmentation. In *Lecture Notes in Computer Science (including Subseries Lecture Notes in Artificial Intelligence and Lecture Notes in Bioinformatics)*; Springer: Berlin/Heidelberg, Germany, 2020; pp. 375–390. [CrossRef]
29. Bergmann, P. MVTec AD—A Comprehensive Real-World Dataset for Unsupervised Anomaly Detection. In Proceedings of the 2019 IEEE/CVF Conference on Computer Vision and Pattern Recognition (CVPR), Long Beach, CA, USA, 15–20 June 2019; pp. 9592–9600.
30. Huang, L.; Nguyen, X.L.; Garofalakis, M.; Jordan, M.I.; Joseph, A.; Taft, N. In-network PCA and anomaly detection. In *Advances in Neural Information Processing Systems*; MIT Press: Cambridge, MA, USA, 2007. [CrossRef]
31. Brauckhoff, D.; Salamatian, K.; May, M. Applying PCA for traffic anomaly detection: Problems and solutions. In Proceedings of the IEEE INFOCOM 2009, Rio de Janeiro, Brazil, 19–25 April 2009. [CrossRef]
32. Jablonski, J.A.; Bihl, T.J.; Bauer, K.W. Principal Component Reconstruction Error for Hyperspectral Anomaly Detection. *IEEE Geosci. Remote Sens. Lett.* **2015**. [CrossRef]
33. Hastie, T.; Tibshirani, R.; Friedman, J. *The Elements of Statistical Learning*; Springer Series in Statistics; Springer: New York, NY, USA, 2001.

34. LeCun, Y.; Bottou, L.; Bengio, Y.; Haffner, P. Gradient-based learning applied to document recognition. *Proc. IEEE* **1998**, *86*. [CrossRef]
35. Simonyan, K.; Zisserman, A. Very deep convolutional networks for large-scale image recognition. In Proceedings of the 3rd International Conference on Learning Representations, ICLR 2015–Conference Track Proceedings, San Diego, CA, USA, 7–9 May 2015.

Article

A Conservative and Implicit Second-Order Nonlinear Numerical Scheme for the Rosenau-KdV Equation

Cui Guo *, Yinglin Wang and Yuesheng Luo

Harbin Engineering University, Harbin 150001, China; wangyl_0598@163.com (Y.W.); luoyuesheng@hrbeu.edu.cn (Y.L.)
* Correspondence: guocui@hrbeu.edu.cn or 2185835@163.com

Abstract: In this paper, for solving the nonlinear Rosenau-KdV equation, a conservative implicit two-level nonlinear scheme is proposed by a new numerical method named the multiple integral finite volume method. According to the order of the original differential equation's highest derivative, we can confirm the number of integration steps, which is just called multiple integration. By multiple integration, a partial differential equation can be converted into a pure integral equation. This is very important because we can effectively avoid the large errors caused by directly approximating the derivative of the original differential equation using the finite difference method. We use the multiple integral finite volume method in the spatial direction and use finite difference in the time direction to construct the numerical scheme. The precision of this scheme is $O(\tau^2 + h^3)$. In addition, we verify that the scheme possesses the conservative property on the original equation. The solvability, uniqueness, convergence, and unconditional stability of this scheme are also demonstrated. The numerical results show that this method can obtain highly accurate solutions. Further, the tendency of the numerical results is consistent with the tendency of the analytical results. This shows that the discrete scheme is effective.

Keywords: multiple integral finite volume method; finite difference method; Rosenau-KdV; conservation; solvability; convergence

1. Introduction

Proposed by Korteweg and de Vries, the Korteweg–de Vries (KdV) equation,

$$u_t + uu_x + u_{xxx} = 0 \tag{1}$$

has been widely studied. It can describe ion–phonon waves, magnetic fluid waves in cold plasma, unidirectional shallow water waves with small amplitude and long waves, and other wave processes in some physical and biological systems.

It has a wide range of physical applications, so there is great interest in this equation. A great many numerical methods have been proposed to obtain the numerical solutions of KdV equations [1–5]. In addition, [6] developed a new integral equation using the negative-order KdV equation and derived multiple soliton solutions, while [7] created various negative-order KdV equations in (3 + 1) dimensions and discussed the solutions for each derived model.

Given the shortcomings of the KdV equation in describing wave–wave and wave–wall interactions, Rosenau [8,9] proposed the Rosenau equation to cope with the compact discrete dynamic system.

$$u_t + u_{xxxxt} + u_x + uu_x = 0 \tag{2}$$

The existence, uniqueness, and regularity of solutions were derived by Park [10]. Since then, several numerical methods have been studied for the Rosenau equation. For example, ref [11] used the Petviashvili iteration method to construct numerical solitary wave solutions; ref [12] applied Galerkin mixed finite element methods to (2) by employing a splitting technique; ref [13] discussed new methods to expand solutions for wave equations like Rosenau-type equations with damping terms; ref [14] constructed an implicit Crank–Nicolson formula of the mixed finite element method for nonlinear fourth-order Rosenau equations; and [15] proposed a meshfree method based on the radial basis function for the Rosenau equation and other higher-order partial differential equations(PDEs). The long-time behavior of solutions was investigated in [16].

To better study nonlinear waves, the viscous term u_{xxx} needs to be included.

$$u_t + u_{xxxxt} + u_x + uu_x + u_{xxx} = 0 \tag{3}$$

Equation (3) is usually called the Rosenau-KdV equation. The authors of [17,18] proposed conservative schemes for the Rosenau-KdV equation based on the finite difference method. The authors of [19] proposed a Crank–Nicolson meshless spectral radial point interpolation (CN-MSRPI) method for the nonlinear Rosenau-KdV equation. The authors of [20] solved the equation by the first-order Lie–Trotter and second-order Strang time-splitting techniques combined with quintic B-spline collocation, while [21] studied numerical solutions by using the subdomain method based on sextic B-spline basis functions. Although various methods have been proposed, we wonder whether there might be a new method with higher accuracy and efficiency that can keep some properties of the original partial differential equation. Further, research on the Rosenau-KdV equation under certain conditions is relatively lacking.

In this paper, we consider the Rosenau-KdV Equation (4) with initial condition

$$u(x,0) = u_0(x), \quad x \in [x_l, x_r] \tag{4}$$

and boundary conditions

$$u(x_l,t) = u(x_r,t) = 0, \quad u_x(x_l,t) = u_x(x_r,t) = 0, \quad t \in [0,T] \tag{5}$$

Here, $u_0(x)$ is a known smooth function, and x_l and x_r are, respectively, the left border and the right border of x.

Theorem 1. *The system (3)–(5) satisfies the following conservative property:*

$$E(t) = \|u\|_{L^2}^2 + \|u_{xx}\|_{L^2}^2 = E(0) = Const. \tag{6}$$

Here, $\|u\|_{L^2}^2 = \int_{x_l}^{x_r} u^2 \, dx$.

Proof. Integrate both sides of Equation (3) from x_l to x_r and apply (5); we thus obtain

$$\int_{x_l}^{x_r} (u_t + u_{xxxxt} + u_{xx} + u_x + uu_x) u \, dx = \frac{1}{2} \frac{\partial}{\partial t} \left(\|u\|_{L^2}^2 + \|u_{xx}\|_{L^2}^2 \right) = 0 \tag{7}$$

Let $E(t) = \|u\|_{L^2}^2 + \|u_{xx}\|_{L^2}^2$. Then we get

$$E(t) = \|u\|_{L^2}^2 + \|u_{xx}\|_{L^2}^2 = E(0) = Const. \tag{8}$$

Hence, the system (3)–(5) meets the conservative property. □

In this paper, we present a two-level implicit nonlinear discrete scheme for the Rosenau-KdV Equations (3)–(5) by using a new method named the multiple integral finite volume method (MIFVM). The remaining contents of this paper are arranged as follows: In Section 2, we introduce MIFVM in detail and propose a numerical scheme. The conservative property of this scheme is also discussed. In Section 3, the solvability of this numerical scheme is derived. Then, in Section 4, we show some prior estimates. According to the prior estimates, we demonstrate the convergence with order $O(\tau^2 + h^3)$ and unconditional stability of this numerical scheme in Section 5. In Section 6, the uniqueness of this numerical solution is verified with the classic theorem. Finally, we verify the effectiveness of the numerical scheme via some numerical experiments in Section 7.

2. A Two-Level Implicit Nonlinear Discrete Scheme and Its Conservative Law

2.1. Notation

Let h and τ be uniform step sizes in the spatial and temporal directions, respectively. Let $x_j = x_l + jh (j = 0, 1, \cdots, J)$, $t_n = n\tau (n = 0, 1, \cdots, N)$, where $h = (x_r - x_l)/J$, $\tau = T/N$. Further, let $u_j = u_j(t) = u(x_l + jh, t)$, $u_j^n = u(x_l + jh, t_n)$, $Z_h^0 = \{u_j | u_0 = u_J = 0, j = 0, 1, \cdots, J\}$, and $\Omega_h = \{x_j | j = 0, 1, \cdots, J\}$. In this paper, we let C denote a generic positive constant independent of h and τ. The difference operators, inner product, and norms we defined are shown below.

$$u_j^{n+\frac{1}{2}} = \frac{u_j^{n+1}+u_j^n}{2}, \left(u_j^n\right)_x = \frac{u_{j+1}^n - u_j^n}{h}, \left(u_j^n\right)_{\bar{x}} = \frac{u_j^n - u_{j-1}^n}{h}, \left(u_j^n\right)_{\hat{x}} = \frac{u_{j+1}^n - u_{j-1}^n}{2h},$$

$$\left(u_j^n\right)_{x\bar{x}} = \left(u_j^n\right)_{\bar{x}x} = \frac{u_{j+1}^n - 2u_j^n + u_{j-1}^n}{h^2}, \left(u_j^n\right)_{xx\hat{x}} = \frac{u_{j+2}^n - 2u_{j+1}^n + 2u_{j-1}^n - u_{j-2}^n}{2h^3},$$

$$\left(u_j^n\right)_{xx\bar{x}\bar{x}} = \frac{u_{j+2}^n - 4u_{j+1}^n + 6u_j^n - 4u_{j-1}^n + u_{j-2}^n}{h^4}, \left(u_j^{n+\frac{1}{2}}\right)_{\hat{t}} = \frac{u_j^{n+1} - u_j^n}{\tau},$$

$$\|u^n\| = \sqrt{(u^n, u^n)}, \|u^n\|_\infty = \max_{x_j \in \Omega_h} |u_j^n|, (u^n, v^n) = \sum_{j=0}^{J} u_j^n v_j^n h$$

However, we should note that if the inner product operates on different functions, there will be different ranges of values of j, for example, $(u_x^n, v_x^n) = \sum_{j=0}^{J-1} \left(u_j^n\right)_x \left(v_j^n\right)_x h$ and $(u_{\bar{x}}^n, v_{\bar{x}}^n) = \sum_{j=1}^{J} \left(u_j^n\right)_{\bar{x}} \left(v_j^n\right)_{\bar{x}} h$.

Lemma 1. *For any two mesh functions $u, v \in Z_h^0$, the following equations hold.*

$$(u_x, v) = -(u, v_{\bar{x}}), (u_{x\bar{x}}, v) = -(u_x, v_x), (u_{\hat{x}}, v) = -(u, v_{\hat{x}})$$

Furthermore, if $\left(u_0^n\right)_{x\bar{x}} = \left(u_J^n\right)_{x\bar{x}} = 0$, then $((u^n)_{xx\bar{x}\bar{x}}, u^n) = \|u_{xx}^n\|^2$.

Lemma 2. *For any mesh function $u \in Z_h^0$, the following equation holds.*

$$\|u_{\hat{x}}\|^2 \leq \|u_x\|^2 \qquad (9)$$

Lemma 3. *For any discrete function $u \in Z_h^0$, we have*

$$(\varphi(u), u) = \sum_{j=1}^{J-1} \frac{1}{3} \left(u_j\right)_{\hat{x}} \left(u_{j-1} + u_j + u_{j+1}\right) u_j h = 0, \qquad (10)$$

where $\varphi(u_j) = \frac{1}{3} \left(u_j\right)_{\hat{x}} \left(u_{j-1} + u_j + u_{j+1}\right)$.

Proof. Because $u \in Z_h^0$, we have

$$(\varphi(u), u) = \frac{1}{6h} \sum_{j=1}^{J-1} \left[u_{j+1} u_j - u_j u_{j-1} + \left(u_{j+1}\right)^2 - \left(u_{j-1}\right)^2 \right] u_j$$
$$= \frac{1}{6h} \left[\sum_{j=1}^{J-2} \left(u_j + u_{j+1}\right) u_j u_{j+1} - \sum_{j=2}^{J-1} \left(u_{j-1} + u_j\right) u_{j-1} u_j \right] = 0 \quad (11)$$

□

2.2. The Multiple Integral Finite Volume Method (MIFVM)

In this paper, we use a method named MIFVM to construct a two-level implicit nonlinear scheme for the Rosenau-KdV Equations (3)–(5). The method uses multiple integrals and combines the finite difference method with the finite volume method. We thus discretize the original PDE into separate spatial and temporal directions.

In the spatial x direction, firstly, by multiple integrals, we turn the original differential Equation (3), with unknown function u and its derivative, into an integral equation with only the unknown function. This is very important because we can effectively avoid the large errors caused by directly approximating the derivative of the original differential equation using the finite difference method. We use the multiple integral finite volume method in the spatial direction and use finite difference in the time direction to construct the numerical scheme. Firstly, in the spatial direction, the number of integration steps m depends on the order of the highest derivative in the x direction of the original PDE. Considering the original Equation (3), the order of the highest derivative in the x direction is four, so $m = 2^4 - 1 = 15$. Now, we define a 15-time integral,

$$\int_{xxxx} u = \int_{xxxx} u(x,t) \stackrel{def}{=} \int_{x_j+\varepsilon_7}^{x_j+\varepsilon_8} dx_{f_2} \int_{x_j+\varepsilon_6}^{x_j+\varepsilon_7} dx_{f_1} \int_{x_j+\varepsilon_5}^{x_j+\varepsilon_6} dx_{e_2} \int_{x_j}^{x_j+\varepsilon_5} dx_{e_1} \int_{x_j-\varepsilon_4}^{x_j} dx_{d_2}$$
$$\int_{x_j-\varepsilon_3}^{x_j-\varepsilon_4} dx_{d_1} \int_{x_j-\varepsilon_2}^{x_j-\varepsilon_3} dx_{c_2} \int_{x_{f_2}}^{x_{d_2}} dx_{c_1} \int_{x_{f_1}}^{x_{f_2}} dx_f \int_{x_{e_1}}^{x_{d_1}} dx_e \int_{x_{d_1}}^{x_{e_1}} dx_d \quad (12)$$
$$\int_{x_{c_1}}^{x_{c_2}} dx_c \int_{x_e}^{x_f} dx_b \int_{x_a}^{x_d} dx_a \int_{x_a}^{x_b} u(x,t) \, dx$$

and we treat original Equation (2) using integral (12). Then, we can get

$$\int_{x_j+\varepsilon_7}^{x_j+\varepsilon_8} dx_{f_2} \int_{x_j+\varepsilon_6}^{x_j+\varepsilon_7} dx_{f_1} \int_{x_j+\varepsilon_5}^{x_j+\varepsilon_6} dx_{e_2} \int_{x_j}^{x_j+\varepsilon_5} dx_{e_1} \int_{x_j-\varepsilon_4}^{x_j} dx_{d_2} \int_{x_j-\varepsilon_3}^{x_j-\varepsilon_4} dx_{d_1} \int_{x_j-\varepsilon_2}^{x_j-\varepsilon_3} dx_{c_2}$$
$$\int_{x_j-\varepsilon_1}^{x_j-\varepsilon_2} dx_{c_1} \int_{x_{f_1}}^{x_{f_2}} dx_f \int_{x_{e_1}}^{x_{e_2}} dx_e \int_{x_{d_1}}^{x_{d_2}} dx_d \int_{x_{c_1}}^{x_{c_2}} dx_c \int_{x_e}^{x_f} dx_b \int_{x_a}^{x_d} dx_a \int_{x_a}^{x_b} u_t \, dx$$
$$+ \int_{x_j+\varepsilon_7}^{x_j+\varepsilon_8} dx_{f_2} \int_{x_j+\varepsilon_6}^{x_j+\varepsilon_7} dx_{f_1} \int_{x_j+\varepsilon_5}^{x_j+\varepsilon_6} dx_{e_2} \int_{x_j}^{x_j+\varepsilon_5} dx_{e_1} \int_{x_j-\varepsilon_4}^{x_j} dx_{d_2} \int_{x_j-\varepsilon_3}^{x_j-\varepsilon_4} dx_{d_1} \int_{x_j-\varepsilon_2}^{x_j-\varepsilon_3} dx_{c_2}$$
$$\int_{x_j-\varepsilon_1}^{x_j-\varepsilon_2} dx_{c_1} \int_{x_{f_1}}^{x_{f_2}} dx_f \int_{x_{e_1}}^{x_{e_2}} dx_e \int_{x_{d_1}}^{x_{d_2}} dx_d \int_{x_{c_1}}^{x_{c_2}} dx_c \int_{x_e}^{x_f} dx_b \int_{x_a}^{x_d} dx_a \int_{x_a}^{x_b} u_{xxxxt} \, dx$$
$$+ \int_{x_j+\varepsilon_7}^{x_j+\varepsilon_8} dx_{f_2} \int_{x_j+\varepsilon_6}^{x_j+\varepsilon_7} dx_{f_1} \int_{x_j+\varepsilon_5}^{x_j+\varepsilon_6} dx_{e_2} \int_{x_j}^{x_j+\varepsilon_5} dx_{e_1} \int_{x_j-\varepsilon_4}^{x_j} dx_{d_2} \int_{x_j-\varepsilon_3}^{x_j-\varepsilon_4} dx_{d_1} \int_{x_j-\varepsilon_2}^{x_j-\varepsilon_3} dx_{c_2} \quad (13)$$
$$\int_{x_j-\varepsilon_1}^{x_j-\varepsilon_2} dx_{c_1} \int_{x_{f_1}}^{x_{f_2}} dx_f \int_{x_{e_1}}^{x_{e_2}} dx_e \int_{x_{d_1}}^{x_{d_2}} dx_d \int_{x_{c_1}}^{x_{c_2}} dx_c \int_{x_e}^{x_f} dx_b \int_{x_a}^{x_d} dx_a \int_{x_a}^{x_b} u_x \, dx$$
$$+ \int_{x_j+\varepsilon_7}^{x_j+\varepsilon_8} dx_{f_2} \int_{x_j+\varepsilon_6}^{x_j+\varepsilon_7} dx_{f_1} \int_{x_j+\varepsilon_5}^{x_j+\varepsilon_6} dx_{e_2} \int_{x_j}^{x_j+\varepsilon_5} dx_{e_1} \int_{x_j-\varepsilon_4}^{x_j} dx_{d_2} \int_{x_j-\varepsilon_3}^{x_j-\varepsilon_4} dx_{d_1} \int_{x_j-\varepsilon_2}^{x_j-\varepsilon_3} dx_{c_2}$$
$$\int_{x_j-\varepsilon_1}^{x_j-\varepsilon_2} dx_{c_1} \int_{x_{f_1}}^{x_{f_2}} dx_f \int_{x_{e_1}}^{x_{e_2}} dx_e \int_{x_{d_1}}^{x_{d_2}} dx_d \int_{x_{c_1}}^{x_{c_2}} dx_c \int_{x_e}^{x_f} dx_b \int_{x_a}^{x_d} dx_a \int_{x_a}^{x_b} uu_x \, dx$$
$$+ \int_{x_j+\varepsilon_7}^{x_j+\varepsilon_8} dx_{f_2} \int_{x_j+\varepsilon_6}^{x_j+\varepsilon_7} dx_{f_1} \int_{x_j+\varepsilon_5}^{x_j+\varepsilon_6} dx_{e_2} \int_{x_j}^{x_j+\varepsilon_5} dx_{e_1} \int_{x_j-\varepsilon_4}^{x_j} dx_{d_2} \int_{x_j-\varepsilon_3}^{x_j-\varepsilon_4} dx_{d_1} \int_{x_j-\varepsilon_2}^{x_j-\varepsilon_3} dx_{c_2}$$
$$\int_{x_j-\varepsilon_1}^{x_j-\varepsilon_2} dx_{c_1} \int_{x_{f_1}}^{x_{f_2}} dx_f \int_{x_{e_1}}^{x_{e_2}} dx_e \int_{x_{d_1}}^{x_{d_2}} dx_d \int_{x_{c_1}}^{x_{c_2}} dx_c \int_{x_e}^{x_f} dx_b \int_{x_a}^{x_d} dx_a \int_{x_a}^{x_b} u_{xxx} \, dx = 0$$

We then use Lagrange interpolation to approximate $u(x_j \pm \varepsilon_i, t)(i = 1, 2, \cdots, 8)$, because they aren't defined on grid nodes. In addition, to obtain a high-precision numerical scheme, the following Lagrange interpolation polynomials are used.

$$\begin{aligned}u(x,t) &= \frac{(x-x_j)(x-x_{j+1})}{2h^2}u_{j-1}(t) - \frac{(x-x_{j-1})(x-x_{j+1})}{h^2}u_j(t) \\ &\quad + \frac{(x-x_{j-1})(x-x_j)}{2h^2}u_{j+1}(t) + O(h^3)\end{aligned} \quad (14)$$

$$\begin{aligned}u(x,t) &= \frac{(x-x_{j-1})(x-x_{j+1})(x-x_{j+2})}{12h^3}u_{j-2}(t) \\ &\quad - \frac{(x-x_{j-2})(x-x_{j+1})(x-x_{j+2})}{6h^4}u_{j-1}(t) \\ &\quad + \frac{(x-x_{j-2})(x-x_{j-1})(x-x_{j+2})}{6h^4}u_{j+1}(t) \\ &\quad - \frac{(x-x_{j-2})(x-x_{j-1})(x-x_{j+1})}{12h^4}u_{j+2}(t) + O(h^4)\end{aligned} \quad (15)$$

and

$$\begin{aligned}u(x,t) &= \frac{(x-x_{j-1})(x-x_j)(x-x_{j+1})(x-x_{j+2})}{24h^4}u_{j-2}(t) \\ &\quad - \frac{(x-x_{j-2})(x-x_j)(x-x_{j+1})(x-x_{j+2})}{6h^4}u_{j-1}(t) \\ &\quad + \frac{(x-x_{j-2})(x-x_{j-1})(x-x_{j+1})(x-x_{j+2})}{4h^4}u_j(t) \\ &\quad - \frac{(x-x_{j-2})(x-x_{j-1})(x-x_j)(x-x_{j+2})}{6h^4}u_{j+1}(t) \\ &\quad + \frac{(x-x_{j-2})(x-x_{j-1})(x-x_j)(x-x_{j+1})}{24h^4}u_{j+2}(t) + O(h^5)\end{aligned} \quad (16)$$

Secondly, in the temporal direction, we use center difference,

$$\left(u_j^{n+\frac{1}{2}}\right)_t = \frac{u_j^{n+1} - u_j^n}{\tau} + O(\tau^2) \quad (17)$$

to approximate the one-order derivative. Then, the numerical scheme will possess two-order accuracy in the temporal direction.

With the 15-time integral, Lagrange interpolation, and center difference, we obtain a series of numerical schemes with eight parameters, $\varepsilon_i (i = 1, 2, \cdots, 8)$. As soon as we identify the eight parameters, we obtain a specific scheme. In fact, finally, we want to obtaina specific scheme that can keep some properties of the original PDE, such as the conservative property.

2.3. A Two-Level Implicit Nonlinear Discrete Scheme

According to the specific steps introduced above, to retain theenergy conservative property of problem (3)–(5), we choose $\varepsilon_1 = -\varepsilon_4 = -\varepsilon_5 = \varepsilon_8 = \sqrt{3}h$ and $\varepsilon_2 = -\varepsilon_3 = -\varepsilon_6 = \varepsilon_7 = \sqrt{3}h/3$. Now, let us substitute the eight parameters and (17) into (13). After simplifying, we obtain a two-level implicit nonlinear discrete scheme for (3)–(5). This is presented below.

$$\begin{aligned}&\frac{1}{9}\left(\left(u_{j-1}^{n+\frac{1}{2}}\right)_{\hat{t}} + 7\left(u_j^{n+\frac{1}{2}}\right)_{\hat{t}} + \left(u_{j+1}^{n+\frac{1}{2}}\right)_{\hat{t}}\right) + \left(u_j^{n+\frac{1}{2}}\right)_{xx\bar{x}\bar{x}\hat{t}} + \left(u_j^{n+\frac{1}{2}}\right)_{\hat{x}} \\ &+ \frac{1}{3}\left(u_j^{n+\frac{1}{2}}\right)_{\hat{x}}\left(u_{j-1}^{n+\frac{1}{2}} + u_j^{n+\frac{1}{2}} + u_{j+1}^{n+\frac{1}{2}}\right) + \left(u_j^{n+\frac{1}{2}}\right)_{x\bar{x}\hat{x}} = 0, \\ &\quad 1 \leq j \leq J-1, \ 0 \leq n \leq N-1\end{aligned} \quad (18)$$

$$u_j^0 = u_0(x_j), \ 1 \leq j \leq J-1 \quad (19)$$

$$u_0^n = u_J^n = 0, \ (u_0^n)_x = \left(u_J^n\right)_x = 0, \ 0 \leq n \leq N-1 \quad (20)$$

2.4. Conservative Law of the Discrete Scheme

Theorem 2. *The two-level implicit nonlinear numerical scheme (18) possesses the following property:*

$$E^n = \frac{7}{9}\|u^n\|^2 + \frac{2h}{9}\sum_{j=0}^{J-1} u_j^n u_{j+1}^n + \|u_{xx}^n\|^2 = E^{n-1} = \ldots = E^0 \quad (21)$$

Proof. Computing the inner product of (18) with $2u^{n+\frac{1}{2}}$ $(i.e. u^{n+1} + u^n)$, we have

$$\frac{7}{9\tau}\left(\|u^{n+1}\|^2 - \|u^n\|^2\right) + \frac{2h}{9\tau}\left(\sum_{j=0}^{J-1} u_j^{n+1} u_{j+1}^{n+1} - \sum_{j=0}^{J-1} u_j^n u_{j+1}^n\right) \\ + \frac{1}{\tau}\left(\|u_{xx}^{n+1}\|^2 - \|u_{xx}^n\|^2\right) + \left(\varphi(u^{n+\frac{1}{2}}), 2u^{n+\frac{1}{2}}\right) = 0 \quad (22)$$

Let $E^n = \frac{7}{9}\|u^n\|^2 + \frac{2h}{9}\sum_{j=0}^{J-1} u_j^n u_{j+1}^n + \|u_{xx}^n\|^2$. Applying Lemma 3, we have

$$E^{n+1} = E^n. \quad (23)$$

Thus, we obtain $E^n = \cdots = E^0$, which proves Theorem 2. It shows that this numerical scheme can retain the conservation property of the original PDE. □

3. Solvability

The following lemmas will be very helpful for proving the solvability of the discrete scheme (17)–(19).

Lemma 4. *Ref [22] Let H be a finite-dimensional inner product space; suppose that $g : H \to H$, is continuous and there exists an $\alpha > 0$ such that $(g(x), x) > 0$ for all $x \in H$ with $\|x\| = \alpha$. Then there is $x^* \in H$ such that $g(x^*) = 0$ and $\|x^*\| \le \alpha$.*

It is a classic theory and comes from the paper *Existence and uniqueness theorems for solutions of nonlinear boundary value problems*. This article was published in the *Proceedings of Symposia in Applied Mathematics* in 1965.

Lemma 5. $2M - E$ *is a positive definite matrix, where E is an identity matrix and*

$$M = \begin{bmatrix} 1 & 0 & 0 & \cdots & 0 & 0 & 0 \\ 0 & 7 & 1 & \cdots & 0 & 0 & 0 \\ 0 & 1 & 7 & \cdots & 0 & 0 & 0 \\ \vdots & \vdots & \vdots & \ddots & \vdots & \vdots & \vdots \\ 0 & 0 & 0 & \cdots & 7 & 1 & 0 \\ 0 & 0 & 0 & \cdots & 1 & 7 & 0 \\ 0 & 0 & 0 & \cdots & 0 & 0 & 1 \end{bmatrix}_{(J+1)\times(J+1)}$$

Proof. We know that

$$2M - E = \begin{bmatrix} 1 & 0 & 0 & \cdots & 0 & 0 & 0 \\ 0 & 13 & 2 & \cdots & 0 & 0 & 0 \\ 0 & 2 & 13 & \cdots & 0 & 0 & 0 \\ \vdots & \vdots & \vdots & \ddots & \vdots & \vdots & \vdots \\ 0 & 0 & 0 & \cdots & 13 & 2 & 0 \\ 0 & 0 & 0 & \cdots & 2 & 13 & 0 \\ 0 & 0 & 0 & \cdots & 0 & 0 & 1 \end{bmatrix}_{(J+1)\times(J+1)}$$

Let $P_i (1 \leq i \leq J+1)$ be ordered principal minor determinants of $2M - E$. Obviously, we have $P_1 = 1, P_2 = 13, P_3 = \begin{vmatrix} 1 & 0 & 0 \\ 0 & 13 & 2 \\ 0 & 2 & 13 \end{vmatrix} = 165$, and $P_J = P_{J+1}$. In addition, from $2M - E$, we have

$$P_i = 13P_{i-1} - 4P_{i-2}, \quad 3 \leq i \leq J.$$

So, when $i = 4$, we have $P_4 = 13P_3 - 4P_2 > P_3$. Similarly, when $5 \leq i \leq J$, we have

$$P_J > P_{J-1} > \cdots > P_5 > P_4$$

Then, we have

$$P_{J+1} = P_J > P_{J-1} > \cdots > P_4 > P_3 > P_2 > P_1 > 0$$

Hence, $2M - E$ is a positive definite matrix. □

Theorem 3. *There is a $u^{n+1} \in Z_h^0$ that satisfies the discrete scheme (18)–(20).*

Proof. Suppose that $u^0, u^1, \ldots, u^{n-1}$ and u^n satisfy (18)–(20) for $n \leq N - 1$. Next, we prove that there is a u^{n+1} that satisfies the discrete scheme (18)–(20).

Let g be an operator on Z_h^0 defined by

$$g(v) = \frac{2}{9}A(v - u^n) + 2(v - u^n)_{xx\bar{x}\,\bar{x}} + \tau v_{\hat{x}} + \frac{\tau}{3}v_{\hat{x}}(v_{j-1} + v_j + v_{j+1}) + \tau v_{x\bar{x}\hat{x}} \quad (24)$$

where

$$A = \begin{bmatrix} 1 & 0 & 0 & \cdots & 0 & 0 & 0 \\ 1 & 7 & 1 & \cdots & 0 & 0 & 0 \\ 0 & 1 & 7 & \cdots & 0 & 0 & 0 \\ \vdots & \vdots & \vdots & \ddots & \vdots & \vdots & \vdots \\ 0 & 0 & 0 & \cdots & 7 & 1 & 0 \\ 0 & 0 & 0 & \cdots & 1 & 7 & 1 \\ 0 & 0 & 0 & \cdots & 0 & 0 & 1 \end{bmatrix}_{(J+1)\times(J+1)}, \quad N = \begin{bmatrix} 0 & 0 & 0 & \cdots & 0 & 0 & 0 \\ 1 & 0 & 0 & \cdots & 0 & 0 & 0 \\ 0 & 0 & 0 & \cdots & 0 & 0 & 0 \\ \vdots & \vdots & \vdots & \ddots & \vdots & \vdots & \vdots \\ 0 & 0 & 0 & \cdots & 0 & 0 & 0 \\ 0 & 0 & 0 & \cdots & 0 & 0 & 1 \\ 0 & 0 & 0 & \cdots & 0 & 0 & 0 \end{bmatrix}_{(J+1)\times(J+1)},$$

$$v = \begin{pmatrix} v_0 \\ v_1 \\ v_2 \\ \vdots \\ v_J \end{pmatrix} \in Z_h^0$$

Obviously, g is continuous, $A = M + N$, and $(Nv, v) = v_0 v_1 + v_{J-1} v_J = 0$. Let $\lambda_0, \lambda_1, \cdots, \lambda_J$ be the eigenvalues of M and let $\lambda_{\min} = \{\lambda_0, \lambda_1, \cdots, \lambda_J\}$. Take the inner product of (24) with v. By Lemma 1 and Lemma 3, we have

35

$$
\begin{aligned}
(g(v), v) &= \tfrac{2}{9}(Av, v) - \tfrac{2}{9}(Au^n, v) + 22\|v_{xx}\|^2 - 22(v_{xx}, u^n_{xx}) \\
&\geq \tfrac{2}{9}(Mv, v) - \tfrac{2}{9}\|Au^n\| \cdot \|v\| + 2\|v_{xx}\|^2 - 2\|v_{xx}\| \cdot \|u^n_{xx}\| \\
&\geq \tfrac{2}{9}(\lambda_0 v_0^2 + \cdots + \lambda_J v_J^2) - \tfrac{1}{9}\|Au^n\|^2 - \tfrac{1}{9}\|v\|^2 + \|v_{xx}\|^2 - \|u^n_{xx}\|^2 \\
&\geq \tfrac{(2\lambda_{\min}-1)}{9}\|v\|^2 - \tfrac{1}{9}\|Au^n\|^2 - \|u^n_{xx}\|^2
\end{aligned}
\tag{25}
$$

From Lemma 2, we can guarantee that $2\lambda_{\min} - 1 > 0$. Therefore, let

$$
\|v\|^2 = \frac{\|Au^n\|^2 + 9\|u^n_{xx}\|^2 + 1}{2\lambda_{\min} - 1}
\tag{26}
$$

For all $v \in Z_h^0$, we have $(g(v), v) > 0$. From Lemma 4, there is a $v^* = \frac{u^n + u^{n+1}}{2} \in Z_h^0$ such that $g(v^*) = 0$. So, there is a $u^{n+1} = 2v^* - u^n$ that satisfies the scheme (18)–(20). □

4. Some Prior Estimates for the Discrete Scheme

Lemma 6. *Suppose that $u_0 \in H_0^2[x_l, x_r]$; then the solution of (3)–(5) satisfies*

$$
\|u\| \leq C, \ \|u_x\| \leq C, \ \|u\|_\infty \leq C, \ \|u_x\|_\infty \leq C
\tag{27}
$$

Proof. From (16), we have

$$
\|u\| \leq C, \ \|u_{xx}\| \leq C
\tag{28}
$$

Then, by the Holder inequality and the Schwarz inequality, we obtain

$$
\begin{aligned}
\|u_x\|^2 &= \int_{x_l}^{x_r} u_x u_x dx = uu_x\big|_{x_l}^{x_r} - \int_{x_l}^{x_r} uu_{xx} dx = -\int_{x_l}^{x_r} uu_{xx} dx \\
&\leq \|u\| \cdot \|u_{xx}\| \leq \tfrac{1}{2}\big(\|u\|^2 + \|u_{xx}\|^2\big)
\end{aligned}
\tag{29}
$$

Thus, $\|u_x\| \leq C$. By the Sobolev inequality we have $\|u\|_\infty \leq C$, $\|u_x\|_\infty \leq C$. □

Lemma 7. *[Discrete Sobolev Inequality] [22]. There are two constants C_1 and C_2 such that*

$$
\|\mathbf{u}^n\|_\infty \leq C_1 \|\mathbf{u}^n\| + C_2 \|\mathbf{u}^n_x\|
\tag{30}
$$

Lemma 8. *Assume that $u \in Z_h^0$; then the solution of the discrete scheme (18)–(20) satisfies*

$$
\|u^n_{xx}\| \leq C, \ \|u^n\| \leq C, \ \|u^n_x\| \leq C, \|u^n\|_\infty \leq C, \|u^n_x\|_\infty \leq C.
\tag{31}
$$

Proof. From (21) we have

$$
\|u^n_{xx}\| \leq C, \|u^n\| \leq C
\tag{32}
$$

By Lemma 1 and the Cauchy–Schwarz inequality, we obtain

$$
\|u_x\|^2 \leq \|u^n\| \cdot \|u^n_{xx}\| \leq \frac{1}{2}\Big(\|u^n_{xx}\|^2 + \|u^n\|^2\Big) \leq C
$$

Applying Lemma 7, we also obtain

$$
\|u^n\|_\infty \leq C, \|u^n_x\|_\infty \leq C.
$$

□

5. Convergence and Stability of the Discrete Scheme

Let $v_j^{n+\frac{1}{2}} = v\left(x_j, t^{n+\frac{1}{2}}\right)$ be the solution of (3)–(5). By substituting this into (18), we obtain the truncation error of scheme (17)–(19)

$$Er_j^{n+\frac{1}{2}} = \frac{1}{9}\left(\left(v_{j-1}^{n+\frac{1}{2}}\right)_{\hat{t}} + 7\left(v_j^{n+\frac{1}{2}}\right)_{\hat{t}} + \left(v_{j+1}^{n+\frac{1}{2}}\right)_{\hat{t}}\right) + \left(v_j^{n+\frac{1}{2}}\right)_{x\bar{x}\bar{x}\hat{t}} + \left(v_j^{n+\frac{1}{2}}\right)_{\hat{x}} \quad (33)$$
$$+ \frac{1}{3}\left(v_j^{n+\frac{1}{2}}\right)_{\hat{x}}\left(v_{j-1}^{n+\frac{1}{2}} + v_j^{n+\frac{1}{2}} + v_{j+1}^{n+\frac{1}{2}}\right) + \left(v_j^{n+\frac{1}{2}}\right)_{x\bar{x}\hat{x}}$$

By Taylor expansion and Lagrange interpolation, we know that $Er_j^{n+\frac{1}{2}} = O(\tau^2 + h^3)$.

Theorem 4. *Suppose $u_0 \in H_0^2[x_l, x_r]$ and $u(x,t) \in C^{5,3}$; then the numerical solution u_j^n of scheme (17)–(19) converges to the solution $v_j^{n+\frac{1}{2}}$ of the initia lboundary value problem (3)–(5) with order $O(\tau^2 + h^3)$ by the norm $\|\cdot\|_\infty$.*

Proof. Let $e_j^{n+\frac{1}{2}} = v_j^{n+\frac{1}{2}} - u_j^{n+\frac{1}{2}}$ and subtract (18) from (33); we then have

$$Er_j^{n+\frac{1}{2}} = \frac{1}{9}\left(\left(e_{j-1}^{n+\frac{1}{2}}\right)_{\hat{t}} + 7\left(e_j^{n+\frac{1}{2}}\right)_{\hat{t}} + \left(e_{j+1}^{n+\frac{1}{2}}\right)_{\hat{t}}\right) + \left(e_j^{n+\frac{1}{2}}\right)_{x\bar{x}\bar{x}\hat{t}} \quad (34)$$
$$+ \left(e_j^{n+\frac{1}{2}}\right)_{\hat{x}} + \left(e_j^{n+\frac{1}{2}}\right)_{x\bar{x}\hat{x}} + (\varphi(v_j^{n+\frac{1}{2}}) - \varphi(u_j^{n+\frac{1}{2}})).$$

Computing the inner product of (34) with $2e^{n+\frac{1}{2}}$ (i.e. $e^{n+1} + e^n$), we have

$$\left(Er_j^{n+\frac{1}{2}}, 2e^{n+\frac{1}{2}}\right) = \frac{7}{9\tau}\|e^{n+1}\|^2 + \frac{2h}{9\tau}\sum_{j=1}^{J-1} e_j^{n+1}e_{j+1}^{n+1} - \frac{7}{9\tau}\|e^n\|^2 - \frac{2h}{9\tau}\sum_{j=1}^{J-1} e_j^n e_{j+1}^n \quad (35)$$
$$+ \frac{1}{\tau}\left(\|e_{xx}^{n+1}\|^2 - \|e_{xx}^n\|^2\right) + \left(\varphi(v^{n+\frac{1}{2}}) - \varphi(u^{n+\frac{1}{2}}), 2e^{n+\frac{1}{2}}\right)$$

From Lemmas 6 and 7 and the Cauchy–Schwarz inequality, we obtain

$$-\left(\varphi(v^{n+\frac{1}{2}}) - \varphi(u^{n+\frac{1}{2}}), 2e^{n+\frac{1}{2}}\right)$$
$$= -\frac{2}{3}h\sum_{j=1}^{J-1}\left(v_{j-1}^{n+\frac{1}{2}} + v_j^{n+\frac{1}{2}} + v_{j+1}^{n+\frac{1}{2}}\right)\left(v_j^{n+\frac{1}{2}}\right)_{\hat{x}} e_j^{n+\frac{1}{2}} + \frac{2}{3}h\sum_{j=1}^{J-1}\left(u_{j-1}^{n+\frac{1}{2}} + u_j^{n+\frac{1}{2}} + u_{j+1}^{n+\frac{1}{2}}\right)\left(u_j^{n+\frac{1}{2}}\right)_{\hat{x}} e_j^{n+\frac{1}{2}}$$
$$= -\frac{2}{3}h\sum_{j=1}^{J-1}\left(e_{j-1}^{n+\frac{1}{2}} + e_j^{n+\frac{1}{2}} + e_{j+1}^{n+\frac{1}{2}}\right)\left(v_j^{n+\frac{1}{2}}\right)_{\hat{x}} e_j^{n+\frac{1}{2}} + \frac{2}{3}h\sum_{j=1}^{J-1}\left(u_{j-1}^{n+\frac{1}{2}} + u_j^{n+\frac{1}{2}} + u_{j+1}^{n+\frac{1}{2}}\right)\left(u_j^{n+\frac{1}{2}}\right)_{\hat{x}} e_j^{n+\frac{1}{2}}$$
$$- \frac{2}{3}h\sum_{j=1}^{J-1}\left(u_{j-1}^{n+\frac{1}{2}} + u_j^{n+\frac{1}{2}} + u_{j+1}^{n+\frac{1}{2}}\right)\left(v_j^{n+\frac{1}{2}}\right)_{\hat{x}} e_j^{n+\frac{1}{2}} \quad (36)$$
$$= -\frac{2}{3}h\sum_{j=1}^{J-1}\left(e_{j-1}^{n+\frac{1}{2}} + e_j^{n+\frac{1}{2}} + e_{j+1}^{n+\frac{1}{2}}\right)\left(v_j^{n+\frac{1}{2}}\right)_{\hat{x}} e_j^{n+\frac{1}{2}} - \frac{2}{3}h\sum_{j=1}^{J-1}\left(u_{j-1}^{n+\frac{1}{2}} + u_j^{n+\frac{1}{2}} + u_{j+1}^{n+\frac{1}{2}}\right)\left(e_j^{n+\frac{1}{2}}\right)_{\hat{x}} e_j^{n+\frac{1}{2}}$$
$$\leq \frac{2}{3}Ch\sum_{j=1}^{J-1}\left(\left|e_{j-1}^{n+\frac{1}{2}}\right| + \left|e_j^{n+\frac{1}{2}}\right| + \left|e_{j+1}^{n+\frac{1}{2}}\right|\right)\left|e_j^{n+\frac{1}{2}}\right| + \frac{2}{3}Ch\sum_{j=1}^{J-1}\left|\left(e_j^{n+\frac{1}{2}}\right)_{\hat{x}}\right|\left|e_j^{n+\frac{1}{2}}\right|$$
$$\leq C\left(\|e^{n+1}\|^2 + \|e^n\|^2 + \|e_{\hat{x}}^{n+1}\|^2 + \|e_{\hat{x}}^n\|^2\right)$$

In addition, we have

$$\left(Er_j^{n+\frac{1}{2}}, 2e^{n+\frac{1}{2}}\right) = \left(Er_j^{n+\frac{1}{2}}, e^{n+1} + e^n\right) \leq \|Er^{n+\frac{1}{2}}\|^2 + \frac{\|e^{n+1}\|^2 + \|e^n\|^2}{2} \quad (37)$$

Substituting (36) and (37) into (35), with Lemma 3, we have

$$\frac{7}{9}\|e^{n+1}\|^2 + \frac{2h}{9}\sum_{j=1}^{J-1} e_j^{n+1} e_{j+1}^{n+1} - \frac{7}{9}\|e^n\|^2 - \frac{2h}{9}\sum_{j=1}^{J-1} e_j^n e_{j+1}^n + \|e_{xx}^{n+1}\|^2 - \|e_{xx}^n\|^2$$
$$\leq \tau\|Er^{n+\frac{1}{2}}\|^2 + C\tau\left(\|e^{n+1}\|^2 + \|e^n\|^2 + \|e_x^{n+1}\|^2 + \|e_x^n\|^2 + \|e_{xx}^n\|^2\|e_{xx}^{n+1}\|^2\right) \quad (38)$$

Let $B^n = \frac{7}{9}\|e^n\|^2 + \frac{2h}{9}\sum_{i=1}^{J-1} e_i^n e_{i+1}^n + \|e_{xx}^n\|^2 + \|e_x^n\|^2$. Obviously, $B^0 = 0$. Then, (38) can be rewritten as

$$B^{n+1} - B^n \leq \tau\|Er^{n+\frac{1}{2}}\|^2 + C\tau\left(B^{n+1} + B^n\right) \quad (39)$$

When τ is sufficiently small that $1 - C\tau > 0$, we have

$$B^{n+1} \leq \frac{1+C\tau}{1-C\tau} B^n + \frac{\tau}{1-C\tau}\|Er^{n+\frac{1}{2}}\|^2 \leq \frac{\tau}{1-C\tau}\sum_{k=0}^n \left(\frac{1+C\tau}{1-C\tau}\right)^{n-k}\|Er^{k+\frac{1}{2}}\|^2$$
$$\leq O^2(\tau^2 + h^3)\sum_{k=1}^{n+1}\left(\frac{1+C\tau}{1-C\tau}\right)^k \quad (40)$$

□

Then we have

$$B^n \leq O^2(\tau^2 + h^3)\sum_{k=1}^n \left(\frac{1+C\tau}{1-C\tau}\right)^k \leq O^2(\tau^2 + h^3)\sum_{k=1}^n \left(1 + \frac{2C\tau}{1-C\tau}\right)^k \leq O^2(\tau^2 + h^3)$$

That is, $\|e^n\| \leq O(\tau^2 + h^3)$, $\|e_x^n\| \leq O(\tau^2 + h^3)$. Using Lemma 8, we have

$$\|e^n\|_\infty \leq O(\tau^2 + h^3) \quad (41)$$

Similarly, we can prove the following theorem.

Theorem 5. *Under the conditions of Theorem 4, the solution u_j^n of discrete scheme (18)–(20) is unconditionally stable by the norm $\|\cdot\|_\infty$.*

6. Uniqueness of the Numerical Solution

Theorem 6. *The solution of the discrete scheme (18)–(20) is unique.*

Proof. We assume that u^n and w^n are two different solutions of (18)–(20). Let $S_j^{n+\frac{1}{2}} = w_j^{n+\frac{1}{2}} - u_j^{n+\frac{1}{2}}$. Then, we have

$$\frac{1}{9}\left(\left(S_{j-1}^{n+\frac{1}{2}}\right)_{\hat{t}} + 7\left(S_j^{n+\frac{1}{2}}\right)_{\hat{t}} + \left(S_{j+1}^{n+\frac{1}{2}}\right)_{\hat{t}}\right) + \left(S_j^{n+\frac{1}{2}}\right)_{xx\bar{x}\hat{t}} + \left(S_j^{n+\frac{1}{2}}\right)_{\hat{x}} + \left(S_j^{n+\frac{1}{2}}\right)_{x\bar{x}\hat{x}}$$
$$+ \frac{1}{3}\left(w_j^{n+\frac{1}{2}}\right)_{\hat{x}}\left(w_{j-1}^{n+\frac{1}{2}} + w_j^{n+\frac{1}{2}} + w_{j+1}^{n+\frac{1}{2}}\right) - \frac{1}{3}\left(u_j^{n+\frac{1}{2}}\right)_{\hat{x}}\left(u_{j-1}^{n+\frac{1}{2}} + u_j^{n+\frac{1}{2}} + u_{j+1}^{n+\frac{1}{2}}\right) = 0 \quad (42)$$

By computing the inner product of (42) with $2S^{n+\frac{1}{2}}$ (*i.e.*$S^{n+1} + S^n$), we obtain

$$\frac{7}{9\tau}\|S^{n+1}\|^2 + \frac{2h}{9\tau}\sum_{j=1}^{J-1} S_j^{n+1} S_{j+1}^{n+1} - \frac{7}{9\tau}\|S^n\|^2 - \frac{2h}{9\tau}\sum_{j=1}^{J-1} S_j^n S_{j+1}^n$$
$$+ \frac{1}{\tau}\left(\|S_{xx}^{n+1}\|^2 - \|S_{xx}^n\|^2\right) + \left(\varphi(w) - \varphi(u), 2e^{n+\frac{1}{2}}\right) = 0 \quad (43)$$

Let $Z^n = \frac{7}{9}\|S^n\|^2 + \frac{2h}{9}\sum_{j=1}^{J-1} S_j^n S_{j+1}^n + \|S_{xx}^n\|^2 + \|S_x^n\|^2$; we know that $Z^0 = 0$. From (43) we obtain

$$Z^{n+1} - Z^n \leq C\tau\left(\|S^{n+1}\|^2 + \|S^n\|^2 + \|S_{xx}^{n+1}\|^2 + \|S_{xx}^n\|^2 + \|S_{\hat{x}}^{n+1}\|^2 + \|S_{\hat{x}}^n\|^2\right) \quad (44)$$

Similarly, while $1 - 2C\tau > 0$, we have

$$Z^{n+1} \leq (1+\beta\tau)Z^n \leq \cdots \leq (1+\beta\tau)^{n+1} Z^0 = 0 \quad (45)$$

Hence, we have $\|S^n\|^2 = 0$, where $\beta = \frac{4C}{1-2C\tau}$. This implies that $u^n = w^n$. The discrete scheme (18)–(20) is thus uniquely solvable. □

7. Results
7.1. Example

We consider the Rosenau-KdV equation

$$u_t + u_{xxxxt} + u_x + uu_x + u_{xxx} = 0, \quad (x,t) \in [-40, 40] \times [0, 10] \quad (46)$$

with initial condition

$$u(x,0) = \left(\frac{35}{312}\sqrt{313} - \frac{35}{24}\right)\text{sech}^4\left[\frac{1}{24}\sqrt{2\sqrt{313} - 26}x\right], x \in [-40, 40] \quad (47)$$

and boundary conditions

$$u(-40, t) = u(40, t) = 0, \quad u_x(-40, t) = u_x(40, t) = 0, \, t \in [0, 10] \quad (48)$$

The exact solution is given by

$$u(x, t) = \left(\frac{35}{312}\sqrt{313} - \frac{35}{24}\right)\text{sech}^4\left\{\frac{\sqrt{2\sqrt{313}-26}}{24}\left[x - \left(\frac{1}{2} + \frac{1}{26}\sqrt{313}\right)t\right]\right\} \quad (49)$$

7.2. Figures, Tables, and Schemes

We discretize the problem (46)–(48) using the numerical scheme (18)–(20).

From Figures 1–6, we can see that the numerical solution is consistent with the exact solution.

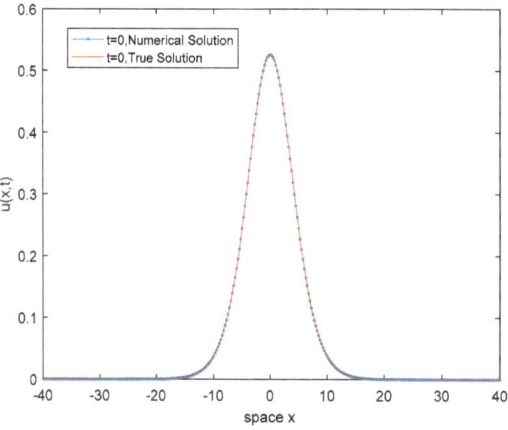

Figure 1. Numerical solution and exact solution with $h = \tau = 1/4$, $t = 0$.

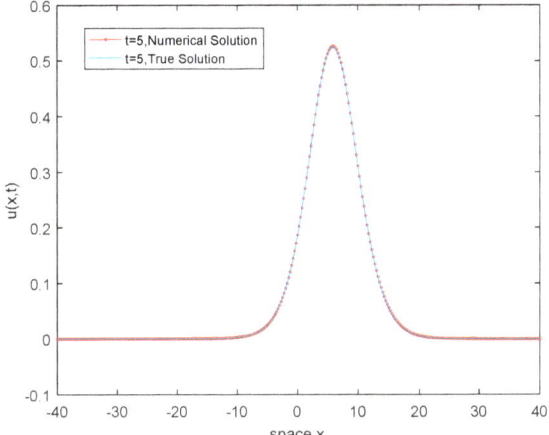

Figure 2. Numerical solution and exact solution with $h = \tau = 1/4$, $t = 5$.

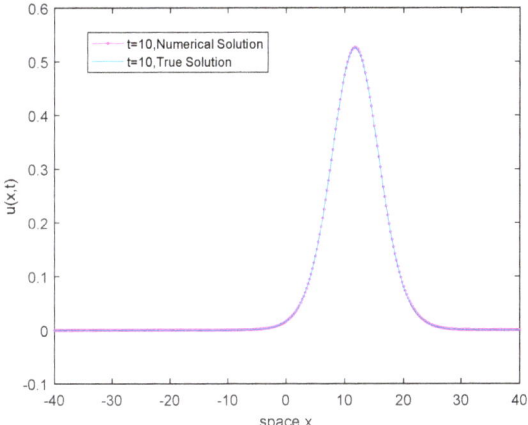

Figure 3. Numerical solution and exact solution with $h = \tau = 1/4$, $t = 10$.

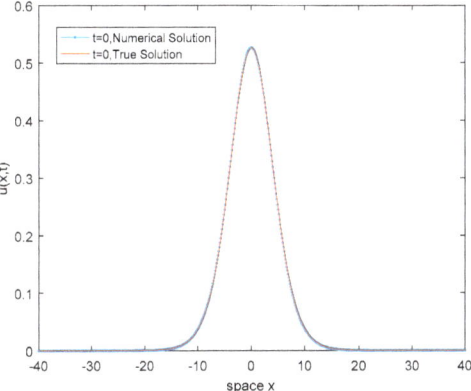

Figure 4. Numerical solution and exact solution with $h = \tau = 1/8$, $t = 0$.

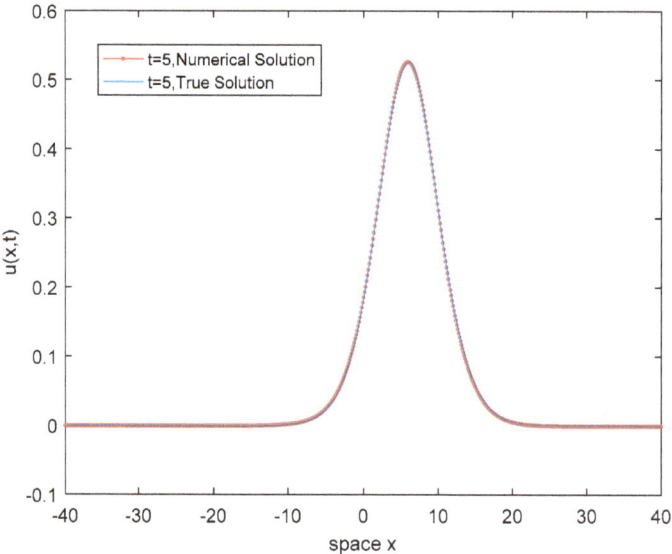

Figure 5. Numerical solution and exact solution with $h = \tau = 1/8$, $t = 5$.

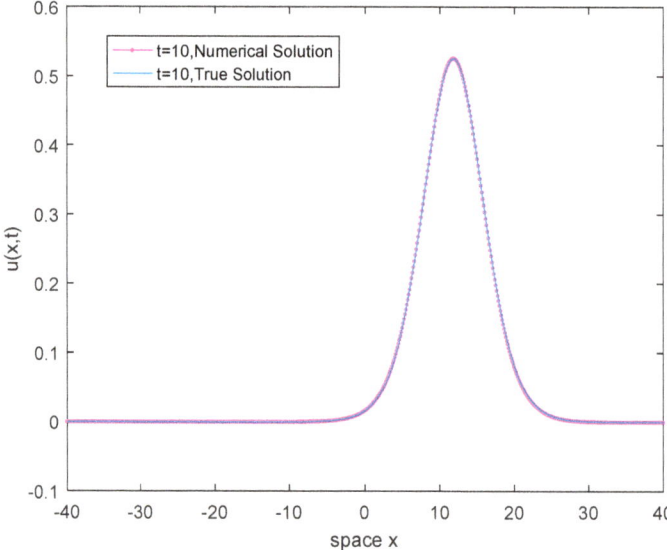

Figure 6. Numerical solution and exact solution with $h = \tau = 1/8$, $t = 10$.

In Table 1, the errors with various h and τ are given. It is obvious that the errors are reducing with decreasing h and τ. Hence, our discrete scheme is reasonable. $\|e^n(h, \tau)\|/\|e^n(h/2, \tau/2)\|$ and $\|e^n(h, \tau)\|_\infty/\|e^n(h/2, \tau/2)\|_\infty$ are given in Table 2, which interprets the convergence rates of the numerical scheme with various h and τ and various norms. From Table 3, we can see that the discrete E_n is conservative. This property is consistent with the original equation. The numerical experiment shows that our discrete scheme is efficient.

Table 1. The errors at different times with various h and τ.

	$h=\tau=1/4$		$h=\tau=1/8$	
	$\|e^n\|_\infty$	$\|e^n\|$	$\|e^n\|_\infty$	$\|e^n\|$
$t=2$	$2.31861708 \times 10^{-4}$	$1.23912354 \times 10^{-3}$	$5.80241511 \times 10^{-5}$	$4.38690718 \times 10^{-4}$
$t=4$	$4.68968494 \times 10^{-4}$	$2.46559087 \times 10^{-3}$	$1.17406720 \times 10^{-4}$	$8.73135975 \times 10^{-4}$
$t=6$	$7.06502002 \times 10^{-4}$	$3.66896388 \times 10^{-3}$	$1.77078291 \times 10^{-4}$	$1.30146495 \times 10^{-3}$
$t=8$	$9.33077015 \times 10^{-4}$	$4.84157574 \times 10^{-3}$	$2.34850462 \times 10^{-4}$	$1.72786004 \times 10^{-3}$
$t=10$	$1.14711186 \times 10^{-3}$	$5.97898965 \times 10^{-3}$	$2.91676007 \times 10^{-4}$	$2.16772647 \times 10^{-3}$

Table 2. The convergence rates with various h and τ and various norms.

	$\|e^n(h,\tau)\|/\|e^n(h/2,\tau/2)\|$			$\|e^n(h,\tau)\|_\infty/\|e^n(h/2,\tau/2)\|_\infty$		
	$\tau=h=1/2$	$\tau=h=1/4$	$\tau=h=1/8$	$\tau=h=1/2$	$\tau=h=1/4$	$\tau=h=1/8$
$t=2$	—	2.81334134	2.82459484	—	3.98408169	3.99595175
$t=4$	—	2.81287147	2.82383379	—	3.96857153	3.99439224
$t=6$	—	2.81210118	2.81910310	—	3.98033046	3.98977197
$t=8$	—	2.81112183	2.80206476	—	3.96316463	3.97306867
$t=10$	—	2.81001148	2.75818454	—	3.97437701	3.93282902

Table 3. Discrete E_n values at different times with various h and τ.

	$h=\tau=1/2$	$h=\tau=1/4$	$h=\tau=1/8$
$t=2$	3.08675012	6.17349199	12.34697937
$t=4$	3.08676651	6.17350095	12.34698364
$t=6$	3.08679087	6.17351432	12.34698593
$t=8$	3.08681918	6.17352996	12.34696462
$t=10$	3.08684844	6.17354622	12.34685552

8. Conclusions

In this paper, a second-order implicit nonlinear discrete scheme for the Rosenau-KdV equation is proposed via the multiple integral finite volume method (MIFVM). The discrete scheme possesses the conservative property of the original equation. The solvability, uniqueness, convergence, and unconditional stability of the scheme were demonstrated in detail. Numerical experiments verified that the discrete scheme given by MIFVM is effective.

Author Contributions: Y.L. analyzed and interpreted the new numerical method, the multiple integral finite volume method. C.G. and Y.W. obtained this conservative nonlinear implicit numerical scheme and demonstrated the existence, uniqueness, convergence, and stability of this numerical scheme. Y.W. proved that the numerical scheme maintained the energy property of the original equation and verified the feasibility of this numerical scheme with a numerical experiment. All authors have read and agreed to the published version of the manuscript.

Funding: This work was supported by the National Natural Science Foundation of China (No. 11526064) and the Fundamental Research Fund for the Central Universities (No. 3072020CF2408).

Institutional Review Board Statement: Not applicable.

Informed Consent Statement: Not applicable.

Data Availability Statement: Not applicable.

Acknowledgments: The authors are grateful to Sun Yue for her work and help. I thank the referees for their valuable work in their paper.

Conflicts of Interest: The authors' interpretation of data or presentation of information wasn't influenced by any personal or other organizations. The interpretation and presentation will not influence the interests of any personal or other organizations, either.

References

1. Yokus, A.; Bulut, H. Numerical simulation of KdV equation by finite difference method. *Indian J. Phys.* **2018**, *92*, 1571–1575. [CrossRef]
2. Başhan, A. A novel approach via mixed Crank–Nicolson scheme and differential quadrature method for numerical solutions of solitons of mKdV equation. *Pramana-J. Phys.* **2019**, *92*, 84. [CrossRef]
3. Başhan, A.; Yağmurlu, N.M.; Uçar, Y.; Esen, A. A new perspective for the numerical solutions of the cmKdV equation via modified cubic B-spline differential quadrature method. *Int. J. Mod. Phys. C* **2018**, *29*, 1850043. [CrossRef]
4. Kong, D.; Xu, Y.; Zheng, Z. Numerical method for generalized time fractional KdV-type equation. *Numer. Methods Partial. Differ. Equ.* **2019**, *36*, 906–936. [CrossRef]
5. Cairone, F.; Anandan, P.; Bucolo, M. Nonlinear systems synchronization for modeling two-phase microfluidics flows. *Nonlinear Dyn.* **2017**, *92*, 75–84. [CrossRef]
6. Wazwaz, A.-M. A new integrable equation that combines the KdV equation with the negative-order KdV equation. *Math. Methods Appl. Sci.* **2018**, *41*, 80–87. [CrossRef]
7. Wazwaz, A.-M. Negative-order KdV equations in (3 + 1) dimensions by using the KdV recursion operator. *Waves Random Complex Media* **2017**, *27*, 768–778. [CrossRef]
8. Rosenau, P. A Quasi-Continuous Description of a Nonlinear Transmission Line. *Phys. Scr.* **1986**, *34*, 827–829. [CrossRef]
9. Rosenau, P. Dynamics of Dense Discrete Systems: High Order Effects. *Prog. Theor. Phys.* **1988**, *79*, 1028–1042. [CrossRef]
10. Park, M.A. On the Rosenau equation. *Matemática Aplic. Comput.* **1990**, *9*, 145–152.
11. Erbay, H.; Erbay, S.; Erkip, A. Numerical computation of solitary wave solutions of the Rosenau equation. *Wave Motion* **2020**, *98*, 102618. [CrossRef]
12. Atouani, N.; Ouali, Y.; Omrani, K. Mixed finite element methods for the Rosenau equation. *J. Appl. Math. Comput.* **2018**, *57*, 393–420. [CrossRef]
13. Michihisa, H. New asymptotic estimates of solutions for generalized Rosenau equations. *Math. Methods Appl. Sci.* **2019**, *42*, 4516–4542. [CrossRef]
14. Shi, D.; Jia, X. Superconvergence analysis of the mixed finite element method for the Rosenau equation. *J. Math. Anal. Appl.* **2020**, *481*, 123485. [CrossRef]
15. Safdari-Vaighani, A.; Larsson, E.; Heryudono, A. Radial Basis Function Methods for the Rosenau Equation and Other Higher Order PDEs. *J. Sci. Comput.* **2018**, *75*, 1555–1580. [CrossRef]
16. Wang, Y.; Feng, G. Large-time behavior of solutions to the Rosenau equation with damped term. *Math. Methods Appl. Sci.* **2017**, *40*, 1986–2004. [CrossRef]
17. Wang, X.; Dai, W. A conservative fourth-order stable finite difference scheme for the generalized Rosenau–KdV equation in both 1D and 2D. *J. Comput. Appl. Math.* **2019**, *355*, 310–331. [CrossRef]
18. Luo, Y.; Xu, Y.; Feng, M. Conservative Difference Scheme for Generalized Rosenau-KdV Equation. *Adv. Math. Phys.* **2014**, *2014*, 986098. [CrossRef]
19. Hussain, M.; Haq, S. Numerical simulation of solitary waves of Rosenau–KdV equation by Crank–Nicolson meshless spectral interpolation method. *Eur. Phys. J. Plus* **2020**, *135*, 98. [CrossRef]
20. Kutluay, S.; Karta, M.; Yağmurlu, N.M. Operator time-splitting techniques combined with quintic B-spline collocation method for the generalized Rosenau–KdV equation. *Numer. Methods Partial. Differ. Equ.* **2019**, *35*, 2221–2235. [CrossRef]
21. Browder, F.E. Existence and uniqueness theorems for solutions of nonlinear boundary value problems. In *Proceedings of the Sum of Squares: Theory and Applications*; American Mathematical Society: Providence, RI, USA, 1965; Volume 17, pp. 24–49.
22. Zhou, Y.L. *Applications of Discrete Functional Analysis to the Finite Difference Method*, 1st ed.; International Academic Publishers: Beijing, China, 1991; pp. 3–20.

Article

Acceleration of Boltzmann Collision Integral Calculation Using Machine Learning

Ian Holloway [1,*,†], Aihua Wood [1,*,†] and Alexander Alekseenko [2,†]

1. Department of Mathematics, Air Force Institute of Technology, WPAFB, OH 45433, USA
2. Department of Mathematics, California State University Northridge, Northridge, CA 91330, USA; alexander.alekseenko@csun.edu
* Correspondence: iholloway@riversideresearch.org (I.H.); aihua.wood@afit.edu (A.W.)
† These authors contributed equally to this work.

Abstract: The Boltzmann equation is essential to the accurate modeling of rarefied gases. Unfortunately, traditional numerical solvers for this equation are too computationally expensive for many practical applications. With modern interest in hypersonic flight and plasma flows, to which the Boltzmann equation is relevant, there would be immediate value in an efficient simulation method. The collision integral component of the equation is the main contributor of the large complexity. A plethora of new mathematical and numerical approaches have been proposed in an effort to reduce the computational cost of solving the Boltzmann collision integral, yet it still remains prohibitively expensive for large problems. This paper aims to accelerate the computation of this integral via machine learning methods. In particular, we build a deep convolutional neural network to encode/decode the solution vector, and enforce conservation laws during post-processing of the collision integral before each time-step. Our preliminary results for the spatially homogeneous Boltzmann equation show a drastic reduction of computational cost. Specifically, our algorithm requires $O(n^3)$ operations, while asymptotically converging direct discretization algorithms require $O(n^6)$, where n is the number of discrete velocity points in one velocity dimension. Our method demonstrated a speed up of 270 times compared to these methods while still maintaining reasonable accuracy.

Keywords: Boltzmann equation; machine learning; collision integral; convolutional neural network

Citation: Holloway, I.; Wood, A.; Alekseenko, A. Acceleration of Boltzmann Collision Integral Calculation Using Machine Learning. *Mathematics* **2021**, *9*, 1384. https://doi.org/10.3390/math9121384

Academic Editor: Ioannis K. Argyros

Received: 31 March 2021
Accepted: 4 June 2021
Published: 15 June 2021

Publisher's Note: MDPI stays neutral with regard to jurisdictional claims in published maps and institutional affiliations.

Copyright: © 2021 by the authors. Licensee MDPI, Basel, Switzerland. This article is an open access article distributed under the terms and conditions of the Creative Commons Attribution (CC BY) license (https://creativecommons.org/licenses/by/4.0/).

1. Introduction

While the Euler and Navier–Stokes equations have for a long time been the work horses in the modeling of fluid dynamics, these equations are inadequate for modeling complex flows, such as rarefied gases, for which the continuum assumption is invalid. Rarefied gas flows have become a topic of increasing interest due to their relevance in practical applications such as hypersonic and space vehicles. To accurately capture the true physics of these non-equilibrium flows, analysis of molecular-level interactions is required. As the governing equation of kinetic theory, the Boltzmann equation is key in understanding these interactions, and therefore also critical in aiding the successful design of these flight vehicles, as well as other applications. Unfortunately, and despite the rapid increase in computing power of recent years, numerical solution of this equation continues to present a major challenge. Among the components of the equation, the main driver of computational complexity is the multi-dimensional collision integral. As a result, a plethora of new mathematical and numerical approaches have been proposed in an effort to reduce the computational cost of solving the Boltzmann collision integral.

Fourier-based spectral methods represent a potent approach to deterministic evaluation of the collision integral [1–5]. These methods use uniform meshes in the velocity space and have complexity of $O(n^6)$ operations, where n is the number of discrete velocity points in one velocity dimension. A discontinuous Galerkin discretization with $O(n^6)$ complexity was proposed in [6]. While these algorithms are suitable for simulation of flows in one and two spatial dimensions, they are difficult to use for three dimensional flows. An additional reduction in complexity is achieved in fast spectral methods by leveraging low rank approximate diagonalization of the weighted convolution form of the collision integral [7–11]. Complexities of the fast spectral methods may vary between $O(M_r M^2 n^3 \log n)$ and $O(M^2 n^3 \log n)$ depending on the form of molecular interaction potential. Numbers M^2 and M_r correspond to the numbers of discrete integration points in angular and radial directions used in diagonalization and usually are significantly smaller than n. In spite of the significant improvement in efficiency, simulation of three dimensional flows of gases with internal energies, multi-component gases, and flows in complex geometries still remains challenging for fast spectral methods.

Other algorithms of lower complexities have also been proposed for either special physics or various representations of the approximate solutions, for example, $O(n^4)$ algorithm for evaluating the collision integral in the case of Maxwell's pseudo-molecules [12], and $O(Mn^3 \log n)$ for hard spheres potentials [13,14]. Simulation of gas mixtures and gases with internal energies, as well as multidimensional models can be found in [14–20], and references therein. Other fast methods include representing the solution as a sum of homogeneous Gaussians [21,22], polynomial spectral discretization [23], utilizing non-uniform meshes [24], and a hyperbolic cross approximation [25]. Additional review of recent results can be found in [26,27].

In this paper, we apply machine learning to accelerate the calculation of the Boltzmann collision integral. The results presented are intended to be an initial demonstration of the viability of machine learning to accelerate solution of the problem at hand, more so than to rigorously prove consistency of machine learning techniques with the discretized Boltzmann equation. For our case study, we consider a class of solutions to the problem of spatially homogeneous relaxation computed using deterministic approach of [6]. The considered class of solutions correspond to hard spheres potential, however, we expect that the results can be replicated for other molecular potentials. We build a deep convolutional neural network to encode/decode the solution vector, and enforce conservation laws during post-processing of the collision integral before each time-step. Our preliminary results for the spatially homogeneous Boltzmann equation show a drastic reduction of computational cost, in the order of $O(n^3)$, compared to $O(n^6)$ for direct discretization algorithm of [6].

Specifically, our model would take the numerical solution as input and return a predicted collision integral at every point of the computational domain as output. This model could then directly replace the collision integral calculation in a time stepping simulation without requiring any other aspect of the simulation to change. Due to the fact that machine learning algorithms are to be trained on a specific set of data, the resulting approximate algorithms will be only applicable to the same classes of problem for which the data was generated. Thus, the proposed approaches are intrinsically applicable to a specific set of problems. For that set of problems, however, the methods provide a significant improvement in speed compared to classical methods. This opens an entire new avenue for addressing the computational complexity associated with solving the Boltzmann equation.

Artificial neural networks and machine learning were previously applied to solution of partial differential equations, see, e.g., [28–30], including solution of kinetic equations [31]. Another data driven approach consists of using low rank tensor approximations of kinetic solutions [32]. Commonly, deep neural networks provide low rank representations of solutions in high dimensional spaces while the governing partial differential equations are used to define penalty functions for network training. It should be noted, however, that a direct implementation of the collision integral in a penalty function in a manner the governing equations are used in physics informed networks, is problematic due to extremely high costs of evaluating the collision integral. As a result, in this paper, we focus on learning the collision operator itself, for a class of solutions. The resulting approximation can be, in principle, combined with approaches of [31,32] to provide an inexpensive physically accurate collision operator. To the authors' knowledge, this is the first attempt at using machine learning to accelerate the calculation of the Boltzmann collision integral. Our early results suggest that the approach has potential to drastically advance the state-of-the-art in simulating complex flows or rarefied gas.

The rest of the paper is organized as follows. Section 2 presents the problem setup and the convolutional network structure that enables the dimension reduction. Conservation considerations are described in Section 3. Section 4 is devoted to error analysis. Our test models and results are shown in Section 5. The paper is concluded in Section 6.

2. Problem Setup

In the kinetic approach the gas is described using the molecular velocity distribution function $f(t, \vec{x}, \vec{v})$ which has the property that $f(t, \vec{x}, \vec{v}) d\vec{x} d\vec{v}$ represents the number of molecules that are contained in the box with the volume $d\vec{x}$ around point $\vec{x}$ whose velocities are contained in a box of volume $d\vec{v}$ around point $\vec{v}$. In this work, we are concerned with the solution of the spatially homogeneous flows that correspond to the assumption that the $f(t, \vec{x}, \vec{v})$ is constant in the $\vec{x}$ variable. In this case, the dynamics of the gas is given by the spatially homogeneous Boltzmann Equation (see, for example [33,34]),

$$\frac{\partial}{\partial t} f(t, \vec{v}) = I[f](t, \vec{v}). \tag{1}$$

Here $I[f]$ is the molecular collision operator

$$I[f](t, \vec{x}, \vec{v}) = \int_{\mathbb{R}^3} \int_{\mathbb{S}^2} (f(t, \vec{v}') f(t, \vec{u}') - f(t, \vec{v}) f(t, \vec{u})) B(|g|, \cos\theta) \, d\sigma \, d\vec{u}, \tag{2}$$

where $\vec{v}$ and $\vec{u}$ are the pre-collision velocities of a pair of particles, $\vec{g} = \vec{v} - \vec{u}$, $\mathbb{S}^2$ is a unit sphere in $\mathbb{R}^3$ centered at the origin, $\vec{w}$ is the unit vector connecting the origin and a point on $\mathbb{S}^2$, θ is the deflection angle defined by the equation $\cos\theta = \vec{w} \cdot \vec{g}/|g|$, $d\sigma = \sin\theta \, d\theta d\varepsilon$, where ε is the azimuthal angle that parametrizes $\vec{w}$ together with the angle θ. Vectors $\vec{v}'$ and $\vec{u}'$ are the post collision velocities of a pair of particles and are computed by

$$\vec{v}' = \vec{v} - \frac{1}{2}(\vec{g} - |g|\vec{w}), \qquad \vec{u}' = \vec{v} - \frac{1}{2}(\vec{g} + |g|\vec{w}).$$

Due to the high computational complexity of the collision integral, use of the Boltzmann equation in practice has been limited.

2.1. Class of Solutions and Solution Collection

The class of solutions for which the training data is constructed consists of solutions to the problem of spatially homogeneous relaxation with the initial data given by two homogeneous Gaussian densities. The initial data is normalized so that the velocity distribution function has unit density, zero bulk velocity and, a set temperature. In the simulations presented in this paper, the value of dimensionless temperature of 0.2 was used. The bulk velocities of the homogeneous Gaussian densities have zero v and w components, thus the solutions are radially symmetric in vw velocity plane.

A collection of solutions is computed by randomly generating macroscopic parameters of density, the u components of the bulk velocity, and temperatures of two homogeneous Gaussian densities and solving (1) until a steady state is reached using the method of [6]. The numerically computed velocity distribution functions are saved at multiple instances in time, each save becoming a data point in the collection. We note that due to normalization of the initial data, the steady state is the same for all computed solutions. All solutions were computed on uniform meshes with dimensions of 41 by 41 by 41 in the velocity domain.

2.2. Dimension Reduction

A key component of finding a faster method of calculating the collision integral is finding low dimensional features that adequately characterize the solution. The true dimensionality of the solution data can be demonstrated using the SVD decomposition. The saved solutions are re-arranged as one dimensional arrays f_j. Then f_j are added as rows to the matrix D_{ij}, where index i runs over all saved solutions and index j runs over all discretization points. This process is schematically depicted in Figure 1.

In Figure 1, singular values of matrix D_{ij}, $i = 1, \ldots, P$, $j = 1, \ldots, M^3$ are shown, where $M = 41$ is the number of velocity points in each velocity dimension in the computed solutions and P is the total number of the considered solution saves, $P \approx 5000$. About 100 of cases of initial data is included in the results in Figure 1. It can be seen that the singular values decrease very fast allowing for low rank approximation $\hat{D}_{ij}$ of the data matrix D_{ij}:

$$\hat{D}_{ij} = \sum_{l=1}^{K} \sigma_l \mu_i^l \xi_j^l. \qquad (3)$$

Here σ_l is the l-th singular value, μ_i^l is the l-th left singular vector and ξ_j^l is the l-th right singular vector of D_{ij}. Vectors ξ_j^l represent orthogonal modes in solutions and σ_l represents the relative importance of these modes in the solution data. A SVD truncation theorem of numerical linear algebra states that the relative L^2 norm of error of approximating D_{ij} with a truncated sum (3) is 0.001 for $K = 21$ and 1.0×10^{-4} for $K = 38$. The relative L^∞ norm of the SVD truncation error is often estimated using the quantity $e_K = (\sum_{i=K+1}^{P} \sigma_i)/\sum_{i=1}^{P} \sigma_i$. Values of e_K for $K = 20$, $K = 36$, and $K = 55$ are 0.0087, 0.00089, and 9.8×10^{-5}, respectively. This suggests that modes corresponding to singular values σ_l, $l > 20$ account for less than 0.01 of the solutions, for $l > 36$ for less than 0.001, and for $l > 55$ for less than 1.0×10^{-4} of the solutions. In other words, the solutions can be approximated accurately with first 55 singular vectors ξ_j^l and these vectors provide a very efficient basis for representing this class of solutions (but not other classes of solutions).

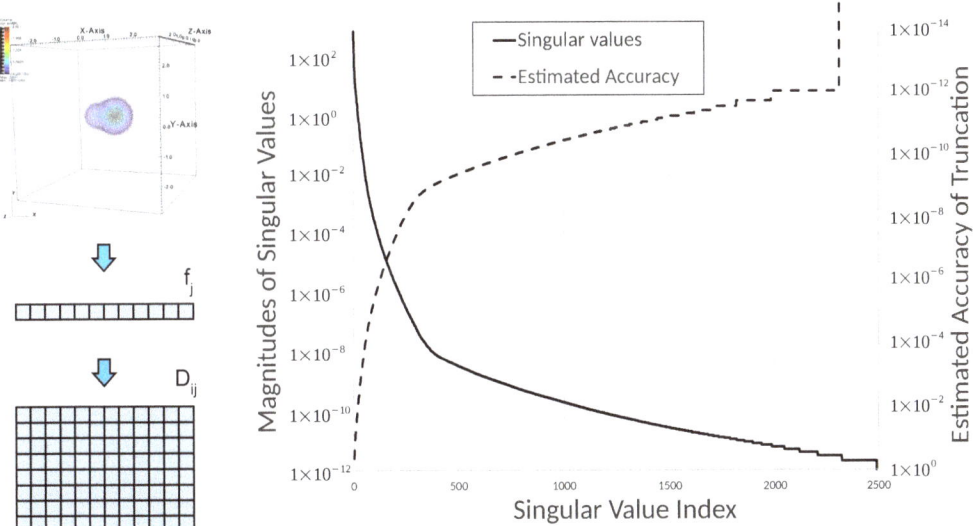

Figure 1. (**left**) Schematic depiction of constructing the solution data matrix: 3D solutions are reshaped into vectors f_j which are then stacked as rows of the solution data matrix D_{ij}. (**right**) Singular values of the solutions data matrix and estimated accuracy plotted on logarithmic scale.

To assess the ability of low rank features to be learned, experiments were conducted applying autoencoders to solution data. An autoencoder is a type of neural network which tries to learn a compressed form of the data on which it is trained. Such a network will contain a constricted layer with only a few output nodes which is meant to be the compressed data. The layers preceding this bottleneck are referred to as the encoder and the layers following are the decoder.

The architecture of autoencoders applied involved convolutional layers, thus making them convolutional autoencoders. As with much of machine learning, there is no specific way that such a network must be constructed. Generally speaking, a 3D convolutional autoencoder will have an encoder with layers arranged as depicted in Figure 2a, and a decoder with layers as depicted in Figure 2b. Figure 2a shows three convolutional filters being applied to the original, resulting in the next three data blocks (color coded to match). These data blocks are then reduced in size by the application of a pooling filter. Another round of convolutional filters is applied, followed by another pooling filter. In Figure 2b, that process is occurring in reverse. The three small data blocks in the beginning are up-sampled to increase their size, then go through a set of convolutional filters. This result is up-sampled again before a final convolutional filter returns the data block to its original dimensions.

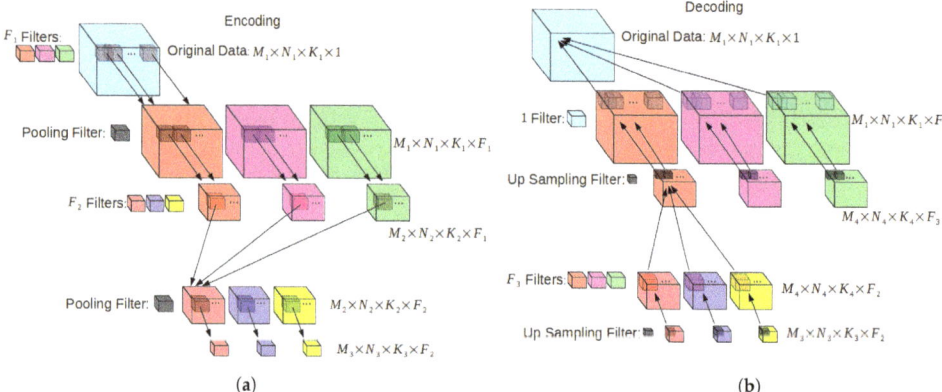

Figure 2. Diagram of convolutional autoencoder. The encoder (**a**) contains successive convolutional and pooling layers which downsize the data. The decoder (**b**) contains successive convolutional and upsampling layers which restore the data to its original size.

The goal is for the convolutional layers to learn key features in the data which are sufficient to reconstruct the data but can be stored in a smaller dimensional space. The compression and decompression is controlled by pooling and upsampling layers. Pooling layers such as max pooling or average pooling replace a block of values in the data with a single value (the max or the average, respectively) which reduces the total number of values being handled. Upsampling layers do the opposite which is to increase the number of values being handled by either repeating values or inserting values which interpolate neighboring values. Alternatively, a network can be allowed to learn its own up-sampling method using transposed convolutional layers.

The convolutional layers are inserted between the pooling and upsampling layers to do the learning. The size of the kernels being applied as well as their activation functions and the number of filters applied are all hyperparameters which must be chosen to get the best results. The requirement for an autoencoder is that at the bottleneck, the number of filters must be such that the total number of values being output is less than the number of values at the input to the network. In Figure 2a this means that $M_3 \times N_3 \times K_3 \times F_2 \ll M_1 \times N_1 \times K_1$.

Samples of the data were taken randomly from the database without discriminating based on initial conditions or the time at which the solution was saved to construct training and test sets. A few different network architectures were constructed and fit to the data using the Keras API included with TensorFlow [35]. The Nadam optimizer proved most effective at training the models. Compared to the Adam and Stochastic Gradient Descent optimizers, Nadam converged the fastest and resulted in the lowest prediction error. A variety of hyperparameter values were explored for the training, some results of this exploration are discussed next.

To demonstrate the performance of the autoencoders, we provide some graphs of the reconstructed solutions to compare to the original solutions. The solutions are 3D data cubes and so can not be easily plotted in totality. Instead we have provide graphs of slices of the domain, taken near the center. The results in Figure 3 come from a network with a bottleneck with dimensions of 8 by 8 by 8 with 8 filters which translates to 8^4 or 4096 variables which is a significant reduction down from 41^3 variables. Both the encoder and decoder portions of the network had three convolutional layers, all of which used the ReLU activation function. The ReLU activation function and its derivative are simple and computationally cheap to use and still grant networks the universal approximator property [36]. Furthermore, we observed ReLU to be easier to train than sigmoid. It has been shown that ReLU is easier to train than sigmoid and sigmoid-like activation functions, because it does not suffer from the vanishing gradient problem [37,38]. Since our autoencoding problem is not a categorical one, the restricted range of sigmoid-like activation functions provides no advantage, and thus ReLU was the preferable activation function.

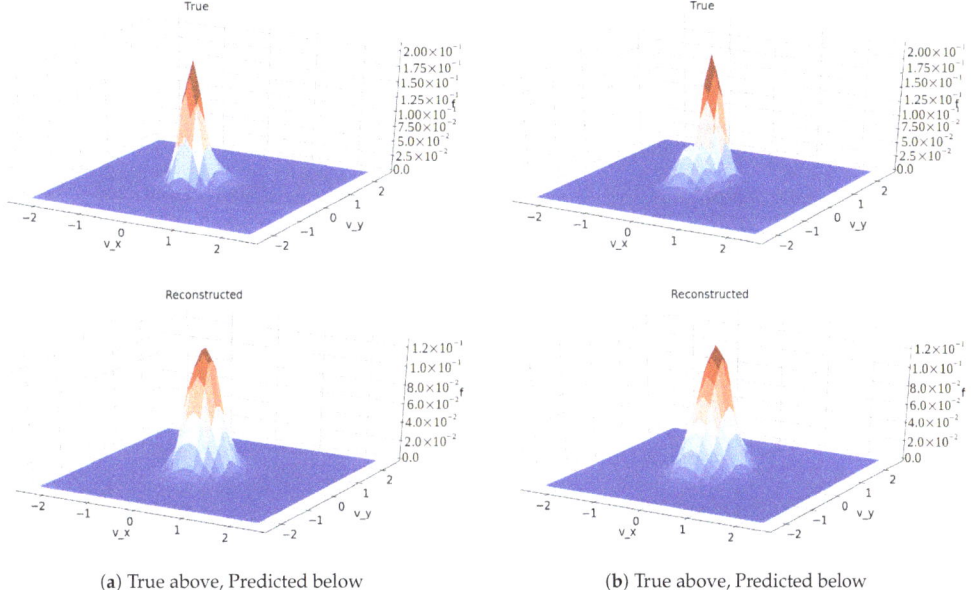

(a) True above, Predicted below   (b) True above, Predicted below

Figure 3. Comparison between true solutions (**top**) and reconstructed (**bottom**) from autoencoder with $8 \times 8 \times 8 \times 8$ bottleneck. Since the solutions are on a 3D grid, only a slice towards the center of the domain is plotted here. The reconstructions are generally good, but with sharp features having been rounded off.

The results in Figure 4 come from a network with a bottleneck with dimensions of 2 by 2 by 2 with 8 filters which translates to 64 variables. The architecture of this network was identical to the previous, less restricted network, other than the bottleneck being tighter.

The results show promise that the convolutional architecture is capable of identifying and capturing important features in this data set, even with a significantly smaller number of variables. This provides hope that machine learning algorithms will be able to compute such solutions with far less computational effort and memory usage than traditional methods. It is most notable that the peaks of the reconstructed graphs have been rounded off and do not reach as high as the true data. Still captured though is the location and general Maxwellian shape, with the reconstructions from the network with the tighter restriction definitely being lossier than those from the less restricted network.

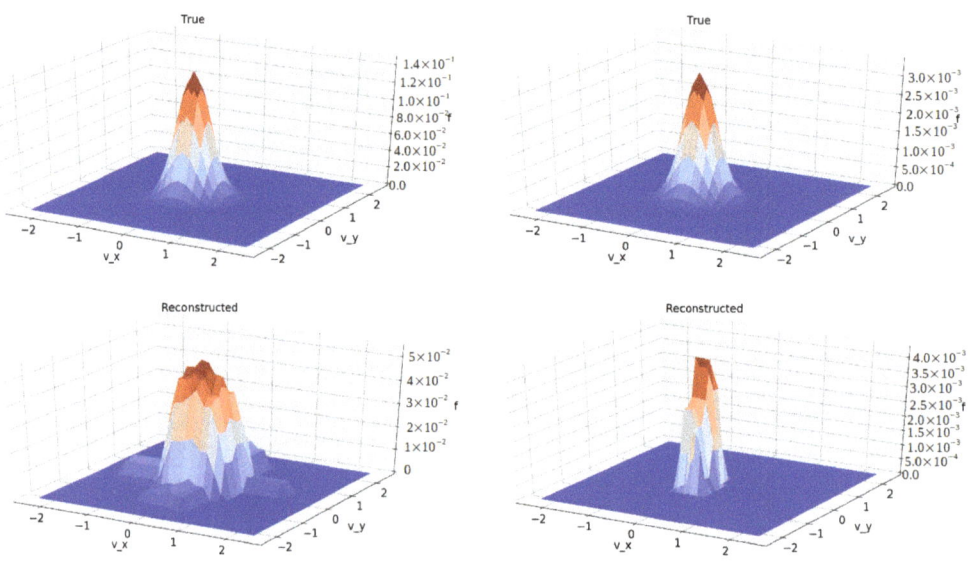

(a) True above, Predicted below   (b) True above, Predicted below

Figure 4. Comparison between true solutions (**top**) and reconstructed (**bottom**) from autoencoder with 2 × 2 × 2 × 8 bottleneck. These are plots of a slice of the domain of the data. Reconstructions were clearly less accurate than those produced by the previous autoencoder, but still demonstrate the Maxwellian's shape and location.

3. Conservation Laws

Exact solutions to the spatially homogeneous Boltzmann equation must adhere to the conservation of mass, momentum, and energy. Each of these quantities is computed from the solution by the application of a linear integral operator. When these operators are applied to the collision integral, the result must be zero in order to ensure conservation.

The present strategy to enforce conservation laws in numerical solution is to post-process the collision integral after it is predicted and before it is used to step in time as shown in Algorithm 1. Many approaches have been proposed to enforce conservation laws in numerical evaluation of the collision integral, see, e.g., [17,39,40]. In this paper, we employ a modification of the Lagrangian multiplier method of [41,42]. The resulting post-processing procedure is schematically described in Algorithm 2. The procedure computes a corrected value of the collision integral that satisfies the discrete conservation laws up to roundoff errors while being as close as possible to the prediction. The difference from the approach of [42] is that values of the predicted collision integral that are small in magnitude are not affected by the procedure. Thus, the procedure avoids creation of small spurious values in the conservative collision integral at the domain boundaries.

In the future, enforcement of the conservation laws can be incorporated into the model and will thus force the training process to account for them. For example, let M be a 5 by m matrix ($5 < m$) which computes the mass, momentum, and energy from the solution. The collision integral must exist in the null space of this matrix. The basis of the null space consists of columns of V_0 from the singular value decomposition of the matrix M,

$$M = U \begin{bmatrix} S & 0 \end{bmatrix} \begin{bmatrix} V_1^T \\ V_0^T \end{bmatrix} \qquad (4)$$

where U, S are 5 by 5 matrices and $V = [V_1, V_0]$ is an m by m matrix. Thus,

$$Q = \sum_{i=6}^{m} \alpha_i v_i, \quad (5)$$

where

$$V_0 = \begin{bmatrix} v_6 & v_7 & \ldots & v_m \end{bmatrix}. \quad (6)$$

A model can then be taught to learn values for the parameters $\{\alpha_i\}$ such that Equation (5) approximates the collision integral. This will ensure that the conservation laws are automatically satisfied.

Algorithm 1 Solve 0-dimensional Boltzmann.

1: **while** $t < t_f$ **do**
2:     Compute $\hat{Q} = \hat{Q}[f(t)]$ using machine learned model
3:     $\hat{Q} = \texttt{enforceConservation}(\hat{Q})$
4:     $f(t + \Delta t) = f(t) + \Delta t * \hat{Q}$
5: **end while**

Algorithm 2 Enforce Conservation.

1: **procedure** ENFORCECONSERVATION(q)
2:     Construct mass, momentum, and energy operator, M
3:     Construct a masking projection operator P that preserves small components of q and nullifies other components
4:     Solve $\min \frac{1}{2}\|q_{\text{corr}} - q\|^2$ s.t. $Mq_{\text{corr}} = 0$, $P(q_{\text{corr}} - q) = 0$
5:     **return** q_{corr}
6: **end procedure**

---

4. Error Propagation

A known weakness of using a machine learned model in place of an analytical one is that errors will be injected into the simulation process. Even if the magnitude of the generated error is small, the way these errors evolve and interact may be significant. It is thus desirable to understand how the errors will behave. In our case, if we define $\hat{Q}$ as the machine-learned model for the collision integral and $\hat{f}$ as the solution computed using that model, then we can define the error functions as

$$e_f = f - \hat{f} \quad (7)$$

and

$$e_Q = Q[f] - \hat{Q}[\hat{f}]. \quad (8)$$

We then have an equation for the evolution of error given by

$$\frac{\partial}{\partial t} e_f + \vec{v} \cdot \nabla_x e_f = e_Q, \quad (9)$$

from which we can derive some approximate error bounds.

If one assumes that the distribution of error is approximately uniform, then the second term can be ignored. This leaves

$$\frac{\partial}{\partial t} e_f = e_Q. \quad (10)$$

Writing e_Q as

$$e_Q = \left(Q[f] - \hat{Q}[f]\right) + \left(\hat{Q}[f] - \hat{Q}[\hat{f}]\right), \quad (11)$$

we see that there are two contributions to the error. The first is the error in the prediction due to the model not being exact, the second is from error accumulated during the time stepping process. If we now assume Lipschitz continuity of the model and that the prediction error is bounded by a constant, then

$$|e_Q| \leq C_1 + C_2|f - \hat{f}| = C_1 + C_2|e_f|. \tag{12}$$

The worst case scenario estimate is

$$\frac{\partial}{\partial t}|e_f| \leq C_1 + C_2|e_f|, \tag{13}$$

which leads to the bound on the solution's error

$$|e_f(t, \vec{x}, \vec{v})| \leq \frac{C_1}{C_2}\left(e^{C_2 t} - 1\right). \tag{14}$$

Depending on the application for which the Boltzmann equation is being solved, it may not be sufficient just to bound the magnitude of the error in the prediction of the solution, but the effect the error has on the moments may also be of interest. It is from the moments that many physical properties of the gas are computed and the errors in the solution may manifest in ways which significantly or insignificantly affect the moments of the solution. In general, a moment of the solution is given by

$$m_i = \int_\Omega q_i f, \tag{15}$$

where q_i is a quantity associated with the definition of the ith moment. Therefore, the error in the moment calculation is

$$e_{m_i} = \int_\Omega q_i f - \int_\Omega q_i \hat{f} = \int_\Omega q_i e_f, \tag{16}$$

which is the corresponding moment of the error. For error in the collision integral prediction satisfying our previous assumptions, the error bound on the moment calculation is

$$|e_{m_i}| \leq \frac{C_1}{C_2}(e^{C_2 t} - 1) \int_\Omega |q_i|. \tag{17}$$

As of yet, it cannot be said what kind of errors should be expected or how they will manifest themselves, other than that the error in the mass, momentum, and energy moments will be exactly zero due to conservation law enforcement. We expect that lower order moments will be less affected by introduced error, however higher order moments could react dramatically to small deviations in the solution.

5. Test Model

Examination of trends in the solution, collision integral pairs in the database led to the conclusion that a second order function should have sufficient flexibility to predict the value at each point of the collision integral. The chosen predictors for each value in the collision integral were the 27 values in the solution in the neighborhood of the index in the collision integral being predicted.

A sparsely connected neural network was then constructed using an expanded feature space which included second order terms computed from the predictors. For a given value in the collision integral, $y_{i^* j^* k^*}$, at indices i^*, j^*, k^*, let $P_{i^* j^* k^*}$ be the set of corresponding features which is

$$P_{i^*j^*k^*} := \{x_{i_1j_1k_1}, x_{i_1j_1k_1}x_{i_2j_2k_2} \mid$$
$$d((i_1,j_1,k_1),(i^*,j^*,k^*)) \le 1, d((i_2,j_2,k_2),(i^*,j^*,k^*)) \le 1\}, \tag{18}$$

where $d((i_1,j_1,k_1),(i^*,j^*,k^*)) = \max(|i_1 - i^*|, |j_1 - j^*|, |k_1 - k^*|)$. We then define $X_{i^*j^*k^*}$ to be a vector of all the elements in $P_{i^*j^*k^*}$. The model to predict $y_{i^*j^*k^*}$ is the support vector machine,

$$y_{i^*j^*k^*} = w_{i^*j^*k^*}^T X_{i^*j^*k^*} + w_{i^*j^*k^*,0}, \tag{19}$$

with parameters $w_{i^*j^*k^*}$, and $w_{i^*j^*k^*,0}$ which correspond to the given $y_{i^*j^*k^*}$. The full network made up of these SVMs is a sparsely connected, single layer network with linear activation functions. In total this architecture has $O(n^3)$ parameters and requires $O(n^3)$ flops to compute the collision integral where n is the number of indices along a single dimension of the discrete mesh. Both the computational complexity and memory requirements are this linear in the size of the mesh. The model was fit to training data using a regularized least squares loss function and achieved overall good predictions.

Comparison of Results

To assess the performance of this machine-learned method for computing the collision integral, the model was used to solve the spatially homogeneous Boltzmann equation

$$\frac{\partial}{\partial t} f(t, \vec{v}) = Q[f](t, \vec{v}) \tag{20}$$

using forward Euler integration in time as shown in Algorithm 1. The solution was integrated starting with initial data $f(0, \vec{v})$ from the database of solutions.

The Python implementation of this method took about 6 min to carry out 667 time steps and about 9 min to carry out 1000 time steps, an estimated $O(10^2)$ times faster than the method of [6]. That method uses a discontinuous Galerkin discretization and takes about 40 h on a single CPU to carry out 1000 time steps. The machine-learned method thus greatly outperformed the method of [6]. The CPU time for both methods to perform one evaluation of the collision operator are summarized in Table 1. The CPU time for the machine-learned method also shows significant improvement compared to times reported in [10] for a fast spectral method.

Table 1. Time to perform one evaluation of the collision operator using machine-learned collision operator and the $O(n^6)$ deterministic method of [6].

	ML Method	Deterministic	Speed Up
Time, s	0.54	147	270×

Solutions computed using the machine-learned collision operator were comparable to the deterministic solutions with better predictions towards the center of the domain than towards the boundary. Figures 5 and 6 show comparisons between solutions achieved using the method in [6] and the present method. As was the case with the autoencoder, these plots are of slices of the domain. The predictions look very similar and trend toward the same steady state over time. The fact that the quality of the prediction is better closer to the center of the domain may be a result of the training data being more diverse towards the center of the domain than towards the boundary. The true solution and collision integral go to zero at the boundary of the domain and so there was not as much information to use to train the model out there. Even still, the magnitude of the difference between the prediction and the true value was small.

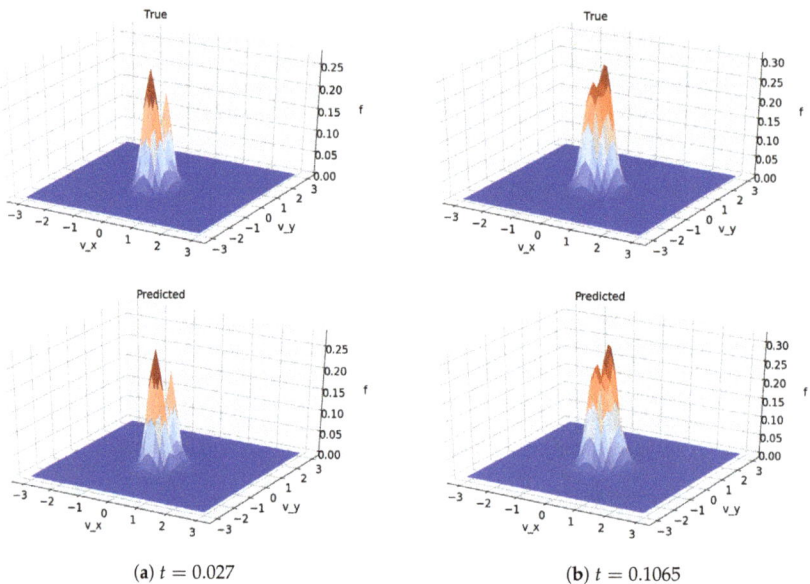

Figure 5. Comparison between solutions computed numerically (**top**) and using the machine learned model (**bottom**). Both comparisons were produced from the same initial data. Time stamps are normalized to the maximum time for which training data existed. These plots are of a single slice of the domain.

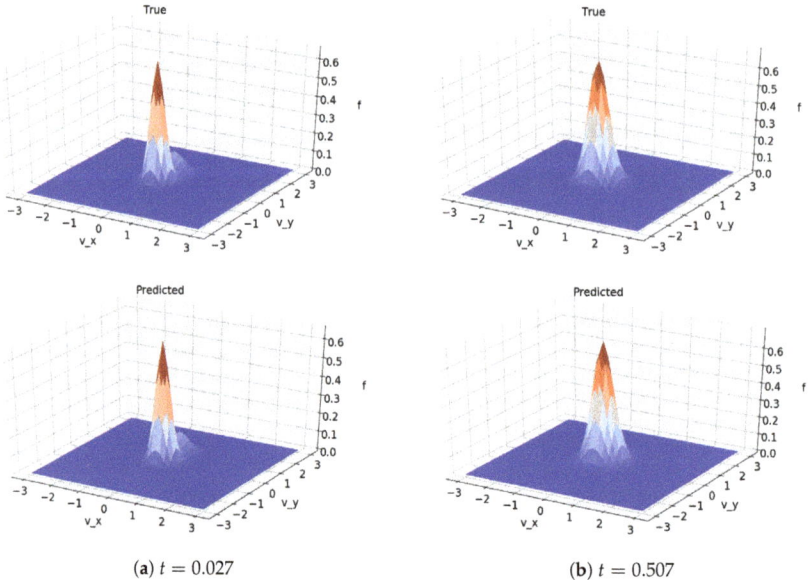

Figure 6. Comparison between solutions computed numerically (**top**) and using the machine learned model (**bottom**). Both comparisons were produced from the same initial data. Time stamps are normalized to the maximum time for which training data existed. These plots are of a single slice of the domain.

Figure 7 demonstrates how the absolute error between the true solution and the predicted solution evolves over time. The total magnitude of the difference remains low throughout the duration of the simulation, being under 10% of the $L1$ norm of the true solution. There is also a consistent behavior among all the error curves, mainly that the difference grows the most during the first few iterations, then starts to flatten out.

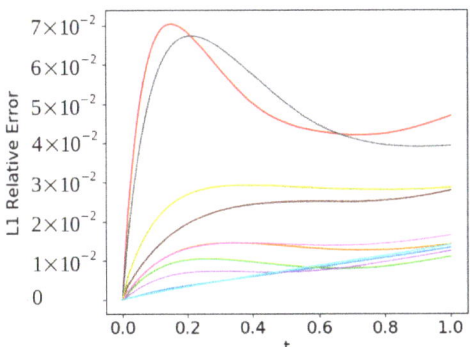

Figure 7. Normalized L_1 error, given by $\|f_{True} - f_{Pred}\|_1 / \|f_{True}\|_1$, shown for 10 different test cases. Time axis is normalized to the maximum time for which training data existed. Error growth is consistently most rapid for $t < 0.2$. For $t > 0.5$ error growth is consistently at a much lower rate.

It is now of interest to see the effect the error has on the moments of the solution. Figure 8 shows a comparison of second and third order moments between the predicted and true solution. The lower order moments corresponding to mass, momentum, and temperature moments are not shown because they match exactly with the analytically computed solution's moments. Generally, the moments of the predicted solutions followed the true moments and headed towards the steady state. They did not however perfectly arrive and remain at the steady state. In some cases the moments of the predicted solution cross over each other, and in other cases they simply fail to meet. The higher order moments exhibited similar behaviors, even more so than the lower order moments, but for many applications moments higher than 3rd order will not be as important.

Solutions remained stable up until and beyond time values for which training data existed. Eventually though there was observed degradation and destabilization of the quality of predictions. Figure 9 demonstrates the long term behavior of the predicted solutions. Training data existed up until the dimensionless time $t = 1.0$, and the simulation was run until $t = 2.0$. The solution has clearly lost its shape by the end of that run; no longer having a nice Gaussian shape. In addition, the moments do not nicely converge to uniform values. The typical behavior was that the moments would tend towards the appropriate steady state early on, but would eventually begin to diverge. We propose that this behavior could be corrected by replacing the machine-learned model with an analytical method once the solution is close to steady state. This would ensure the appropriate long term behavior, and would still run much faster than using a direct discretization method for the full duration. Additionally, we are confident a more advanced architecture can be developed which achieves better accuracy and will likely still be faster than the true method. Even a model architecture that requires ten times the computational work of this simple model would still be tremendously time saving.

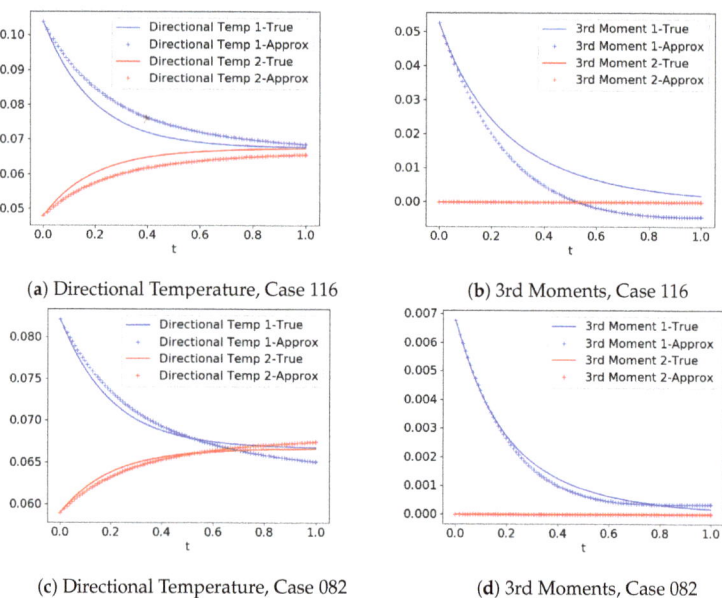

(a) Directional Temperature, Case 116

(b) 3rd Moments, Case 116

(c) Directional Temperature, Case 082

(d) 3rd Moments, Case 082

Figure 8. Comparison of moments between numerically computed and machine learning computed solutions. The red and blue curves correspond to different coordinate directions. The case numbers merely serve to differentiate the runs. Time axis is normalized to the maximum time for which training data existed.

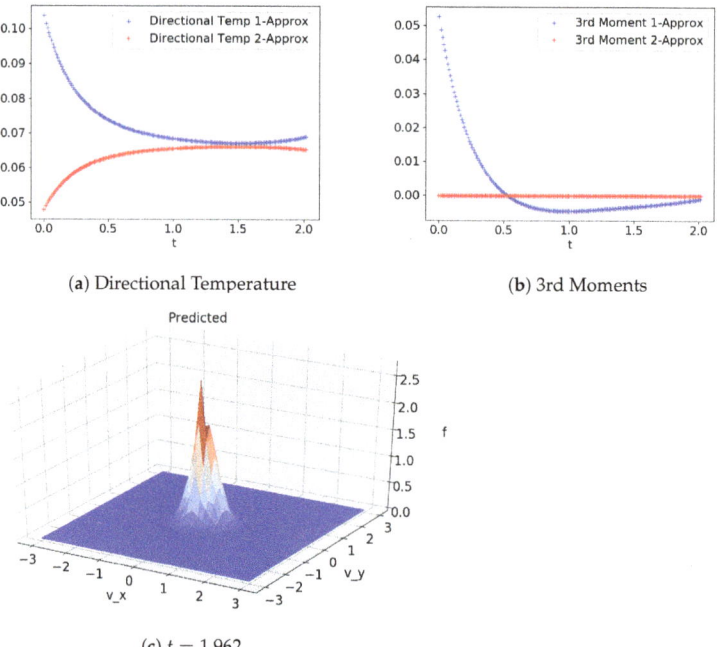

(a) Directional Temperature

(b) 3rd Moments

(c) $t = 1.962$

Figure 9. Long term behavior of predicted solution, Case 116. Key moments and domain slice are shown. Divergence of the directional temperature is seen around $t = 1.75$, and the plot of the solution is visibly degraded. Time axis is normalized to the maximum time for which training data existed.

6. Conclusions

In an effort to advance the state-of-the-art in simulating complex flows, we have conducted investigations into the ability of machine learning to calculate the Boltzmann collision integral more quickly than traditional methods. Our experiments show that the machine-learned models are capable of finding low dimensional features that can encode the solutions with good accuracy. Consequently, predictions of the collision integral are able to be computed in much less time than is required for analytical methods. In the spatially homogeneous case, this approach shows greatly accelerated integration time. The resulting approximate solutions and key moments are generally similar to those of the true solutions. A key weakness exhibited by the method is the long term degradation of the solutions. An ideal numerical method will convergence to the correct steady state solution, however we have observed that the method presented here does not. Future work will involve improving accuracy even further by implementing more sophisticated machine learning methods. This will include investigation of different model architectures, and incorporating the enforcement of conservation laws into the model and the training process.

Author Contributions: Conceptualization, I.H. and A.W.; Data curation, A.A.; Funding acquisition, A.W.; Methodology, I.H., A.W., and A.A.; Project administration, A.W.; Software, I.H. and A.A.; Writing–review & editing, I.H., A.W. and A.A. All authors have read and agreed to the published version of the manuscript.

Funding: The first author was supported in part by an appointment to the Student Research Participation Program at the U.S. Air Force Institute of Technology administered by the Oak Ridge Institute for Science and Education through an interagency agreement between the U.S. Department of Energy and USAFIT. The second author was supported by AFOSR grant F4FGA08305J005. The third author was supported by AFOSR grant F4FGA08305J005 and the NSF DMS-1620497 grant.

Institutional Review Board Statement: Not applicable.

Informed Consent Statement: Not applicable.

Data Availability Statement: Fortran and Python source code, and training data used in this study are available upon request. If interested please reach out to the corresponding author.

Acknowledgments: The authors would like to thank Robert Martin for motivating and inspiring discussions at the early stages of this work.

Conflicts of Interest: The authors declare no conflict of interest.

References

1. Pareschi, L.; Perthame, B. A Fourier spectral method for homogeneous Boltzmann equations. *Transp. Theory Stat. Phys.* **1996**, *25*, 369–382. [CrossRef]
2. Ibragimov, I.; Rjasanow, S. Numerical solution of the Boltzmann equation on the uniform grid. *Computing* **2002**, *69*, 163–186. [CrossRef]
3. Kirsch, R.; Rjasanow, S. A weak formulation of the Boltzmann equation based on the Fourier transform. *J. Stat. Phys.* **2007**, *129*, 483–492. [CrossRef]
4. Gamba, I.M.; Tharkabhushanam, S.H. Shock and boundary structure formation by spectral-Lagrangian methods for the inhomogeneous Boltzmann transport equation. *J. Comput. Math.* **2010**. [CrossRef]
5. Munafò, A.; Haack, J.R.; Gamba, I.M.; Magin, T.E. A spectral-Lagrangian Boltzmann solver for a multi-energy level gas. *J. Comput. Phys.* **2014**, *264*, 152–176. [CrossRef]
6. Alekseenko, A.; Limbacher, J. Evaluating high order discontinuous Galerkin discretization of the Boltzmann collision integral in $\mathcal{O}(N^2)$ operations using the discrete Fourier transform. *Kinet. Relat. Model.* **2019**, *12*, 703. [CrossRef]
7. Mouhot, C.; Pareschi, L. Fast Algorithms for Computing the Boltzmann Collision Operator. *Math. Comput.* **2006**, *75*, 1833–1852. [CrossRef]
8. Wu, L.; Liu, H.; Zhang, Y.; Reese, J.M. Influence of intermolecular potentials on rarefied gas flows: Fast spectral solutions of the Boltzmann equation. *Phys. Fluids* **2015**, *27*, 082002. [CrossRef]
9. Mouhot, C.; Pareschi, L.; Rey, T. Convolutive decomposition and fast summation methods for discrete-velocity approximations of the Boltzmann equation. *ESAIM M2AN* **2013**, *47*, 1515–1531. [CrossRef]
10. Gamba, I.M.; Haack, J.R.; Hauck, C.D.; Hu, J. A fast spectral method for the Boltzmann collision operator with general collision kernels. *SIAM J. Sci. Comput.* **2017**, *39*, B658–B674. [CrossRef]

11. Wu, L.; White, C.; Scanlon, T.J.; Reese, J.M.; Zhang, Y. Deterministic numerical solutions of the Boltzmann equation using the fast spectral method. *J. Comput. Phys.* **2013**, *250*, 27–52. [CrossRef]
12. Bobylev, A.; Rjasanow, S. Difference scheme for the Boltzmann equation based on Fast Fourier Transfrom. *Eur. J. Mech.-B/Fluids* **1997**, *16*, 293–306.
13. Bobylev, A.; Rjasanow, S. Fast deterministic method of solving the Boltzmann equation for hard spheres. *Eur. J. Mech.-B/Fluids* **1999**, *18*, 869–887. [CrossRef]
14. Filbet, F.; Mouhot, C.; Pareschi, L. Solving the Boltzmann Equation in $N \log_2 N$. *SIAM J. Sci. Comput.* **2006**, *28*, 1029–1053. [CrossRef]
15. Kloss, Y.Y.; Tcheremissine, F.G.; Shuvalov, P.V. Solution of the Boltzmann equation for unsteady flows with shock waves in narrow channels. *Comput. Math. Math. Phys.* **2010**, *50*, 1093–1103. [CrossRef]
16. Morris, A.; Varghese, P.; Goldstein, D. Monte Carlo solution of the Boltzmann equation via a discrete velocity model. *J. Comput. Phys.* **2011**, *230*, 1265–1280. [CrossRef]
17. Varghese, P.L. Arbitrary post-collision velocities in a discrete velocity scheme for the Boltzmann equation. In Proceedings of the 25th International Symposium on Rarefied Gas Dynamics, Saint-Petersburg, Russia, 21–28 July 2006; Ivanov, M., Rebrov, A., Eds.; Publishing House of Siberian Branch of RAS: Novosibirsk, Russia, 2007; pp. 227–232.
18. Dimarco, G.; Loubère, R.; Narski, J.; Rey, T. An efficient numerical method for solving the Boltzmann equation in multidimensions. *J. Comput. Phys.* **2018**, *353*, 46–81. [CrossRef]
19. Jaiswal, S.; Alexeenko, A.A.; Hu, J. A discontinuous Galerkin fast spectral method for the full Boltzmann equation with general collision kernels. *J. Comput. Phys.* **2019**, *378*, 178–208. [CrossRef]
20. Wu, L.; Zhang, J.; Reese, J.M.; Zhang, Y. A fast spectral method for the Boltzmann equation for monatomic gas mixtures. *J. Comput. Phys.* **2015**, *298*, 602–621. [CrossRef]
21. Alekseenko, A.; Grandilli, A.; Wood, A. An ultra-sparse approximation of kinetic solutions to spatially homogeneous flows of non-continuum gas. *Results Appl. Math.* **2020**, *5*, 100085. [CrossRef]
22. Alekseenko, A.; Nguyen, T.; Wood, A. A deterministic-stochastic method for computing the Boltzmann collision integral in $\mathcal{O}(MN)$ operations. *Kinet. Relat. Model.* **2018**, *11*, 1211. [CrossRef]
23. Grohs, P.; Hiptmair, R.; Pintarelli, S. Tensor-product discretization for the spatially inhomogeneous and transient Boltzmann equation in 2D. *SMAI J. Comput. Math.* **2017**, *3*, 219–248. [CrossRef]
24. Heintz, A.; Kowalczyk, P.; Grzhibovskis, R. Fast numerical method for the Boltzmann equation on non-uniform grids. *J. Comput. Phys.* **2008**, *227*, 6681–6695. [CrossRef]
25. Fonn, E.; Grohs, P.; Hiptmair, R. Hyperbolic cross approximation for the spatially homogeneous Boltzmann equation. *IMA J. Numer. Anal.* **2015**, *35*, 1533–1567. [CrossRef]
26. Dimarco, G.; Pareschi, L. Numerical methods for kinetic equations. *Acta Numer.* **2014**, *23*, 369–520. [CrossRef]
27. Narayan, A.; Klöckner, A. Deterministic numerical schemes for the Boltzmann equation. *arXiv* **2009**, arXiv:0911.3589.
28. Raissi, M.; Perdikaris, P.; Karniadakis, G. Physics-informed neural networks: A deep learning framework for solving forward and inverse problems involving nonlinear partial differential equations. *J. Comput. Phys.* **2019**, *378*, 686–707. [CrossRef]
29. Weinan, E.; Han, J.; Jentzen, A. Deep Learning-Based Numerical Methods for High-Dimensional Parabolic Partial Differential Equations and Backward Stochastic Differential Equations. *Commun. Math. Stat.* **2017**, *5*, 349–380. [CrossRef]
30. Sirignano, J.; Spiliopoulos, K. DGM: A deep learning algorithm for solving partial differential equations. *J. Comput. Phys.* **2018**, *375*, 1339–1364. [CrossRef]
31. Lou, Q.; Meng, X.; Karniadakis, G.E. Physics-informed neural networks for solving forward and inverse flow problems via the Boltzmann-BGK formulation. *arXiv* **2020**, arXiv:2010.09147.
32. Boelens, A.M.; Venturi, D.; Tartakovsky, D.M. Tensor methods for the Boltzmann-BGK equation. *J. Comput. Phys.* **2020**, *421*, 109744. [CrossRef]
33. Kogan, M. *Rarefied Gas Dynamics*; Plenum Press: New York, NY, USA, 1969.
34. Cercignani, C. *Rarefied Gas Dynamics: From Basic Concepts to Actual Caclulations*; Cambridge University Press: Cambridge, UK, 2000.
35. Chollet, F.; et al. Keras. 2015. Available online: https://keras.io (accessed on 15 October 2019).
36. Hanin, B. Universal Function Approximation by Deep Neural Nets with Bounded Width and ReLU Activations. *Mathematics* **2019**, *7*, 992. [CrossRef]
37. Maas, A.L.; Hannun, A.Y.; Ng, A.Y. Rectifier nonlinearities improve neural network acoustic models. *Proc. Icml. Citeseer* **2013**, *30*, 3.
38. Krizhevsky, A.; Sutskever, I.; Hinton, G. ImageNet Classification with Deep Convolutional Neural Networks. *Neural Inf. Process. Syst.* **2012**, *25*. [CrossRef]
39. Tcheremissine, F.G. Solution to the Boltzmann kinetic equation for high-speed flows. *Comput. Math. Math. Phys.* **2006**, *46*, 315–329. [CrossRef]
40. Aristov, V. *Direct Methods for Solving the Boltzmann Equation and Study of Nonequilibrium Flows*; Fluid Mechanics and Its Applications; Kluwer Academic Publishers: Dordrecht, The Netherlands, 2001.
41. Gamba, I.M.; Tharkabhushanam, S.H. Spectral-Lagrangian methods for collisional models of non-equilibrium statistical states. *J. Comput. Phys.* **2009**, *228*, 2012–2036. [CrossRef]
42. Zhang, C.; Gamba, I.M. A Conservative Discontinuous Galerkin Solver for the Space Homogeneous Boltzmann Equation for Binary Interactions. *SIAM J. Numer. Anal.* **2018**, *56*, 3040–3070. [CrossRef]

Article

Nanofluid Flow on a Shrinking Cylinder with Al$_2$O$_3$ Nanoparticles

Iskandar Waini [1,2], Anuar Ishak [2,*] and Ioan Pop [3]

[1] Fakulti Teknologi Kejuruteraan Mekanikal dan Pembuatan, Universiti Teknikal Malaysia Melaka, Melaka 76100, Malaysia; iskandarwaini@utem.edu.my
[2] Department of Mathematical Sciences, Faculty of Science and Technology, Universiti Kebangsaan Malaysia UKM, Bangi 43600, Malaysia
[3] Department of Mathematics, Faculty of Mathematics and Computer Science, Babeș-Bolyai University, 400084 Cluj-Napoca, Romania; ipop@math.ubbcluj.ro
* Correspondence: anuar_mi@ukm.edu.my

Abstract: This study investigates the nanofluid flow towards a shrinking cylinder consisting of Al$_2$O$_3$ nanoparticles. Here, the flow is subjected to prescribed surface heat flux. The similarity variables are employed to gain the similarity equations. These equations are solved via the bvp4c solver. From the findings, a unique solution is found for the shrinking strength $\lambda \geq -1$. Meanwhile, the dual solutions are observed when $\lambda_c < \lambda < -1$. Furthermore, the friction factor $Re_x^{1/2} C_f$ and the heat transfer rate $Re_x^{-1/2} Nu_x$ increase with the rise of Al$_2$O$_3$ nanoparticles φ and the curvature parameter γ. Quantitatively, the rates of heat transfer $Re_x^{-1/2} Nu_x$ increase up to 3.87% when φ increases from 0 to 0.04, and 6.69% when γ increases from 0.05 to 0.2. Besides, the profiles of the temperature $\theta(\eta)$ and the velocity $f'(\eta)$ on the first solution incline for larger γ, but their second solutions decline. Moreover, it is noticed that the streamlines are separated into two regions. Finally, it is found that the first solution is stable over time.

Keywords: heat transfer; prescribed heat flux; similarity solutions; dual solutions; stability analysis

Citation: Waini, I.; Ishak, A.; Pop, I. Nanofluid Flow on a Shrinking Cylinder with Al$_2$O$_3$ Nanoparticles. *Mathematics* **2021**, *9*, 1612. https://doi.org/10.3390/math9141612

Academic Editor: Aihua Wood

Received: 4 May 2021
Accepted: 6 July 2021
Published: 8 July 2021

Publisher's Note: MDPI stays neutral with regard to jurisdictional claims in published maps and institutional affiliations.

Copyright: © 2021 by the authors. Licensee MDPI, Basel, Switzerland. This article is an open access article distributed under the terms and conditions of the Creative Commons Attribution (CC BY) license (https://creativecommons.org/licenses/by/4.0/).

1. Introduction

The fluid flow toward a stagnation point on a fixed surface was first introduced by Hiemenz [1] in 1911. The axisymmetric flow was then studied by Homann [2]. Ariel [3] followed by examining the flow with the hydromagnetic effects. The flow on a shrinking sheet was reported by Wang [4] and Kamal et al. [5]. Different from the aforementioned studies, which considered the flow over a flat plate, Wang [6] discussed the fluid flow over a circular cylinder. This was then followed by several researchers, including Ishak et al. [7] and Awaludin et al. [8], who studied the flow over a shrinking cylinder subject to a prescribed surface heat flux. They found that the increment of the curvature parameter delayed the boundary layer separation from the surface of the cylinder. Muthtamilselvan and Prakash [9] studied the unsteady flow and heat transfer of a nanofluid over a moving surface with prescribed heat and mass fluxes, and stated that the heat flux condition is important in a microelectromechanical (MEM) condensation application. Several researchers [10–16] have also considered this type of surface heating condition in their studies.

Nanoparticles and structures have been used by humans in fourth century AD, by the Romans, which demonstrated one of the most interesting examples of nanotechnology in the ancient world [17]. The term nanofluid, a mixture of the base fluid and nanoparticles, was initiated by Choi and Eastman [18]. It seems that Pak and Cho [19] were the first who introduced the thermophysical correlations for the nanofluid. Several studies have considered these nanofluid correlations [20–25]. The nanofluid correlations introduced by Pak and Cho [19] were improved by Ho et al. [26]. They reported that the numerical predictions from the existing nanofluid correlations are contradicted with the experimental results. The dispersion of nanoparticles in the base fluid was observed to result in a marked reduction, instead of an enhancement. Therefore, they have introduced the new correlations of the Al_2O_3-water nanofluid through a least-square curve fitting from the experimental results. They concluded that these new correlations give more accurate predictions with the experimental data. It should be noted that the studies of the nanofluid employing these nanofluid correlations are very limited. Among them, Sheremet et al. [27] employed these correlations to study the natural convective heat transfer and fluid flow of Al_2O_3-water nanofluid in an inclined wavy-walled cavity under the effect of non-uniform heating. They found that the heat transfer rate and fluid flow rate are non-monotonic functions of the cavity inclination angle and undulation number. Similarly, these correlations have been considered by Waini et al. [28] to examine the impact of Dufour and Soret diffusions on Al_2O_3-water nanofluid flow over a moving thin needle. They reported that the skin friction coefficient and the heat transfer coefficients increase, but the mass transfer coefficient decreases in the presence of Al_2O_3 nanoparticles. This concept has been upgraded by considering two or more types of nanoparticles that dispersed simultaneously into the base fluid and is called 'hybrid nanofluid'. Some works on such fluids can be found in references [29–31]. Additionally, Takabi and Salehi [32] and Devi and Devi [33] introduced the hybrid nanofluid thermophysical models, which were widely used by many researchers [34–43] in the boundary layer problems. Furthermore, Waini et al. [44–47] scrutinized the temporal stability of the hybrid nanofluid flow.

In this study, the stagnation point flow towards a shrinking cylinder with the Al_2O_3 nanoparticle subjected to prescribed surface heat flux is investigated. Different from the previous studies, the present study examines the flow and thermal behavior of the Al_2O_3/water nanofluid by employing the correlations introduced by Ho et al. [26]. Most importantly, this is the first attempt to study the flow towards a stagnation region of a shrinking cylinder by considering these correlations. Moreover, the dual solutions and their stability are also reported in this study. The finding from this study can contribute to foresee the flow and thermal behaviors in industrial applications.

2. Mathematical Formulation

Consider the nanofluid flow on a shrinking cylinder with Al_2O_3 nanoparticles as shown in Figure 1.

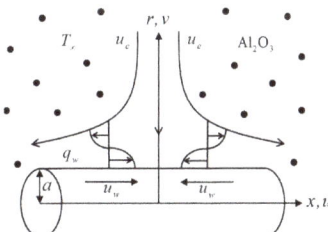

Figure 1. The flow configuration.

Here, $u_e(x) = c_1 x/L$ denotes the external flow velocity with $c_1 > 0$. The surface velocity is represented by $u_w(x) = c_2 x/L$ where c_2 is a constant. Besides, $q_w(x) = T_0 x/L$ is the prescribed heat flux where T_0 and T_∞ correspond to the reference and the ambient temperatures, respectively.

Accordingly, the governing equations are [7,8]:

$$\frac{\partial (ru)}{\partial x} + \frac{\partial (rv)}{\partial r} = 0 \qquad (1)$$

$$u\frac{\partial u}{\partial x} + v\frac{\partial u}{\partial r} = u_e\frac{du_e}{dx} + \frac{\mu_{nf}}{\rho_{nf}}\left(\frac{\partial^2 u}{\partial r^2} + \frac{1}{r}\frac{\partial u}{\partial r}\right) \qquad (2)$$

$$u\frac{\partial T}{\partial x} + v\frac{\partial T}{\partial r} = \frac{k_{nf}}{(\rho C_p)_{nf}}\left(\frac{\partial^2 T}{\partial r^2} + \frac{1}{r}\frac{\partial T}{\partial r}\right) \qquad (3)$$

Subject to:

$$\begin{array}{l} u = u_w(x), v = 0, k_{nf}\frac{\partial T}{\partial r} = -q_w(x) \text{ at } r = a \\ u \to u_e(x), T \to T_\infty \text{ as } r \to \infty \end{array} \qquad (4)$$

where (u, v) are the corresponding velocity components and T is the temperature. Further, Table 1 provides the properties of water and Al_2O_3 [22]. Here, Prandtl number, Pr is taken as Pr = 6.2. Meanwhile, the nanofluid thermophysical models are given by [19,26]:

$$\begin{array}{l} \mu_{nf} = \mu_f(1 + 4.93\varphi + 222.4\varphi^2), \quad k_{nf} = k_f(1 + 2.944\varphi + 19.672\varphi^2), \\ \rho_{nf} = (1-\varphi)\rho_f + \varphi\rho_s, \quad (\rho C_p)_{nf} = (1-\varphi)(\rho C_p)_f + \varphi(\rho C_p)_s \end{array} \qquad (5)$$

where μ, k, ρ, and (ρC_p) denote the dynamic viscosity, the thermal conductivity, the density, and the heat capacity, respectively with φ is the Al_2O_3 nanoparticle volume fractions and the subscript s represents its solid component. Meanwhile, the subscripts f and nf correspond to fluid and nanofluid, respectively. Note that these thermophysical models were also considered by Sheremet et al. [27] and Waini et al. [28].

Table 1. Thermophysical properties.

Properties	Nanoparticle	Base Fluid
	Al_2O_3	water
C_p (J/kgK)	765	4179
ρ (kg/m^3)	3970	997.1
k (W/mK)	40	0.613

Consider the following dimensionless variables [7,8]:

$$\psi = \left(\frac{c_1 \nu_f}{L}\right)^{1/2} axf(\eta), T = T_\infty + \frac{q_w}{k_f}\left(\frac{\nu_f L}{c_1}\right)^{1/2}\theta(\eta), \eta = \left(\frac{c_1}{\nu_f L}\right)^{1/2}\frac{r^2 - a^2}{2a} \qquad (6)$$

With the stream function ψ, the characteristic length L, and the fluid kinematic viscosity ν_f. Here, $u = (\partial \psi/\partial r)/r$ and $v = -(\partial \psi/\partial x)/r$. So that:

$$u = \frac{c_1 x}{L} f'(\eta), v = -\frac{a}{r}\left(\frac{c_1 \nu_f}{L}\right)^{1/2} f(\eta) \qquad (7)$$

On using Equations (6) and (7), the continuity equation, i.e., Equation (1), is identically satisfied. Now, Equations (2) and (3) become:

$$\frac{\mu_{nf}/\mu_f}{\rho_{nf}/\rho_f}[2\gamma f'' + (1+2\gamma\eta)f'''] + 1 - f'^2 + f f'' = 0 \qquad (8)$$

$$\frac{1}{\Pr}\frac{k_{nf}/k_f}{(\rho C_p)_{nf}/(\rho C_p)_f}[2\gamma\theta' + (1+2\gamma\eta)\theta''] + f\theta' - f'\theta = 0 \qquad (9)$$

Subject to:

$$\begin{array}{l} f'(0) = \lambda, f(0) = 0, \theta'(0) = -\frac{k_f}{k_{nf}}, \\ f'(\eta) \to 1, \theta(\eta) \to 0 \text{ as } \eta \to \infty \end{array} \qquad (10)$$

The physical parameters appearing in Equations (8)–(10) are the stretching/shrinking parameter λ, the curvature parameter γ, and the Prandtl number Pr, given as:

$$\lambda = \frac{c_2}{c_1}, \quad \gamma = \left(\frac{\nu_f L}{c_1 a^2}\right)^{1/2}, \quad \Pr = \frac{(\mu C_p)_f}{k_f} \tag{11}$$

Note that, $\lambda < 0$ and $\lambda > 0$ signify the shrinking and stretching sheets, while $\lambda = 0$ is for the static sheet. Here, by taking $\varphi = \lambda = \gamma = 0$, Equation (8) reduces to the Hiemenz flow, see White [48]. The local Nusselt number Nu_x and the skin friction coefficients C_f are:

$$Nu_x = -\frac{x k_{nf}}{k_f(T_w - T_\infty)}\left(\frac{\partial T}{\partial r}\right)_{r=a}, \quad C_f = \frac{\mu_{nf}}{\rho_f u_e^2}\left(\frac{\partial u}{\partial r}\right)_{r=a} \tag{12}$$

On using Equation (6), one obtains

$$Re_x^{-1/2} Nu_x = \frac{1}{\theta(0)}, \quad Re_x^{1/2} C_f = \frac{\mu_{nf}}{\mu_f} f''(0) \tag{13}$$

where $Re_x = u_e x / \nu_f$ is the local Reynolds number.

3. Stability Analysis

This temporal stability analysis was first introduced by Merkin [49] and then followed by Weidman et al. [50]. Firstly, consider the new variables as follows [8]:

$$\psi = \left(\frac{c_1 \nu_f}{L}\right)^{1/2} a x f(\eta, \tau), \; T = T_\infty + \frac{q_w}{k_f}\left(\frac{\nu_f L}{c_1}\right)^{1/2} \theta(\eta, \tau), \; \eta = \left(\frac{c_1}{\nu_f L}\right)^{1/2} \frac{r^2 - a^2}{2a}, \; \tau = \frac{c_1}{L} t \tag{14}$$

where τ is the dimensionless time variable. Then, the unsteady form of Equations (2) and (3) are employed. On using Equation (14), one obtains:

$$\frac{\mu_{nf}/\mu_f}{\rho_{nf}/\rho_f}\left[2\gamma \frac{\partial^2 f}{\partial \eta^2} + (1+2\gamma\eta)\frac{\partial^3 f}{\partial \eta^3}\right] + 1 - \left(\frac{\partial f}{\partial \eta}\right)^2 + f\frac{\partial^2 f}{\partial \eta^2} - \frac{\partial^2 f}{\partial \eta \partial \tau} = 0 \tag{15}$$

$$\frac{1}{\Pr}\frac{k_{nf}/k_f}{(\rho C_p)_{nf}/(\rho C_p)_f}\left[2\gamma \frac{\partial \theta}{\partial \eta} + (1+2\gamma\eta)\frac{\partial^2 \theta}{\partial \eta^2}\right] + f\frac{\partial \theta}{\partial \eta} - \theta\frac{\partial f}{\partial \eta} - \frac{\partial \theta}{\partial \tau} = 0 \tag{16}$$

Subject to:

$$\begin{array}{c}\frac{\partial f}{\partial \eta}(0,\tau) = \lambda, \; f(0,\tau) = 0, \; \frac{\partial \theta}{\partial \eta}(0,\tau) = -\frac{k_f}{k_{nf}}, \\ \frac{\partial f}{\partial \eta}(\infty,\tau) = 1, \; \theta(\infty,\tau) = 0\end{array} \tag{17}$$

To investigate the temporal stability, the following perturbation functions are employed [50]:

$$f(\eta,\tau) = f_0(\eta) + e^{-\alpha\tau} F(\eta), \quad \theta(\eta,\tau) = \theta_0(\eta) + e^{-\alpha\tau} G(\eta) \tag{18}$$

where $F(\eta)$ and $G(\eta)$ are comparatively small compared to $f_0(\eta)$ and $\theta_0(\eta)$, and α denotes the eigenvalue. On using Equation (18), Equations (15) and (16) respectively become:

$$\frac{\mu_{nf}/\mu_f}{\rho_{nf}/\rho_f}[2\gamma F'' + (1+2\gamma\eta)F'''] - 2f_0'F' + f_0''F + f_0 F'' + \alpha F' = 0 \tag{19}$$

$$\frac{1}{\Pr}\frac{k_{nf}/k_f}{(\rho C_p)_{nf}/(\rho C_p)_f}[2\gamma G' + (1+2\gamma\eta)G''] + f_0 G' + \theta_0' F - f_0' G - \theta_0 F' + \alpha G = 0 \tag{20}$$

The boundary conditions then become:

$$F'(0) = 0, \quad F(0) = 0, G'(0) = 0; \\ F'(\infty) = 0, \quad G(\infty) = 0 \quad (21)$$

Without loss of generality, following Harris et al. [51], we fix the value of $F''(0)$ as $F''(0) = 1$ to obtain the smallest eigenvalues α in Equations (19) and (20).

4. Results and Discussion

The solutions of Equations (8)–(10) are attained by utilizing the package bvp4c in MATLAB software [52]. The effects of various physical parameters are then examined and presented in tabular and graphical forms.

By taking $\varphi = \lambda = \gamma = 0$, we obtain $f''(0) = 1.232588$, which is in agreement with what is reported by Wang [4] and Awaludin et al. [8]. The values of $f''(0)$ and $1/\theta(0)$ for several values of λ when $\varphi = \gamma = 0$ are also provided in Table 2 for future reference. Further, the values of $Re_x^{-1/2} Nu_x$ and $Re_x^{1/2} C_f$ when $Pr = 6.2$ with various values of φ, γ, and λ are given in Table 3. The values of $Re_x^{-1/2} Nu_x$ and $Re_x^{1/2} C_f$ are intensified with the rise of γ and φ. Quantitatively, a 3.87% increment of $Re_x^{-1/2} Nu_x$ is observed when φ increases from 0 to 0.04. Moreover, it is noticeable that the values of $Re_x^{-1/2} Nu_x$ increase up to 6.69% when γ increases from 0.05 to 0.2. Meanwhile, the values of $Re_x^{1/2} C_f$ reduce, but the values of $Re_x^{-1/2} Nu_x$ increase when λ increases from -0.5 to 0.5. It is seen that the nanoparticle volume fractions, the curvature, and the stretching/shrinking parameters can be utilized to control the heat transfer rate.

Table 2. Values of $f''(0)$ and $1/\theta(0)$ for regular fluid ($\varphi = 0$) under different λ when $\gamma = 0$ (flat plate).

λ	Wang [4]	Awaludin et al. [8]	Present Results	
	$f''(0)$	$f''(0)$	$f''(0)$	$1/\theta(0)$
-1	1.32882		1.328817	-2.359393
-0.5	1.49567		1.495670	0.314542
0	1.232588	1.232588	1.232588	1.573433
0.1	1.14656	1.146561	1.146561	1.767533
0.2	1.051130	1.051130	1.051130	1.949500
0.5	0.7133	0.713295	0.713295	2.438276
1	0	0	0	3.120727
2	-1.88731	-1.887307	-1.887307	4.203068
5	-10.26475	-10.264749	-10.264749	6.491300

Table 3. Values of $Re_x^{-1/2} Nu_x$ and $Re_x^{1/2} C_f$ for φ, γ, and λ when $Pr = 6.2$.

φ	γ	λ	$Re_x^{-1/2} Nu_x$	$Re_x^{1/2} C_f$
0	0	0	1.573433	1.232588
0.02			1.610281	1.382684
0.04			1.634333	1.625081
0.04	0.05		1.673416	1.667025
	0.1		1.711566	1.708036
	0.2		1.785416	1.787623
	0.1	-0.5	0.354240	2.110589
		-0.2	1.242946	1.914480
		0.5	2.645346	0.979397

Next, the results in graphical forms are provided to have a better insight into the effect of the physical parameters. The variations of the local Nusselt number $Re_x^{-1/2} Nu_x$ and the skin friction coefficient $Re_x^{1/2} C_f$ against λ when $\varphi = 0.02$ and $Pr = 6.2$ for several values of γ are shown in Figures 2 and 3. Larger γ gives higher values of $Re_x^{1/2} C_f$ and $Re_x^{-1/2} Nu_x$

on the first solution compared to the flat plate case ($\gamma = 0$). Besides, a unique solution is found when $\lambda \geq -1$. Meanwhile, two solutions are observed for the limited range of λ when the sheet is shrunk ($\lambda_c < \lambda < -1$). The similarity solutions also terminate in this region at $\lambda = \lambda_c$ (critical value). Here, the critical values are respectively $\lambda_c = -1.24657$, -1.32099, and -1.38801 for $\gamma = 0, 0.1$ and 0.2. The velocity $f'(\eta)$ and temperature $\theta(\eta)$ profiles for $\varphi = 0, 0.02$, and 0.04 when $\Pr = 6.2$, $\lambda = -1.24$, and $\gamma = 0.1$ are given in Figures 4 and 5. The reduction of $f'(\eta)$ and $\theta(\eta)$ are observed for both branches with the rising of φ. Physically, the addition of the nanoparticles makes the fluid more viscous and thus, slows down the flow. Consequently, the fluid velocity decreases. Also, the added nanoparticles dissipate energy in the form of heat and consequently exert more energy, which enhances the temperature. However, in this study, we discover that the temperature decreases as φ increases. This behavior is due to the prescribed heat flux on the shrinking surface of the cylinder.

Further, Figures 6 and 7 show the effect of γ on $f'(\eta)$ and $\theta(\eta)$ when $\varphi = 0.02$, $\lambda = -1.24$ and $\Pr = 6.2$. The profiles of $f'(\eta)$ and $\theta(\eta)$ on the first solution incline for larger γ. However, the profiles on the second solution decline. Besides, the negative values of $\theta(\eta)$ are noticed in Figures 5 and 7. The definition of the curvature parameter γ is inversely proportional to the radius of the cylinder, see Equation (11). Thus, the radius of the cylinder decreases as γ increases. Hence, the fluid velocity amplifies due to less resistance occurring between the surface of the cylinder and the fluid. Consequently, the fluid temperature increases for cumulative γ. Since the Kelvin temperature of substances is defined as the average kinetic energy of the particles of substances, as velocity enhances with γ, the kinetic energy increases, and consequently intensifies the temperature [16].

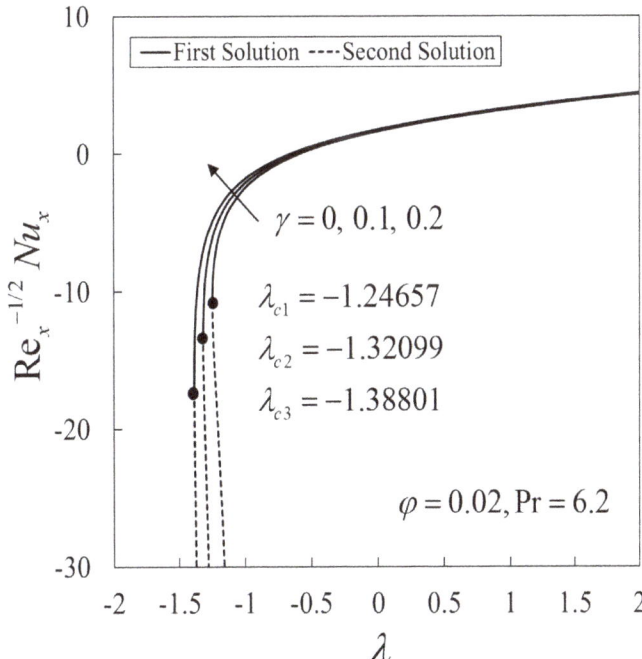

Figure 2. Local Nusselt number $Re_x^{-1/2} Nu_x$ against λ for various values of γ.

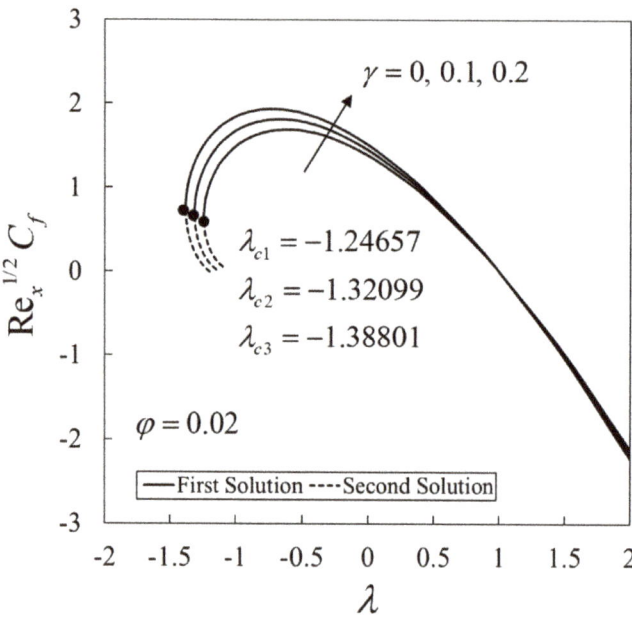

Figure 3. Skin friction coefficient $Re_x^{1/2}C_f$ against λ for various values of γ.

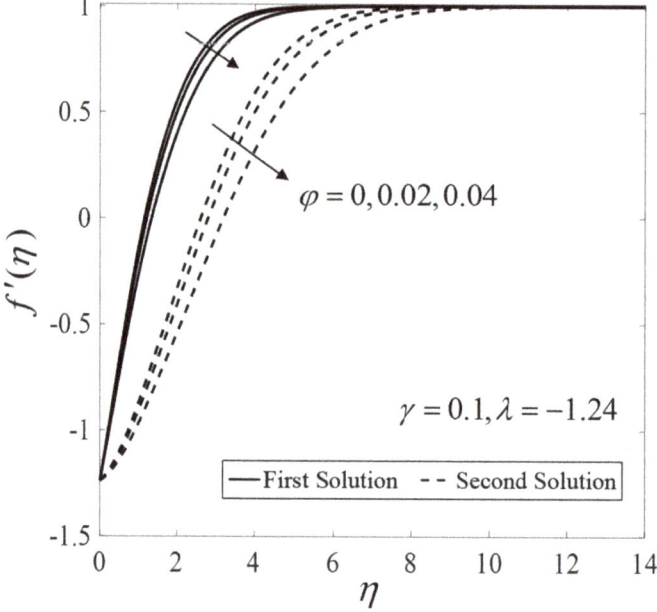

Figure 4. Velocity profiles $f'(\eta)$ for various values of φ.

Figures 8 and 9 display the streamlines when $\lambda = -1.24$ (shrinking sheet), $\varphi = 0.02$, and $\gamma = 0.1$ for the first and the second solutions, respectively. Here, the streamlines are plotted for several values of $\overline{\psi} = \psi/a(c_1\nu_f/L)^{1/2}$. The streamlines are separated into two regions by the horizontal line for both solutions. It is notable that the horizontal line that separates the flow is nearer to the shrinking sheet for the first solution. Besides, the reverse rotating flow occurs in the lower region. Meanwhile, the flow pattern on the upper region behaves as the normal stagnation point.

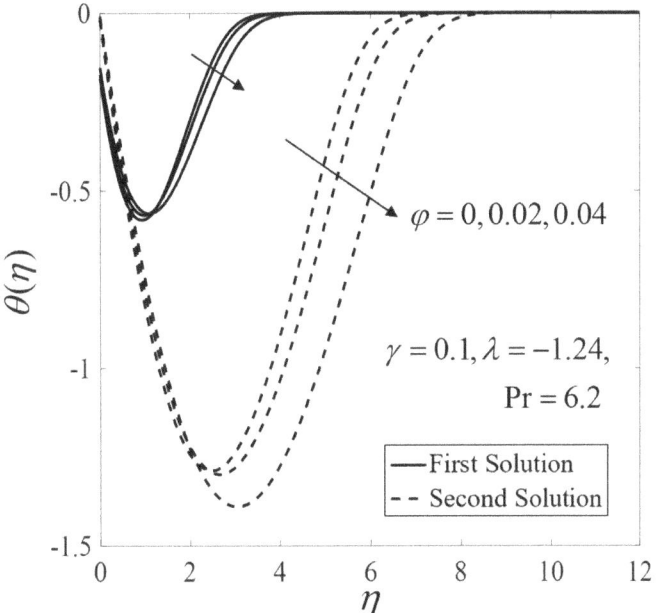

Figure 5. Temperature profiles $\theta(\eta)$ for various values of φ.

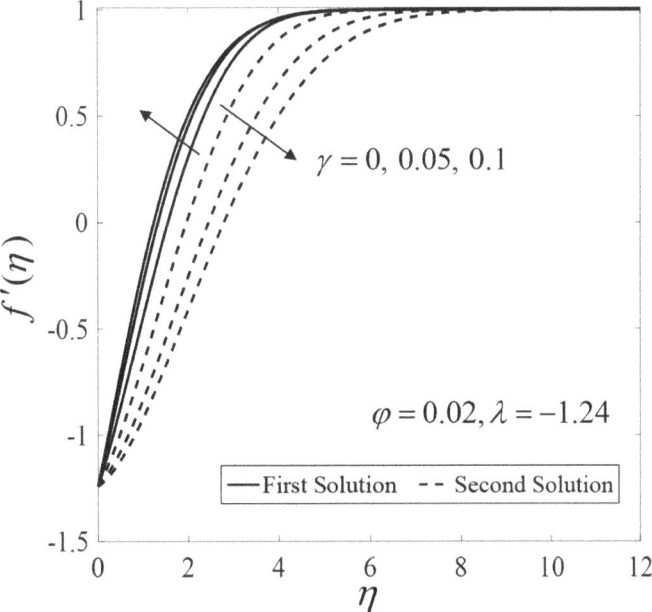

Figure 6. Velocity profiles $f'(\eta)$ for various values of γ.

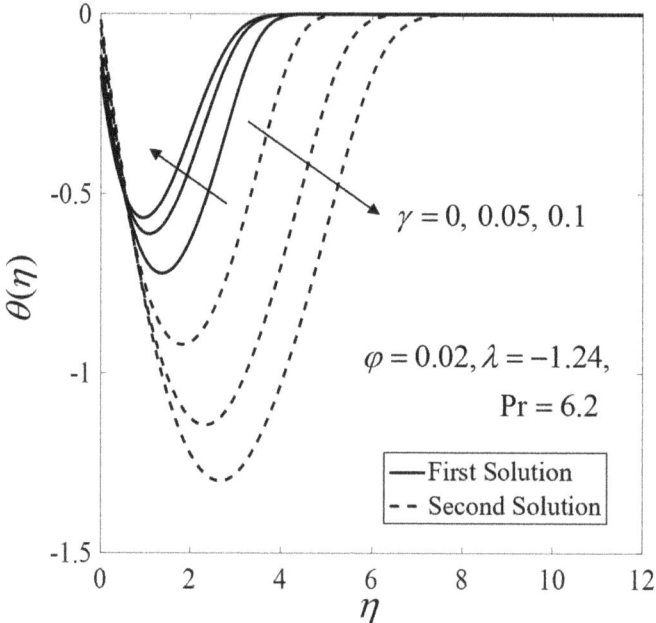

Figure 7. Temperature profiles $\theta(\eta)$ for various values of γ.

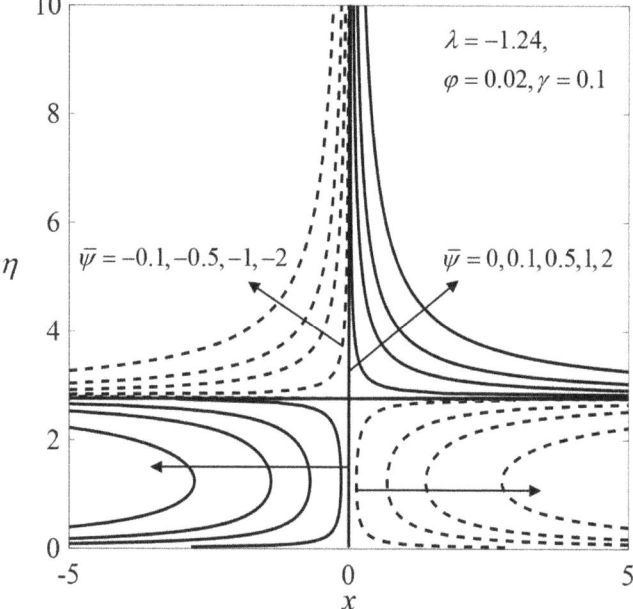

Figure 8. Streamlines for the first solution.

The variation of α against λ when $\varphi = 0.02$ and $\gamma = 0.1$ is described in Figure 10. For positive values of α, it is noted that $e^{-\alpha\tau} \to 0$ as time evolves ($\tau \to \infty$). In contrast, negative values of α, $e^{-\alpha\tau} \to \infty$ as $\tau \to \infty$ show a growth of disturbance as time evolves.

These behaviors show that the first solution is stable, while the second solution is unstable in the long run.

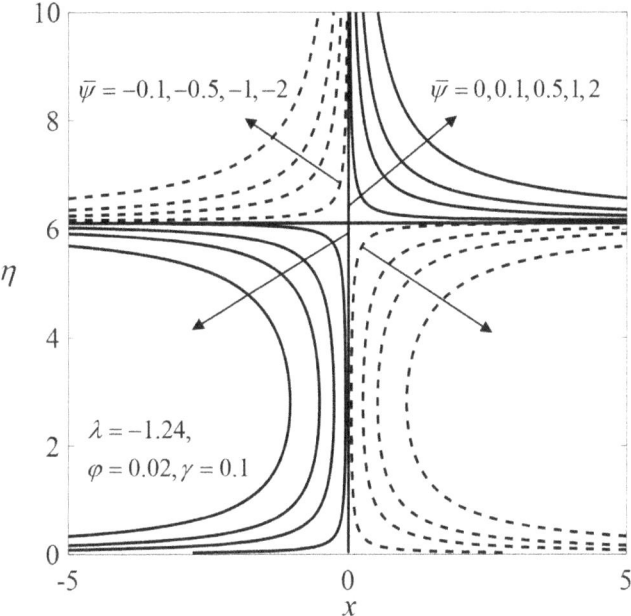

Figure 9. Streamlines for the second solution.

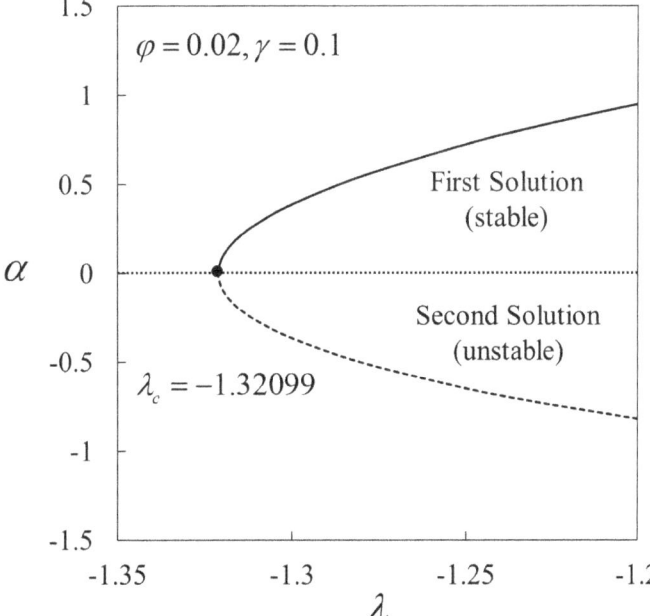

Figure 10. Smallest eigenvalues α against λ.

5. Conclusions

This study examined the stagnation point flow on a shrinking cylinder filled with Al_2O_3 nanoparticles. The surface of the cylinder is subjected to prescribed surface heat flux. The correlations of Al_2O_3/water nanofluid introduced by Ho et al. [25] were employed. Findings revealed two solutions to be observed for the limited range of λ when the sheet is shrunk ($\lambda_c < \lambda < -1$). The similarity solutions terminated in this region at $\lambda = \lambda_c$. Meanwhile, a unique solution was found when $\lambda \geq -1$. The skin friction coefficient $Re_x^{1/2} C_f$ and the local Nusselt number $Re_x^{-1/2} Nu_x$ were intensified with the rising of the nanoparticle volume fraction φ and the curvature parameter γ. Quantitatively, the values of $Re_x^{-1/2} Nu_x$ increased up to 3.87% when φ is increased from 0 to 0.04, and 6.69% when γ is increased from 0.05 to 0.2. Furthermore, Al_2O_3/water nanofluid produced higher values of $Re_x^{1/2} C_f$ and $Re_x^{-1/2} Nu_x$ compared to water. Moreover, the rising of φ tended to reduce the velocity $f'(\eta)$ and the temperature $\theta(\eta)$ for both branches. Besides, the profiles on the first solution incline when larger values of γ are applied. Finally, the temporal stability analysis showed that the first solution is stable while the second solution is unstable over time.

Author Contributions: Conceptualization, I.P.; funding acquisition, A.I.; methodology, I.W.; Project administration, A.I.; supervision, A.I. and I.P.; validation, I.P.; writing—original draft, I.W.; writing—review and editing, A.I., I.P. All authors have read and agreed to the published version of the manuscript.

Funding: This research was funded by Universiti Kebangsaan Malaysia (Project Code: DIP-2020-001).

Institutional Review Board Statement: Not applicable.

Informed Consent Statement: Not applicable.

Data Availability Statement: Not applicable.

Acknowledgments: We acknowledge the Universiti Teknikal Malaysia Melaka and the Universiti Kebangsaan Malaysia (DIP-2020-001) for financial supports

Conflicts of Interest: The authors declare no conflict of interest.

References

1. Hiemenz, K. Die Grenzschicht an einem in den gleichförmigen Flüssigkeitsstrom eingetauchten geraden Kreiszylinder. *Dinglers Polytech. J.* **1911**, *326*, 321–410.
2. Homann, F. Der Einflub grober Zähigkeit bei der Strömung um den Zylinder und um die Kugel. *Z. Angew. Math. Mech.* **1936**, *16*, 153–164. [CrossRef]
3. Ariel, P.D. Hiemenz flow in hydromagnetics. *Acta Mech.* **1994**, *103*, 31–43. [CrossRef]
4. Wang, C.Y. Stagnation flow towards a shrinking sheet. *Int. J. Nonlinear Mech.* **2008**, *43*, 377–382. [CrossRef]
5. Kamal, F.; Zaimi, K.; Ishak, A.; Pop, I. Stability analysis of MHD stagnation-point flow towards a permeable stretching/shrinking sheet in a nanofluid with chemical reactions effect. *Sains Malays.* **2019**, *48*, 243–250. [CrossRef]
6. Wang, C.Y. Fluid flow due to a stretching cylinder. *Phys. Fluids* **1988**, *31*, 466. [CrossRef]
7. Ishak, A. Mixed convection boundary layer flow over a vertical cylinder with prescribed surface heat flux. *J. Phys. A Math. Theor.* **2009**, *42*, 195501. [CrossRef]
8. Awaludin, I.S.; Ahmad, R.; Ishak, A. On the stability of the flow over a shrinking cylinder with prescribed surface heat flux. *Propuls. Power Res.* **2020**, *9*, 181–187. [CrossRef]
9. Muthtamilselvan, M.; Prakash, D. Unsteady hydromagnetic slip flow and heat transfer of nanofluid over a moving surface with prescribed heat and mass fluxes. *Proc. Inst. Mech. Eng. Part C J. Mech. Eng. Sci.* **2015**, *229*, 703–715. [CrossRef]
10. Bachok, N.; Ishak, A. Flow and heat transfer over a stretching cylinder with prescribed surface heat flux. *Malays. J. Math. Sci.* **2010**, *4*, 159–169.
11. Bhattacharyya, K. Heat transfer in boundary layer stagnation-point flow towards a shrinking sheet with non-uniform heat flux. *Chin. Phys. B* **2013**, *22*, 074705. [CrossRef]
12. Qasim, M.; Khan, Z.H.; Khan, W.A.; Ali Shah, I. MHD boundary layer slip flow and heat transfer of ferrofluid along a stretching cylinder with prescribed heat flux. *PLoS ONE* **2014**, *9*, e83930.
13. Mabood, F.; Lorenzini, G.; Pochai, N.; Ibrahim, S.M. Effects of prescribed heat flux and transpiration on MHD axisymmetric flow impinging on stretching cylinder. *Contin. Mech. Thermodyn.* **2016**, *28*, 1925–1932. [CrossRef]
14. Megahed, A.M. Carreau fluid flow due to nonlinearly stretching sheet with thermal radiation, heat flux, and variable conductivity. *Appl. Math. Mech.* **2019**, *40*, 1615–1624. [CrossRef]

15. Waini, I.; Ishak, A.; Pop, I. On the stability of the flow and heat transfer over a moving thin needle with prescribed surface heat flux. *Chin. J. Phys.* **2019**, *60*, 651–658. [CrossRef]
16. Giri, S.S.; Das, K.; Kundu, P.K. Homogeneous–heterogeneous reaction mechanism on MHD carbon nanotube flow over a stretching cylinder with prescribed heat flux using differential transform method. *J. Comput. Des. Eng.* **2020**, *7*, 337–351.
17. Bayda, S.; Adeel, M.; Tuccinardi, T.; Cordani, M.; Rizzolio, F. The history of nanoscience and nanotechnology: From chemical–physical applications to nanomedicine. *Molecules* **2020**, *25*, 112. [CrossRef] [PubMed]
18. Choi, S.U.S.; Eastman, J.A. Enhancing thermal conductivity of fluids with nanoparticles. In Proceedings of the ASME International Mechanical Engineering Congress & Exposition, San Francisco, CA, USA, 12–17 November 1995; pp. 99–105.
19. Pak, B.C.; Cho, Y.I. Hydrodynamic and heat transfer study of dispersed fluids with submicron metallic oxide. *Exp. Heat Transf.* **1998**, *11*, 151–170. [CrossRef]
20. Khanafer, K.; Vafai, K.; Lightstone, M. Buoyancy-driven heat transfer enhancement in a two-dimensional enclosure utilizing nanofluids. *Int. J. Heat Mass Transf.* **2003**, *46*, 3639–3653. [CrossRef]
21. Tiwari, R.K.; Das, M.K. Heat transfer augmentation in a two-sided lid-driven differentially heated square cavity utilizing nanofluids. *Int. J. Heat Mass Transf.* **2007**, *50*, 2002–2018. [CrossRef]
22. Oztop, H.F.; Abu-Nada, E. Numerical study of natural convection in partially heated rectangular enclosures filled with nanofluids. *Int. J. Heat Fluid Flow* **2008**, *29*, 1326–1336. [CrossRef]
23. Mohebbi, R.; Rashidi, M.M. Numerical simulation of natural convection heat transfer of a nanofluid in an L-shaped enclosure with a heating obstacle. *J. Taiwan Inst. Chem. Eng.* **2017**, *72*, 70–84. [CrossRef]
24. Gavili, A.; Isfahani, T. Experimental investigation of transient heat transfer coefficient in natural convection with Al_2O_3-nanofluids. *Heat Mass Transf.* **2020**, *56*, 901–911. [CrossRef]
25. Turkyilmazoglu, M. Single phase nanofluids in fluid mechanics and their hydrodynamic linear stability analysis. *Comput. Methods Programs Biomed.* **2020**, *187*, 105171. [CrossRef] [PubMed]
26. Ho, C.J.; Liu, W.K.; Chang, Y.S.; Lin, C.C. Natural convection heat transfer of alumina-water nanofluid in vertical square enclosures: An experimental study. *Int. J. Therm. Sci.* **2010**, *49*, 1345–1353. [CrossRef]
27. Sheremet, M.A.; Trîmbiţaş, R.; Groşan, T.; Pop, I. Natural convection of an alumina-water nanofluid inside an inclined wavy-walled cavity with a non-uniform heating using Tiwari and Das' nanofluid model. *Appl. Math. Mech.* **2018**, *39*, 1425–1436. [CrossRef]
28. Waini, I.; Ishak, A.; Pop, I. Dufour and Soret effects on Al_2O_3-water nanofluid flow over a moving thin needle: Tiwari and Das model. *Int. J. Numer. Methods Heat Fluid Flow* **2021**, *31*, 766–782. [CrossRef]
29. Turcu, R.; Darabont, A.; Nan, A.; Aldea, N.; Macovei, D.; Bica, D.; Vekas, L.; Pana, O.; Soran, M.L.; Koos, A.A.; et al. New polypyrrole-multiwall carbon nanotubes hybrid materials. *J. Optoelectron. Adv. Mater.* **2006**, *8*, 643–647.
30. Jana, S.; Salehi-Khojin, A.; Zhong, W.H. Enhancement of fluid thermal conductivity by the addition of single and hybrid nano-additives. *Thermochim. Acta* **2007**, *462*, 45–55. [CrossRef]
31. Suresh, S.; Venkitaraj, K.P.; Selvakumar, P.; Chandrasekar, M. Synthesis of Al_2O_3-Cu/water hybrid nanofluids using two step method and its thermo physical properties. *Colloids Surf. A Physicochem. Eng. Asp.* **2011**, *388*, 41–48. [CrossRef]
32. Takabi, B.; Salehi, S. Augmentation of the heat transfer performance of a sinusoidal corrugated enclosure by employing hybrid nanofluid. *Adv. Mech. Eng.* **2014**, *6*, 147059. [CrossRef]
33. Devi, S.P.A.; Devi, S.S.U. Numerical investigation of hydromagnetic hybrid Cu-Al_2O_3/water nanofluid flow over a permeable stretching sheet with suction. *Int. J. Nonlinear Sci. Numer. Simul.* **2016**, *17*, 249–257. [CrossRef]
34. Jamshed, W.; Aziz, A. Cattaneo–Christov based study of TiO_2–CuO/EG Casson hybrid nanofluid flow over a stretching surface with entropy generation. *Appl. Nanosci.* **2018**, *8*, 685–698. [CrossRef]
35. Subhani, M.; Nadeem, S. Numerical analysis of micropolar hybrid nanofluid. *Appl. Nanosci.* **2019**, *9*, 447–459. [CrossRef]
36. Hassan, M.; Faisal, A.; Ali, I.; Bhatti, M.M.; Yousaf, M. Effects of Cu–Ag hybrid nanoparticles on the momentum and thermal boundary layer flow over the wedge. *Proc. Inst. Mech. Eng. Part E J. Process Mech. Eng.* **2019**, *233*, 1128–1136. [CrossRef]
37. Maskeen, M.M.; Zeeshan, A.; Mehmood, O.U.; Hassan, M. Heat transfer enhancement in hydromagnetic alumina–copper/water hybrid nanofluid flow over a stretching cylinder. *J. Therm. Anal. Calorim.* **2019**, *138*, 1127–1136. [CrossRef]
38. Khashi'ie, N.S.; Arifin, N.M.; Nazar, R.; Hafidzuddin, E.H.; Wahi, N.; Pop, I. Magnetohydrodynamics (MHD) axisymmetric flow and heat transfer of a hybrid nanofluid past a radially permeable stretching/shrinking sheet with Joule heating. *Chin. J. Phys.* **2020**, *64*, 251–263. [CrossRef]
39. Zainal, N.A.; Nazar, R.; Naganthran, K.; Pop, I. MHD mixed convection stagnation point flow of a hybrid nanofluid past a vertical flat plate with convective boundary condition. *Chin. J. Phys.* **2020**, *66*, 630–644. [CrossRef]
40. Ghalambaz, M.; Rosca, N.C.; Rosca, A.V.; Pop, I. Mixed convection and stability analysis of stagnation-point boundary layer flow and heat transfer of hybrid nanofluids over a vertical plate. *Int. J. Numer. Methods Heat Fluid Flow* **2020**, *30*, 3737–3754. [CrossRef]
41. Mahanthesh, B.; Shehzad, S.A.; Ambreen, T.; Khan, S.U. Significance of Joule heating and viscous heating on heat transport of MoS2–Ag hybrid nanofluid past an isothermal wedge. *J. Therm. Anal. Calorim.* **2021**, *143*, 1221–1229. [CrossRef]
42. Bilal, M.; Khan, I.; Gul, T.; Tassaddiq, A.; Alghamdi, W.; Mukhtar, S.; Kumam, P. Darcy-forchheimer hybrid nano fluid flow with mixed convection past an inclined cylinder. *Comput. Mater. Contin.* **2021**, *66*, 2025–2039. [CrossRef]
43. Lund, L.A.; Omar, Z.; Dero, S.; Chu, Y.; Khan, I.; Nisar, K.S. Temporal stability analysis of magnetized hybrid nanofluid propagating through an unsteady shrinking sheet: Partial slip conditions. *Comput. Mater. Contin.* **2021**, *66*, 1963–1975. [CrossRef]

44. Waini, I.; Ishak, A.; Pop, I. Hiemenz flow over a shrinking sheet in a hybrid nanofluid. *Results Phys.* **2020**, *19*, 103351. [CrossRef]
45. Waini, I.; Ishak, A.; Pop, I. Flow towards a stagnation region of a vertical plate in a hybrid nanofluid: Assisting and opposing flows. *Mathematics* **2021**, *9*, 448. [CrossRef]
46. Waini, I.; Ishak, A.; Pop, I. Hybrid nanofluid flow towards a stagnation point on an exponentially stretching/shrinking vertical sheet with buoyancy effects. *Int. J. Numer. Methods Heat Fluid Flow* **2021**, *31*, 216–235. [CrossRef]
47. Waini, I.; Ishak, A.; Pop, I. Hybrid nanofluid flow towards a stagnation point on a stretching/shrinking cylinder. *Sci. Rep.* **2020**, *10*, 9296. [CrossRef]
48. White, F.M. *Viscous Fluid Flow*, 3rd ed.; McGraw-Hill: New York, NY, USA, 2006.
49. Merkin, J.H. On dual solutions occurring in mixed convection in a porous medium. *J. Eng. Math.* **1986**, *20*, 171–179. [CrossRef]
50. Weidman, P.D.; Kubitschek, D.G.; Davis, A.M.J. The effect of transpiration on self-similar boundary layer flow over moving surfaces. *Int. J. Eng. Sci.* **2006**, *44*, 730–737. [CrossRef]
51. Harris, S.D.; Ingham, D.B.; Pop, I. Mixed convection boundary-layer flow near the stagnation point on a vertical surface in a porous medium: Brinkman model with slip. *Transp. Porous Media* **2009**, *77*, 267–285. [CrossRef]
52. Shampine, L.F.; Gladwell, I.; Thompson, S. *Solving ODEs with MATLAB*; Cambridge University Press: Cambridge, UK, 2003.

Article

Meta-Heuristic Optimization Methods for Quaternion-Valued Neural Networks

Jeremiah Bill [1], Lance Champagne [1,*], Bruce Cox [1] and Trevor Bihl [2]

[1] Air Force Institute of Technology, Department of Operational Sciences, WPAFB, OH 45433, USA; jeremiah.bill@afit.edu (J.B.); bruceacox1@gmail.com (B.C.)
[2] Air Force Research Laboratory, Sensors Directorate, WPAFB, OH 45433, USA; trevor.bihl.2@us.af.mil
* Correspondence: lance.champagne@afit.edu

Citation: Bill, J.; Champagne, L.; Cox, B.; Bihl, T. Meta-Heuristic Optimization Methods for Quaternion-Valued Neural Networks. *Mathematics* **2021**, *9*, 938. https://doi.org/10.3390/math9090938

Academic Editor: Alessandro Niccolai

Received: 31 March 2021
Accepted: 17 April 2021
Published: 23 April 2021

Publisher's Note: MDPI stays neutral with regard to jurisdictional claims in published maps and institutional affiliations.

Copyright: © 2021 by the authors. Licensee MDPI, Basel, Switzerland. This article is an open access article distributed under the terms and conditions of the Creative Commons Attribution (CC BY) license (https://creativecommons.org/licenses/by/4.0/).

Abstract: In recent years, real-valued neural networks have demonstrated promising, and often striking, results across a broad range of domains. This has driven a surge of applications utilizing high-dimensional datasets. While many techniques exist to alleviate issues of high-dimensionality, they all induce a cost in terms of network size or computational runtime. This work examines the use of quaternions, a form of hypercomplex numbers, in neural networks. The constructed networks demonstrate the ability of quaternions to encode high-dimensional data in an efficient neural network structure, showing that hypercomplex neural networks reduce the number of total trainable parameters compared to their real-valued equivalents. Finally, this work introduces a novel training algorithm using a meta-heuristic approach that bypasses the need for analytic quaternion loss or activation functions. This algorithm allows for a broader range of activation functions over current quaternion networks and presents a proof-of-concept for future work.

Keywords: multilayer perceptrons; quaternion neural networks; metaheuristic optimization; genetic algorithms

1. Introduction

Over the last several decades, the explosive growth in artificial intelligence and machine learning (AI/ML) research has driven a need for more efficient data representations and machine learning training methods. As machine learning applications have expanded into new and exciting domains, the scale of data processed through enterprise systems has grown to an almost incomprehensible level. While computational resources have grown commensurately with this increase in data, inefficiencies in current neural network architectures continue to hamper progress on difficult optimization problems.

This work examines the use of hypercomplex numbers in neural networks, with a particular emphasis on the use of quaternions in neural network architectures. This work demonstrates that quaternion data representations can reduce the total number of trainable neural network parameters by a factor of four, resulting in improvements in both computer memory allocations and computational runtime. Additionally, this work presents a novel, gradient-free, quaternion genetic algorithm that enables the use of several loss and activation functions previously unavailable due to differentiability requirements.

The remainder of this article is organized as follows: Section 2 provides a review of neural networks, the quaternion number system, quaternion neural networks, and metaheuristic optimization techniques. Section 3 describes the methodology used to develop a quaternion neural network and a novel quaternion genetic training algorithm. Section 4 presents the network results, comparing the quaternion genetic algorithm performance to two analogous real-valued networks. Additionally, a multidimensional input/multidimensional output network is presented for predicting the Lorenz attractor chaotic dynamical system. Finally, Section 5 provides conclusions, recommendations, and proposals for future work.

2. Background and Related Work

2.1. Neural Networks and Multi-Layer Perceptrons

Statistical learning processes have received increasing attention in recent years with the proliferation of large datasets, ever-increasing computing power, and simplified data exploration tools. In 1957, Frank Rosenblatt proposed a neural structure called the perceptron [1]. A perceptron is composed of several threshold logic units (TLUs), each of which takes a weighted sum of input values and uses the resulting sum as the input to a non-linear activation function. While each TLU computes a linear combination of the inputs based on the network weights, the use of a non-linear activation function allows the perceptron to estimate a number of non-linear functions by adjusting the weights of each input.

Stacking multiple layers of perceptrons together so that the output of one perceptron forms the input to a subsequent perceptron allows for the estimation of a vast set of linear and non-linear problems. In fact, two contemporaries, Cybenko [2] and Hornik et al. [3] both independently showed that a network with a single hidden layer and sigmoidal activation functions is able to approximate any nonlinear function to an arbitrary degree of accuracy. This network structure is called the multilayer perceptron (MLP) and it forms the most basic deep neural network (DNN). This result (called the Universal Approximation Theorem) has provided the theoretical justification that has driven neural network research to the present day. A representation of an MLP is shown in Figure 1, and [4] provides an overview of MLPs and other common neural network structures.

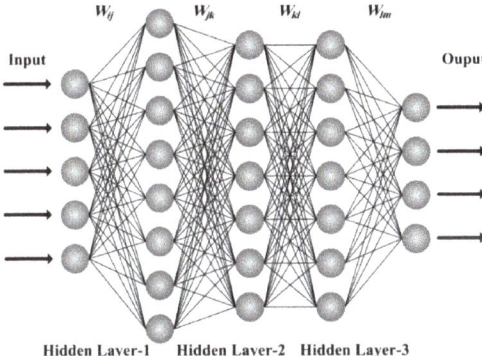

Figure 1. Representation of a basic MLP [5].

2.1.1. The Backpropagation Algorithm

Although artificial neural networks have existed since the mid-20th century, researchers found them to be computationally expensive to train and impractical for most applications. As a result, neural network research was largely stagnant until 1986, when Rumelhart et al. [6] introduced the backpropagation algorithm for training a neural network. The algorithm developed by Rumelhart et al. extended several key ideas that Werbos [7] presented in his unpublished doctoral dissertation.

The backpropagation algorithm has proven to be a straightforward, easy-to-understand, and easy-to-implement algorithm that has enabled efficient implementations of neural networks across a wide-range of problem sets. Examples of custom architectures include convolutional neural networks (CNNs) for processing image data, recurrent neural networks (RNNs) for processing sequence data, and generative adversarial networks (GANs) which have been used in recent years to create deep fakes and very convincing counterfeit data [8].

2.1.2. Shortfalls

Despite artificial neural networks achieving state-of-the-art results in a breathtaking array of problem domains, ANNs are not without their shortfalls. For example, ANNs often require a vast amount of training data. Standard machine learning datasets such as the ImageNet dataset for computer vision often contain several million datapoints [9]. Consequently, training an ANN requires a large amount of computer resources, in terms of both RAM and processing time. Additionally, the backpropagation algorithm requires a significant amount of low-level computational power in order to perform the matrix multiplications for each forward and backward pass. While GPUs have proven to be particularly well-suited for this task [10], many of the current large-scale ANN research applications require prohibitive amounts of computer memory and GPU hours.

Finally, MLPs can struggle to maintain any sort of spatial relationships that are present within the training data. A simple example of this is seen in color image processing. In general, each of the three color channels of an RGB image are processed separately in an MLP since the 3-dimensional matrix representation of the image must first be flattened into a vector for the network forward pass step. This results in the loss of the spatial relationship between the red, green, and blue pixel intensities at each pixel.

Many spatial dependency issues can be alleviated using more advanced ANN architectures such as convolutional neural networks, which preserve spatial relationships within the data using successive convolutional layers to transform the input data [11]. However, every CNN must contain at least one fully-connected layer prior to the output layer which flattens the output of the final convolution into a 1-dimensional real-valued vector. Thus, even with a CNN, there is some spatial information that is lost when the output of the final convolution is flattened.

On the other hand, Yin et al. [12] highlight the fact that this spatial hierarchy between pixel intensity values can be maintained when using higher-dimensional number systems such as quaternions as opposed to real numbers, and their result is a significant motivation for this paper. Matsui et al. [13] demonstrated similar experimental results on a 3-dimensional affine transformation problem, showing that quaternion-valued deep neural networks were able to recover the spatial relationships between 3-dimensional coordinates. Section 2.2 provides a brief summary of hypercomplex number systems, along with a review of their use and success in advanced neural network applications.

2.2. The Quaternions

The quaternion numbers (denoted by $\mathbb{H}$) are a four-dimensional extension of the complex numbers. Complex numbers have the form $x + iy$, consisting of a real part x and an imaginary part y, and can be thought of as an isomorphism of $\mathbb{R}^2$. That is, the complex numbers contain two copies of the real number line, allowing a single complex number to encode twice as much information as a single real number. Complex numbers are particularly useful for describing motion in 2-dimensional space, since there is a very succinct analogue between complex multiplication and rotations in the plane [14].

Quaternions are referred to as hypercomplex numbers. Each quaternion $\mathbf{q}$ consists of a real part and three imaginary parts, so that the quaternions form an isomorphism with $\mathbb{R}^4$ with basis elements 1, $\mathbf{i}$, $\mathbf{j}$, and $\mathbf{k}$:

$$\mathbf{q} = r + x\mathbf{i} + y\mathbf{j} + z\mathbf{k}. \qquad (1)$$

Quaternions form a generalization of the complex numbers, where the three imaginary components $\mathbf{i}$, $\mathbf{j}$, and $\mathbf{k}$ follow the same construct as $\mathbf{i}$ in $\mathbb{C}$:

$$\mathbf{i}^2 = \mathbf{j}^2 = \mathbf{k}^2 = -1. \qquad (2)$$

However, the three imaginary basis components must also satisfy the following rules:

$$\mathbf{jk} = -\mathbf{kj} = \mathbf{i} \qquad (3)$$
$$\mathbf{ki} = -\mathbf{ik} = \mathbf{j} \qquad (4)$$
$$\mathbf{ij} = -\mathbf{ji} = \mathbf{k}. \qquad (5)$$

These rules clearly demonstrate that quaternion multiplication is non-commutative. However, since the multiplication of any two basis elements is plus or minus another basis element, the quaternions under these rules form a non-abelian group, denoted Q_8. The group Q_8, along with the operations of addition and multiplication form a division algebra, which is an algebraic structure similar to a field where multiplication is non-commutative.

The 4-dimensional structure of each quaternion number indicates that quaternions are capable of encoding four copies of the real number line into a single quaternion number, analogous to the two copies of $\mathbb{R}$ encoded in the complex numbers. Quaternions were discovered by the Irish mathematician Sir William Rowan Hamilton in 1843 [15], hence why the set of quaternions is referred to as $\mathbb{H}$ and the quaternion notion of multiplication, described below, is referred to as the *Hamilton Product*. For an in-depth review of quaternions and their applications, see [16].

2.2.1. Quaternion Algebra

The quaternions form a division algebra, meaning that the set of quaternions along with the operations of addition and multiplication follow 8 of the 9 field axioms (all but commutativity). Quaternion addition is defined using the element-wise addition operation. For two quaternions $\mathbf{q}_1, \mathbf{q}_2 \in \mathbb{H}$, where:

$$\mathbf{q}_1 = r_1 + x_1 \mathbf{i} + y_1 \mathbf{j} + z_1 \mathbf{k}$$

and

$$\mathbf{q}_2 = r_2 + x_2 \mathbf{i} + y_2 \mathbf{j} + z_2 \mathbf{k}.$$

The sum $\mathbf{q}_1 + \mathbf{q}_2$ is defined as,

$$\mathbf{q}_1 + \mathbf{q}_2 := (r_1 + r_2) + (x_1 + x_2)\mathbf{i} + (y_1 + y_2)\mathbf{j} + (z_1 + z_2)\mathbf{k}. \tag{6}$$

Quaternion multiplication, referred to as the Hamilton Product, can easily be derived using the basis multiplication rules in Equations (3)–(5) and the distributive property. In reduced form, the Hamilton Product of two quaternions $\mathbf{q}_1$ and $\mathbf{q}_2$ is defined as:

$$\begin{aligned}\mathbf{q}_1 * \mathbf{q}_2 :=& (r_1 r_2 - x_1 x_2 - y_1 y_2 - z_1 z_2) \\ &+ (r_1 x_2 + x_1 r_2 + y_1 z_2 - z_1 y_2)\mathbf{i} \\ &+ (r_1 y_2 - x_1 z_2 + y_1 r_2 + z_1 x_2)\mathbf{j} \\ &+ (r_1 z_2 + x_1 y_2 - y_1 x_2 + z_1 r_2)\mathbf{k}. \end{aligned} \tag{7}$$

2.2.2. Quaternion Conjugates, Norms, and Distance

The notion of a quaternion conjugate is analogous to that of complex conjugates in $\mathbb{C}$. The conjugate of a quaternion $\mathbf{q} = r + x\mathbf{i} + y\mathbf{j} + z\mathbf{k}$ is given by $\mathbf{q}^* = r - x\mathbf{i} - y\mathbf{j} - z\mathbf{k}$. The norm of a quaternion is equivalent to the Euclidean norm in $\mathbb{R}$ and is given by:

$$||\mathbf{q}|| := \sqrt{\mathbf{q}\mathbf{q}^*} = \sqrt{r^2 + x^2 + y^2 + z^2}. \tag{8}$$

With this quaternion norm, one can also define a notion of distance $d(\mathbf{q}, \mathbf{p})$ between two quaternions $\mathbf{q}$ and $\mathbf{p}$ as:

$$d(\mathbf{q}, \mathbf{p}) := ||\mathbf{q} - \mathbf{p}||. \tag{9}$$

2.2.3. Quaternionic Matrices

Since the set of quaternions $\mathbb{H}$ form a division algebra under addition and the Hamilton product, they also form a non-commutative ring under the same operations. Hence, quaternionic matrix operations can be defined as for matrices over an arbitrary ring. Given any two quaternionic matrices $A, B \in \mathbb{H}^{M \times N}$, the sum $A + B$ is defined element-wise:

$$(A + B)_{ij} := A_{ij} + B_{ij}. \tag{10}$$

Similarly, for any quaternionic matrix $A \in \mathbb{H}^{M \times N}$ and $B \in \mathbb{H}^{N \times P}$, the product $AB \in \mathbb{H}^{M \times P}$ is defined as:

$$(AB)(m, p) := \sum_{n=1}^{N} A(m, n) B(n, p), \quad \forall m = 1, \ldots, M, p = 1, \ldots, P. \tag{11}$$

As with matrix multiplication over an arbitrary ring, quaternionic matrix multiplication is non-commutative. Additionally, great care must be taken to ensure the proper execution of the Hamilton product when multiplying each row of A with each column of B, since the Hamilton product itself is non-commutative.

2.3. Quaternion-Valued Neural Networks (QNNs)

Many practical applications of machine learning techniques involve data that are multidimensional. With the mathematical machinery described in Section 2.2, the quaternions provide a succinct and efficient way of representing multidimensional data. Additionally, when applied to neural network architectures, quaternions have been shown to preserve spatial hierarchies and interrelated data components that are often separated and distorted in real-valued MLP architectures. This section provides a brief review of QNN research, starting with a brief note on some of the issues in QNN construction stemming from quaternionic analysis and quaternion calculus. Then, the development of QNNs is traced chronologically from early works to the state of the art.

2.3.1. A Note on Quaternion Calculus and Quaternionic Analysis

There are very few analytic functions of a quaternion variable. To account for this, quaternion networks generally utilize "split" activation functions, where a real-valued activation function is applied to each quaternion coefficient. For example, the split quaternion sigmoid function [17] for a quaternion $\mathbf{q} = r + x\mathbf{i} + y\mathbf{j} + z\mathbf{k}$ is given by:

$$\sigma(\mathbf{q}) = \sigma(r) + \sigma(x)\mathbf{i} + \sigma(y)\mathbf{j} + \sigma(z)\mathbf{k}, \tag{12}$$

where $\sigma(\cdot)$ is the real-valued sigmoid function. Similar definitions hold for any real-valued activation function, and many QNNs utilize these split activation functions even when quaternionic functions, such as the quaternion-valued hyperbolic tangent function, are available. Research has indicated that true quaternionic activation functions can improve performance over split activation functions [18], but they require special considerations since their analyticity can only be defined over a localized domain, and the composition of two locally analytic quaternion functions is generally not locally analytic [19], providing limited utility in deep neural networks. Additionally, many complex and quaternion-valued elementary transcendental functions, including the hyperbolic tangent, are unbounded and contain singularities [20] that make neural network training difficult.

These issues, along with the non-commutativity of quaternions, also affect the gradient descent algorithm employed in many quaternion networks. Generally speaking, the non-commutativity of quaternions precludes the development of a general product rule and a quaternion chain rule to compute quaternion derivatives and partial derivatives. Thus, quaternion networks must employ split loss functions and the partial derivatives used in the backpropagation algorithm are calculated using a similar "split" definition. The split partial derivative used in training a Quaternion Multilayer Perceptron (QMLP) network, first defined by [17], is given by:

$$\frac{\partial E}{\partial W^l} = \frac{\partial E}{\partial W_r^l} + \frac{\partial E}{\partial W_x^l}\mathbf{i} + \frac{\partial E}{\partial W_y^l}\mathbf{j} + \frac{\partial E}{\partial W_z^l}\mathbf{k}, \tag{13}$$

where E is the loss function and W^l is the weight matrix at layer l. Some researchers refer to this as a "channelwise" [18] or vectorized implementation.

Researchers have made several advances in quaternion calculus, dubbed the generalized Hamilton-Real (G$\mathbb{H}$R) calculus [21], with novel product and chain rules. However, as of this writing, the G$\mathbb{H}$R calculus and the associated learning algorithms implementing the G$\mathbb{H}$R product and chain rules have yet to be applied to any real-world machine learning dataset with a deep quaternion network.

This work proposes a genetic algorithm to train a quaternion-valued neural network with fully quaternion activation functions at each layer of the network. The genetic algorithm circumvents the need for the convoluted calculus rules that one must employ in traditional QNNs due to the non-commutativity of quaternions and the locally analytic nature of the activation functions, allowing for a broader range of available activation functions. While not yet proven in the quaternion domain, this approach has a strong theoretical basis that is supported in both the complex- and real-valued domains ([2,3,20]).

2.3.2. Quaternion Neural Networks

The QMLP was first introduced by Arena et al. [17] in 1994, as noted in Section 2.3.1. The initial QMLP used split sigmoid activation functions and a version of the mean square error (MSE) loss function E, formed by substituting quaternions into the real-valued MSE equation. For a network with $l = 1, \ldots, M$ layers and $1 < n < N_l$ nodes per layer, the output of each node n in each layer l is computed as:

$$y_n^l = \sigma(S_n^l), \tag{14}$$

where σ is any split sigmoidal activation function and S_n^l is the linear combination of network weights, biases, and the output of the $l-1$ layer computed as in a normal MLP:

$$S_n^l = \sum_{m=0}^{N_{l-1}} w_{nm}^l * y_m^{l-1} + b_n^l. \tag{15}$$

For each S_n^l, the weights, biases, and y-values are all quaternions. Thus, $*$ represents the Hamilton Product. The loss function E is given by:

$$E = \frac{1}{N}\sum_{n=1}^{N}(\mathbf{t}_n - \mathbf{y}_n^{(M)})^2, \tag{16}$$

where $\mathbf{t}$ represents the target (truth) data and $\mathbf{y}^{(M)}$ represents the neural network output at the Mth layer.

The authors also introduced a simple learning algorithm using the split or "channelwise" partial derivatives discussed in Section 2.3.1, where the gradient Δ_n^l at the output layer is simply the output error of the network $(\mathbf{t}_n - \mathbf{y}_n^{(M)})$ and the error at each prior layer l is calculated using the formula:

$$\Delta_n^l = \sum_{n=1}^{N^{l+1}} w_{hn}^{*l+1} * (\Delta_n^{l+1} \cdot \sigma'(S_n^{l+1})), \tag{17}$$

where w_{hn}^{*l+1} represents the quaternion conjugate of the weight connecting node h in the lth layer to node n in the $l+1$st layer. Additionally, $(\cdot)$ represents the componentwise product, not the Hamilton Product between the gradient at the $l+1$st layer and the channelwise partial derivative of $\sigma(\cdot)$. Using this gradient rule, the biases at each layer are updated according to the normal backpropagation process:

$$b_n^l = b_n^l + \epsilon \Delta_n^l, \qquad (18)$$

where ϵ is the learning rate. Note, however, that the weights are updated using the rule:

$$w_{nm}^l = w_{nm}^l + \epsilon \Delta_n^l * S_m^{*l-1}, \qquad (19)$$

where S_m^{*l-1} represents the conjugate of the input to the lth layer S_m^{l-1}.

Although the quaternion backpropagation algorithm bears similarities to the real-valued backpropagation algorithm, it is unique in several ways. The first is the use of split derivatives in the weight and bias update step. Although the use of split derivatives may seem like a trick to bypass a true quaternion derivative definition, it builds on [22], which proved that split activation functions and derivatives in the complex domain could universally approximate complex-valued functions. While unproven in the quaternion domain, Arena et al. demonstrated the effectiveness of this network on a small function approximation problem, where a quaternion network was used to approximate a quaternion-valued function. Additionally, the weight update and the gradients leverage the quaternion conjugate, which improves training performance.

Since the introduction of the QMLP and its associated training algorithm, researchers have used QMLPs for a variety of tasks. In particular, QMLPs have been used as autoencoders [23], for color image processing [24], text processing [25], and polarized signal processing [26]. Another natural application of quaternions is in robotic control [27], since quaternions can compactly represent 3-dimensional rotation and motion through space. Parcollet et al. [28] note that in every scenario, QMLPs always outperform real-valued MLPs when processing 3- or 4-dimensional signals. These simple networks have driven further research in more advanced network architectures such as convolutional neural networks and recurrent neural networks, both of which have shown promise in the quaternion domain for advanced image processing [29], speech recognition [30], and other tasks.

2.4. Metaheuristic Optimization Techniques

Whereas the backpropagation algorithm discussed in Section 2.1.1 has dominated nearly all neural network research since it was first introduced, recent work has shown that heuristic search methods can also effectively train neural networks at a scale comparable to gradient descent and backpropagation. Metaheuristic optimization encompasses a broad range of optimization techniques that do not provide guarantees of algorithmic closure or convergence, but have shown empirically to perform well in a variety of complex optimization tasks. In contrast to gradient-based methods such as the backpropagation algorithm, many metaheuristics do not require any gradient information.

Perhaps the most famous application of a metaheuristic approach in training neural networks is the NeuroEvolution through Augmenting Topologies (NEAT) [31] process, which uses a genetic algorithm to simultaneously train and grow neural networks through an evolutionary process. NEAT has proven to be a very effective neural network training tool, and subsequent variants of NEAT have successfully evolved neural networks with millions of weight and bias parameters [32]. More recently, researchers with Uber's OpenAI Labs have shown that even basic Genetic Algorithms can compete with backpropagation in training large networks with up to four million parameters [33]. Several other metaheuristic implementations have shown promise in training neural networks and optimizing the hyperparameters of neural networks. See [34] for a full review of metaheuristic optimization in neural network design.

Metaheuristic optimization methods have also been applied to a limited number of search problems in the quaternion domain. A quaternion variant of the Firefly Algorithm [35] demonstrated comparable performance to the real-valued Firefly Algorithm in optimizing nonlinear test functions. In addition, [36] introduced a quaternion-based Harmony Search algorithm, demonstrating the algorithm's performance on a similar range of nonlinear test functions. The hypothesis of both approaches is that the search space in the hypercomplex domain is smoother than the search space in $\mathbb{R}$. While not proven, [37] summarizes the approach. Additionally, Khuat et al. [38] introduced a quaternion genetic algorithm with multi-parent crossover that was used to optimize a similar set of nonlinear test functions. Finally, [39] used the Harmony Search algorithm introduced in [36] to fine-tune the hyperparameters of a neural network. However, as of this writing, quaternion metaheuristic search methods have yet to be applied to more complex tasks, such as optimizing a large number of weights and biases in a quaternion neural network.

Given the difficulties in defining globally analytic quaternion loss functions, activation functions, and quaternion partial derivatives, metaheuristic optimization provides an ideal method of training quaternion neural networks. Section 3 outlines a novel quaternion genetic algorithm for training the weights and biases of quaternion neural networks. The algorithm does not require gradient information and makes no assumptions on the analyticity of the activation functions of the network at each layer, allowing for a broader range of quaternion activation functions than have been available in prior works.

3. Methodology

This section describes the test methodology employed in comparing the performance of real-valued MLPs to quaternion-valued MLPs in several multidimensional function approximation tasks. First, Section 3.1 describes the test functions selected for use in the study. Section 3.2 outlines the structure of the neural networks, including an overview of the neurons, layers, and total trainable parameters of each network. Section 3.3 details the genetic algorithm used to train the real- and quaternion-valued networks. Finally, Section 3.4 presents a description of the evaluation strategy and key comparison metrics.

3.1. Test Functions

Demonstrating the ability of a neural network to approximate an arbitrary nonlinear function is a crucial step in the development of any ANN structure. Cybenko's Universal Approximation Theorem [2], discussed in Section 2.1, provides the theoretical underpinning for all modern ANN research and has legitimized many of the ANN applications to date. While still unproven for the quaternion domain, this research demonstrates that quaternion neural networks with elementary transcendental activation functions and a genetic training algorithm can effectively approximate arbitrary nonlinear functions, using the Ackley function and the Lorenz attractor chaotic system as test cases.

3.1.1. The Ackley Function

The Ackley function is a non-convex test function that is often used to test global optimization algorithms. It was first introduced by David Ackley [40] and has since been included in a standard library of optimization test functions. In three dimensions, the function is characterized by an elevated eggcrate-like surface, with a global minimum in the center of the function that sinks down to zero. The Ackley function is a good test case for quaternion networks since it can easily be defined in any number of dimensions. A vector representation of the function is given in Equation (20), where a, b, and c are constants and n represents the dimensionality of the vector $\mathbf{x}$. Additionally, a three-dimensional plot of the Ackley function is shown in Figure 2.

$$f(\mathbf{x}) = -a \exp\left(-b\sqrt{\frac{1}{n}\sum_{i=1}^{n} x_i^2}\right) - \exp\left(\frac{1}{n}\sum_{i=1}^{n} \cos(c \cdot x_i)\right) + a + \exp(1) \qquad (20)$$

This research uses a 4-dimensional Ackley function, with the a, b, and c coefficient values set to 20.0, -0.2, and 2π, respectively. The function's x, y, and z values are generated over the range $[-5, 5]$, using a meshgrid with a spacing of 0.5 between each point. With three-dimensional input, and this results in 9261 data points. The coordinate values are then translated from $\mathbb{R}$ into $\mathbb{H}$ by taking the coordinates of each point and casting them into the three imaginary parts of a quaternion. For example, the point $(-5, -5, -5) \Rightarrow \mathbf{q}_1 = 0r - 5\mathbf{i} - 5\mathbf{j} - 5\mathbf{k}$.

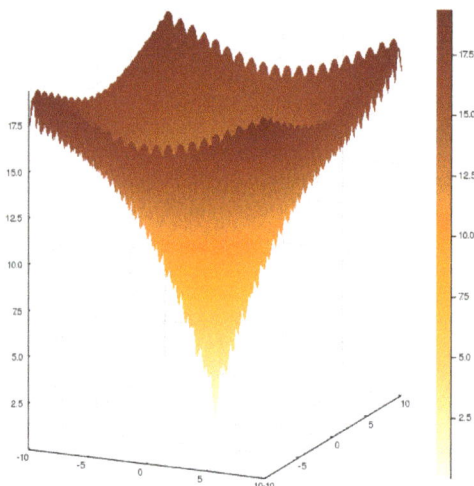

Figure 2. 3D Ackley function.

Finally, the data is split into a training set and a test set. The purpose of this split is to ensure that the neural networks are producing functions with good generalization capabilities. The data points are randomly shuffled and 80% of the data points are retained as training data while 20% of the data points are split into the test set.

3.1.2. The Lorenz Attractor Chaotic System

The Lorenz attractor is a deterministic system of differential equations that was first presented by Edward Lorenz [41]. The Lorenz attractor is a chaotic system, meaning that while it is deterministic, the system never cycles and never reaches a steady state. Additionally, the system is very sensitive to initial conditions. When represented as a set of 3-dimensional coordinates, the Lorenz attractor produces a mesmerizing graph often referred to as the Lorenz butterfly. A static representation of this is shown in Figure 3.

The Lorenz attractor is governed by the following system of differential equations:

$$\frac{dx}{dt} = \sigma(y - x) \tag{21}$$

$$\frac{dy}{dt} = \rho x - y - xz \tag{22}$$

$$\frac{dz}{dt} = xy - \beta z \tag{23}$$

where σ, ρ, and β are constants. For this experiment (and in Figure 3), $\sigma = 10$, $\rho = 28$, and $\beta = \frac{8}{3}$. Quaternions are naturally well-suited to predicting chaotic time series, including the Lorenz attractor, since the problem involves both a multidimensional input and a multidimensional output. Split quaternion neural networks have proven quite successful at chaotic time series prediction based on small training datasets ([42–45]).

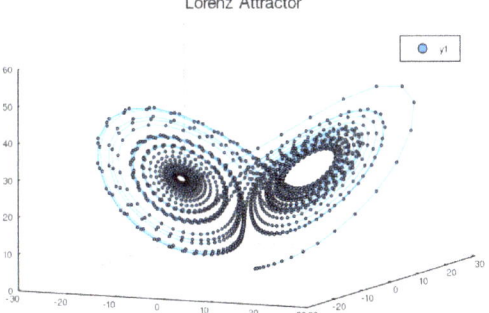

Figure 3. Lorenz Attractor.

The data for the Lorenz attractor was again split 80%/20% between training and test datasets. Additionally, both the inputs and the outputs were cast into the quaternion domain. This allowed for a direct output error calculation using the quaternion distance metric defined in Section 2.2.2. The full details of the loss function, the activation functions, the neurons, and the layers of the networks used in both experiments are discussed in Section 3.2.

3.2. MLP Network Topologies

3.2.1. Function Approximation

The function approximation experiment focused on the relative performance of real-valued network architectures to quaternion networks with pure quaternion activation functions. The comparison experiment operated on three distinct network architecture and training algorithm combinations. The first is the quaternion multilayer perceptron trained with a genetic algorithm (from here on referred to as QMLP+GA). This network consists of an input layer, two hidden layers, and an output layer.

Between each layer of the network, a "normalization" step was added, where the output of each layer is individually normalized. Since the training data-points were encoded into quaternion values, the input and output layer require a single node each. The two hidden layers of the network contain 3 nodes each, resulting in a total of 22 trainable weights and biases for the network. The pure-quaternion hyperbolic tangent (*tanh*) function was selected as the nonlinear activation function for the input layer and both hidden layers. The *tanh* function in the quaternion domain is defined as:

$$tanh(\mathbf{q}) = \frac{e^{2\mathbf{q}} - 1}{e^{2\mathbf{q}} + 1}, \qquad \mathbf{q} \in \mathbb{H}. \tag{24}$$

To determine the loss at the output layer, the final output is first mapped from $\mathbb{H}$ into $\mathbb{R}$ using the norm defined in Section 2.2.2. This mapping allows for the use of any real-valued loss function, and the mean absolute error (MAE) loss function was selected due to its simplicity. The MAE is given by:

$$\frac{1}{N} \sum_{i=1}^{N} |\hat{y} - y|, \tag{25}$$

where N is the number of data-points, $\hat{y}$ is the predicted value, and y is the truth or target value.

To provide a baseline comparison for the QMLP+GA network, an equivalent real-valued network is constructed and trained using the same genetic algorithm as the QMLP+GA. Finally, an identical MLP is constructed and trained using the gradient descent (GD) algorithm. These two variants are referred to as the MLP+GA network and the

MLP+GD network, respectively. The layers, neurons per layer, and total parameters of each of the three networks are summarized in Table 1.

Table 1. Neural network topologies for the Ackley Function approximation.

Network	Input	Hidden 1	Hidden 2	Output	Parameters
QMLP+GA	1	3	3	1	22
MLP+GA	3	9	9	3	136
MLP+GD	3	9	9	3	136

The real-valued hyperbolic tangent was used as the activation function on the input layer and both hidden layers, with a MAE loss function. However, since the hyperbolic tangent is globally analytic in $\mathbb{R}$, the normalization layers from the QMLP were removed. The learning rate η for the gradient descent algorithm was set to $\eta = 0.03$. The real-valued MLPs contained a total of 136 trainable weight and bias parameters, a six-fold increase over the QMLP.

3.2.2. Chaotic Time Series Prediction

Chaotic time series prediction of the Lorenz attractor requires multidimensional input data as well as multidimensional output data. It is a notoriously difficult problem, especially considering the system's sensitivity to initial conditions. In contrast with the function approximation experiment, the time series prediction experiment focused on the ability of quaternion networks to learn complex multidimensional nonlinearities. To that end, the time series prediction experiment centered on optimizing a set of quaternion network hyperparameters and did not consider any equivalent real-valued networks.

To test the predictive capabilities of a simple QMLP+GA network, a set of 500 time series inputs were generated using a fixed-timestep 4th-order Runge–Kutta Ordinary Differential Equation (ODE) solver. The first 400 time series formed the training dataset, while the last 100 were held out for the test set. The starting point for each time series was randomly generated using a uniform $U[-10.0, 10.0]$ distribution for the x- and y-coordinates and a uniform $U[0.0, 10.0]$ distribution for the z-coordinates. Initial tests focused on relatively short time series inputs. Each series was generated over a range of 20 timesteps, and the first 10 values of each series formed the input training data, while the last 10 values formed the target values for training.

Figure 4 illustrates the sensitivity of the Lorenz system to initial starting conditions. Several initial starting points were generated using the distributions defined above for the x-, y-, and z-coordinates. Each system was then solved for 500 timesteps, starting at the initial position in 3-space. While each curve exhibits the characteristic "butterfly" shape, the individual coordinates of each series at each time step are drastically different.

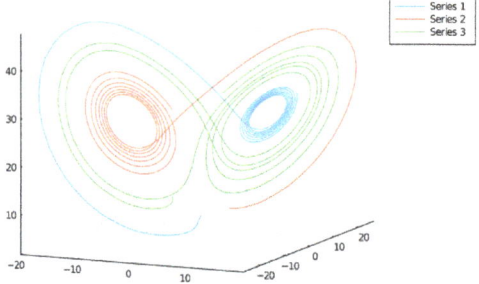

Figure 4. Impact of initial conditions on the Lorenz system.

Initial experiments showed that simple, smaller networks performed better with the genetic algorithm then larger networks. A 4-layer network was constructed for the time series prediction experiment. The structure of the network closely resembles an autoencoder network, where large input layers are scaled down throughout the network before being scaled back up for the output layer. This structure proved successful over several rounds of experimentation in predicting the 10-step ahead x, y, and z coordinates for the test set data. As a final experiment, a QMLP was created to predict the Lorenz coordinates 50 steps ahead based on in input time series of 25 steps. The layers, neurons per layer, and total parameters of each network are summarized in Table 2.

Table 2. Neural network topology for chaotic prediction.

Network	Input	Hidden 1	Hidden 2	Output	Parameters
QMLP+GA	10	3	3	10	85
QMLP+GA	25	5	10	50	740

Before processing through the network, the training and test datasets were cast into the quaternion domain using a vectorized approach. For an input vector τ_i, the corresponding quaternion input vector was constructed using the following approach:

$$\tau_i = \begin{bmatrix} \vec{x}_1 \\ \vec{x}_2 \\ \vdots \\ \vec{x}_{10} \end{bmatrix} \implies \tau_{q_i} = \begin{bmatrix} 0.0 + x_1\mathbf{i} + y_1\mathbf{j} + z_1\mathbf{k} \\ 0.0 + x_2\mathbf{i} + y_2\mathbf{j} + z_2\mathbf{k} \\ \vdots \\ 0.0 + x_{10}\mathbf{i} + y_{10}\mathbf{j} + z_{10}\mathbf{k} \end{bmatrix}. \tag{26}$$

Additionally, the target values were cast into quaternions. At each iteration, a quaternionic form of the MAE measured the fitness of each solution. Only the imaginary components of each input and target vector contained coordinate information, so this experiment introduced a $QMAE_{imag}$ calculation, defined in Equation (27) below.

$$\begin{aligned} QMAE_{imag} &:= \frac{1}{N} \sum_{i=1}^{N} ||\hat{y}_{\mathbf{q_i}} - y_{\mathbf{q_i}}||_{imag} \\ &= \frac{1}{N} \sum_{i=1}^{N} ||(\hat{x}_i\mathbf{i} + \hat{y}_i\mathbf{j} + \hat{z}_i\mathbf{k}) - (x_i\mathbf{i} + y_i\mathbf{j} + z_i\mathbf{k})|| \\ &= \frac{1}{N} \sum_{i=1}^{N} \left(\sqrt{(\hat{x}_i - x_i)^2 + (\hat{y}_i - y_i)^2 + (\hat{z}_i - z_i)^2} \right)^2. \end{aligned} \tag{27}$$

Since this experiment did not consider any real-valued networks, several quaternion activation functions were utilized during testing that are not available as activation functions in the real-domain. In particular, Ref. [46] notes that quaternionic functions with local analytic conditions are isomorphic to analytic complex functions. Additionally, Ref. [20] demonstrate that hyperbolic and inverse hyperbolic trigonometric functions are universal approximators in the complex domain. This experiment explored the use of several quaternionic elementary transcendental functions and found the inverse hyperbolic tangent, defined in [47], to provide the best performance:

$$\operatorname{arctanh}(p) := \frac{\ln(1+p) - \ln(1-p)}{2}. \tag{28}$$

Whereas the Lorenz prediction QMLP+GA networks required a slightly different network structure than the Ackley function approximation networks, both networks employed an identical genetic algorithm in the training phase. This approach eliminated the need for differentiability of both the loss function and the activation functions of the network. Additionally, it eliminated the need for a quaternion partial derivative calculation, which is a notoriously difficult problem. Section 3.3 describes the details of the algorithm, while Section 4 discusses the results and performance of the algorithm in both experiments.

3.3. Quaternion Genetic Algorithm

This section describes the quaternion genetic algorithm that was developed to train the QMLP-GA. A simple change of the underlying data type from quaternions to real-valued inputs, weights, and biases enabled the training of the MLP-GA with an identical algorithm. This research took a similar approach to Uber's OpenAI Labs genetic algorithm training process [33], opting for a very basic algorithm with minimal enhancements to demonstrate the proof-of-concept. Based on the success of this approach in Uber's experiments as well as in the quaternion domain presented here, a more advanced algorithm incorporating any of the many algorithmic improvements would likely improve on the baseline results discussed in Section 4.

A general diagram of the genetic algorithm process flow is shown in Figure 5. A genetic algorithm is a population-based search method, operating on a population of solutions to iteratively find improving solutions. In this case, an individual neural network, defined by its weights and biases, represents a single solution. To initialize the algorithm, a population of $N = 20$ distinct neural networks was instantiated, with all weights and biases randomly generated following a uniform distribution over $[-1, 1]$.

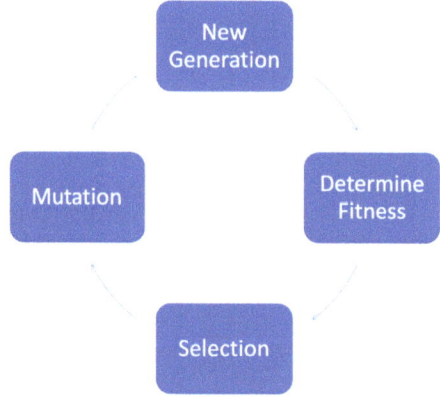

Figure 5. Genetic algorithm/genetic programming process.

After instantiation, the algorithm measures the fitness of each solution. For each neural network, the entire training dataset is processed through the network, capturing the total MAE for each network. The networks are then rank-ordered based on the lowest MAE value.

In the selection step, the n best solutions are retained as the "parents" for the next generation of the algorithm. In this research, $n = 5$ networks were retained as the parent generation in each iteration of the algorithm. While many advanced selection techniques exist, this work employed a simple rank selection, which selected the five best networks from each generation.

Finally, to generate a new population of solutions, the genetic algorithm performs a random mutation step, where a parent solution is randomly selected from the $n = 5$ best parent solutions. Then, the algorithm creates a "child" solution by mutating roughly half of the weights and biases of the parent solution with random noise. In this case, the generating distribution for the random noise was the standard normal distribution, $\mathcal{N}(0, 1)$. This process repeats for $N - n = 20 - 5 = 15$ times to create a new generation of solutions.

This process is commonly referred to as a genetic program, where generations are created solely through the mutation process. Often, genetic algorithms will include an additional crossover step prior to mutation, where new child solutions are created using a selection of features from separate parent solutions. Crossover was omitted from this algorithm, since mutation alone provided a good baseline performance, reiterating the fact that the most simple genetic algorithms are competitive to the popular backpropagation algorithm. A summary of the algorithm is shown in Algorithm 1. Additional details of the genetic algorithm along with a brief comparison of the computational effort required for the genetic algorithm versus classic gradient descent are provided in Appendix A.

Algorithm 1 Quaternion Genetic Algorithm

1: Instantiate $\mathcal{P}_m$ parent networks, $m \in N = \{1, \ldots, 20\}$, input mutation function ψ.
2: **for** $i \in N$ **do**
3:      Evaluate population fitness F_i
4: **end for**
5: **for** $g = 1$ to G generations **do**
6:      Sort population $\leftarrow F_i$
7:      Select best parents $\mathcal{P}_n^{g-1}$, $n = 1, \ldots, 5$
8:      **for** $j = n+1$ to N **do**
9:           Generate $k = \mathbf{UniformInt}(1, n)$
10:          $\mathcal{P}_j^g = \psi(\mathcal{P}_k^{g-1})$.
11:     **end for**
12: **end for**
13: Return final population $\mathcal{P}_m^G$ for $m \in N$.

3.4. Evaluation and Analysis Strategy

Each of the networks described in Section 3.2 processed the training data from the Ackley function and the Lorenz attractor system. At each training epoch, the algorithms either recorded the MAE of the overall system in the case of the gradient descent network, or the MAE or (QMAE) of the best solution for the genetic algorithm networks. Additionally, several computational metrics were recorded including memory allocations and computational runtime. Finally, each of the trained models processed the test data, recording the test set percentage error for each instance. Section 4 contains a discussion of network performance in each problem instance for each network in regards to these metrics.

4. Results

All computations presented here were performed on a desktop workstation running Windows 10 Enterprise with 64 GB of RAM and dual Intel Xeon Silver 4108 CPUs. Each CPU contained eight physical cores running at 1.80 GHz. Coding was performed in Julia 1.5.3 using the Quaternion.jl package and Flux.jl [48] for the MLP+GD network.

4.1. Function Approximation Results

The focus of the function approximation test was twofold. First, the function approximation task served as a proof-of-concept for the QMLP-GA. While quaternion neural networks and metaheuristic neural network training algorithms both exist separately in the literature, this work demonstrates the first use of metaheuristics to effectively train quaternion neural networks. Second, this experiment demonstrated some of the computational benefits that quaternions provide.

In keeping with these two goals, the three neural networks employed default parameters and very basic training algorithm implementations. No attempt was made to tune the hyperparameters of any of the models; instead, the results speak for themselves. The training set error for each of the three networks versus epoch is shown in Figure 6.

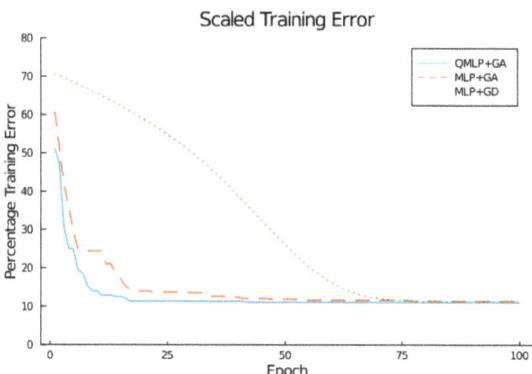

Figure 6. The training set mean absolute error for each network.

The QMLP+GA initialized using the random uniform weight initialization scheme described in Section 3 had the lowest initial prediction error, at roughly 50% in the first epoch. In contrast, the MLP+GA started with a nearly 60% initial error, while the MLP+GD was above 70%. The genetic algorithm improved rapidly, showing significantly faster initial algorithmic improvement versus the gradient descent algorithm. Both GA-trained networks showed rapid improvements over the first 25 training epochs, while the MLP+GD network searched for nearly 75 epochs before catching up to the GA-trained networks. The MLP+GD eventually caught up to the other two networks, but the prediction error remained slightly higher for the gradient descent network throughout the entire training process.

Table 3 shows the test set performance for each of the three networks across several measures of merit. In each column, the best results are highlighted in bold text. The quaternion network had the fastest overall runtime, resulting in the lowest test set error with the fewest number of trainable parameters. The real-valued MLP had a similar performance and required less overall system memory throughout the runtime of the algorithm, but required nearly six times the number of trainable parameters. Finally, the gradient descent-trained MLP had the worst performance in every category. While the test set error was comparable to the other two networks, the MLP+GD took more than 50 times as long to run with over 70 times as much memory allocated to store the gradient and error information for the backpropagation process.

These results, while cursory, clearly demonstrate the viability of quaternion networks trained with genetic algorithms. The quaternion network showed the fastest overall improvement, lowest final error, and lowest computational cost (in terms of runtime) when compared to two comparable networks. Additionally, the two GA-trained networks outperformed the gradient descent network across all measures of merit. These results validate the use of genetic algorithms in neural network training and show that quaternion networks can easily outperform equivalent real-valued networks involving multidimensional input data.

Table 3. Neural network comparison results.

Network	Runtime (s)	Memory (GB)	Parameters	Test Error
QMLP+GA	**17.421**	10.238	**22**	**11.01**%
MLP+GA	18.069	**9.497**	136	11.15%
MLP+GD	955.040	778.027	136	11.23%

4.2. Time Series Prediction Results

While the function approximation results demonstrate a viable proof-of-concept for quaternion neural networks, the chaotic time series prediction task illustrates the power of QNNs in the difficult task of predicting noisy systems. Additionally, chaotic time series prediction provides a natural multidimensional input + multidimensional output test that is almost tailor made for quaternion networks. In each of the figures displayed in this section, the orange graph represents the true chaotic time series, while the blue graph represents the predicted values. The final prediction results presented in Figure 7 are far from current state-of-the-art results using deep recurrent neural networks (RNNs) or long-short term memory (LSTM) networks, yet they illustrate the ability of simple QNNs to learn complex nonlinearities over time.

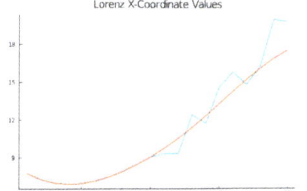 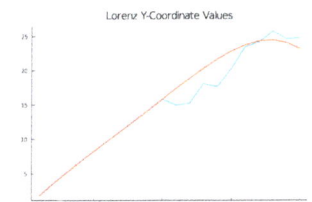 

Figure 7. 10-step ahead predicted coordinate values.

This experiment utilized two distinct QNN network topologies. The first network predicted the Lorenz attractor for 10 timesteps in the future based on an input time series of 10 timesteps. The second network predicted the Lorenz attractor for 50 timesteps in the future based on an input of 25 observations. The structure of each network is listed in Table 2, while the results for both networks are listed in Table 4. The test error percentage listed in Table 4 was measured using the mean absolute percentage error (MAPE) for time series forecasting, defined in Equation (29), where e_t is the unscaled prediction error for observation t and y_t is the target value at t:

$$MAPE = mean\left(\left|100\frac{e_t}{y_t}\right|\right). \qquad (29)$$

Early tests indicated that smaller networks performed better with the genetic algorithm. The final two networks contained comparatively few nodes in each layer and were structured as autoencoder networks, which perform a type of downsampling and subsequent upsampling as information passes through the network. Each network was trained for 50,000 epochs, which equated to roughly 28 min for the 10-step prediction network and around 4 h for the 50-step prediction network.

Table 4. Lorenz prediction results.

Prediction Steps	Runtime (s)	Memory (GB)	Params	Test Error
10	1668.565	947.304	85	10.89%
50	14769.069	2.815 (TB)	740	9.59%

The test set error listed in Table 4 indicates that on average, individual predicted values were off by about 11%. The actual versus predicted x-, y-, and z-coordinates for one of the test set time series are shown in Figure 7, while two 3-dimensional path predictions are shown in Figure 8. While the test error is relatively high, the QMLP+GA performs remarkably well on future predictions, especially in the long sweeping sections of the Lorenz attractor curves. The errors understandably grow and compound in the two "wings" of the curve, where the graph circles closely around each pole of the attractor.

The final experiment tested the ability of the QMLP to predict long sequences based on a relatively short input. The network was trained over 50,000 epochs to predict 50 observations based on an input sequence of length 25. Table 4 summarizes several measures of merit for the network, while the x-, y-, and z-coordinate results for a representative test set sequence are shown in Figure 9.

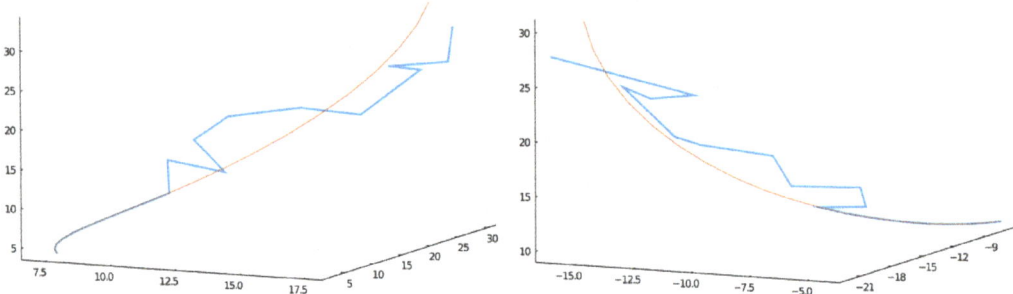

Figure 8. 10-step ahead path predictions.

In each coordinate direction, there is some clear noise at each prediction step, but the network accurately predicts the general motion of each variable. The motion of each prediction path is even more evident in the 3-dimensional plots shown in Figure 10, which shows two path predictions for two series from the test set data. As with the 10-step prediction model, the 50-step model makes the best predictions along the long sweeping arcs of the system, with errors compounding near the two "wings" of the attractor.

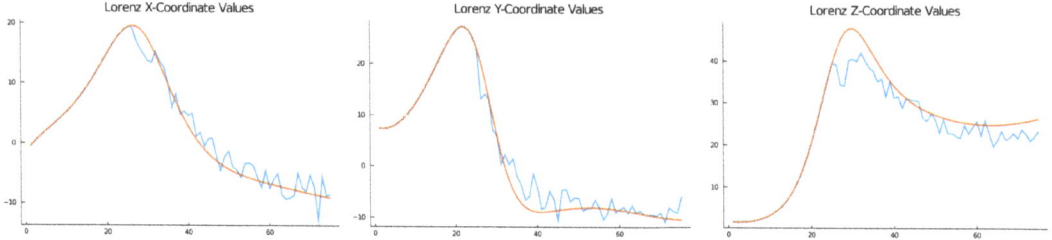

Figure 9. 50-step ahead predicted coordinate values.

Finally, the unscaled training error plots for both networks are shown in Figure 11. The genetic algorithm showed similar performance in both time series prediction tasks as it did in the function approximation task, with dramatic initial improvements and slow but consistent improvements as the iterations progressed. Surprisingly, the 50-step prediction experiment resulted in a lower test set prediction error than the 10-step prediction network, likely due to the scale of each predicted value.

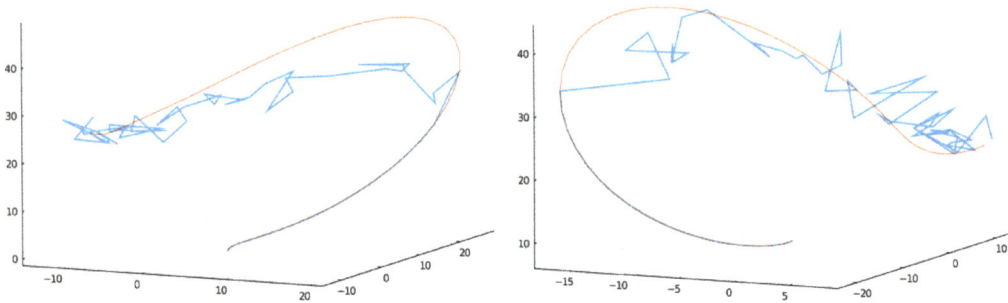

Figure 10. 50-step ahead path predictions.

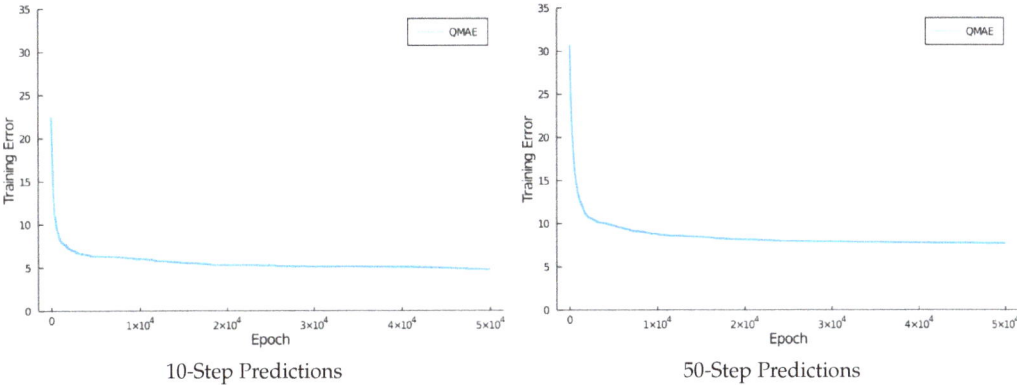

|10-Step Predictions | 50-Step Predictions|

Figure 11. Unscaled QMAE training error.

5. Discussion

In the Ackley function approximation experiment, all three networks utilized a random uniform weight initialization scheme. However, the quaternion network had between a 10–20% lower initial prediction error than the real-valued networks. This is likely due to the fact that the quaternion network employed six times fewer weight and bias parameters than the real-valued networks. The quaternion network maintained the lowest training set error across the entire 100-epoch training period, resulting in the best test set performance. The larger networks constructed in the second experiment demonstrated similar training characteristics and test set performance.

The genetic algorithm removes the need for expensive gradient calculations, resulting in better memory performance and more than 50x faster runtime in the first experiment versus the real-valued gradient descent algorithm. Given the difficulty of calculating quaternion gradients, the improvement over a quaternion gradient descent algorithm would likely be even greater. However, a genetic programming approach does come with some drawbacks. In the naive approach presented here, the algorithm would sometimes stall for several iterations while searching for an improving solution. There are many existing techniques designed to mitigate this stalling, but the literature on genetic algorithms is much less developed compared to comparable work on gradient descent optimization.

Despite this, the genetic algorithm opened the aperture on viable activation functions and loss functions for use with quaternion networks. This is perhaps the most significant contribution of this research. The results from [46] indicate that any locally analytic complex-valued activation function can be extended and used in the quaternion domain, but this work presents the first successful implementation of inverse hyperbolic trigonometric functions in quaternion networks. The success of the inverse hyperbolic tangent function in the chaotic time series prediction task demonstrates the value of using gradient free optimization methods in the quaternion domain.

The quaternions and quaternion neural networks are relatively unexplored compared to real analysis and real-valued neural networks. While certain applications in image processing and other domains have driven research in the quaternions and QNNs, there is still room for significant improvement in both the theoretical and practical aspects of quaternions. Going forward, the following lines of research will be crucial for continued innovation in the quaternion domain.

First, a solid foundation of quaternionic analysis is crucial to theoretically sound QNN research. While a handful of researchers have published works on quaternionic analysis, the corpus is quite thin. Research in novel quaternion activation functions, quaternion differentiability, quaternion analytic conditions, and novel quaternion training algorithms could significantly enhance both the current understanding of quaternion optimization as well as quaternion implementations of common machine learning models. Additionally, the quaternion Universal Approximation Theorem for either split or pure quaternion activation functions is an outstanding problem that is vital for establishing the legitimacy of quaternion networks from a theoretical point of view. Proving either variant of the Universal Approximation Theorem would be a substantial contribution to the field.

Finally, this research simply provided a proof-of-concept for GA-trained quaternion neural networks. The two examples presented were limited in scope and future work should build on these results to demonstrate the viability of GA-trained networks in large-scale optimization problems. In particular, quaternions are particularly well suited to the fields of image processing and robotic control, both of which have a plethora of neural network-related application opportunities. The authors intend to build on this proof-of-concept in future work by examining the scalability of quaternion GAs to large machine learning datasets and an in-depth comparative analysis of real-valued versus quaternion-valued neural networks using a design of experiments (DoE) hyperparameter-tuning approach. Finally, the authors intend to apply GA-trained QNNs to problem domains for which quaternions are particularly well suited, including 3D optimal satellite control and reinforcement learning for autonomous flight models.

Author Contributions: Conceptualization, J.B. and L.C.; methodology, J.B. and L.C.; software, J.B.; validation, J.B., L.C. and B.C.; formal analysis, J.B. and L.C.; investigation, J.B.; resources, L.C.; data curation, J.B.; writing—original draft preparation, J.B. and L.C.; writing—review and editing, J.B., L.C., B.C. and T.B.; visualization, J.B.; supervision, L.C.; project administration, L.C.; funding acquisition, L.C., B.C. and T.B. All authors have read and agreed to the published version of the manuscript.

Funding: This research was funded in part by Air Force Research Laboratory, WPAFB, OH, USA.

Institutional Review Board Statement: Not applicable.

Informed Consent Statement: Not applicable.

Data Availability Statement: Not applicable.

Conflicts of Interest: This report was prepared as an account of work sponsored by an agency of the United States Government. Neither the United States Government nor any agency thereof, nor any of their employees, make any warranty, express or implied, or assume any legal liability or responsibility for the accuracy, completeness, or usefulness of any information, apparatus, product, or process disclosed, or represent that its use would not infringe privately owned rights. Reference herein to any specific commercial product, process, or service by trade name, trademark, manufacturer, or otherwise does not necessarily constitute or imply its endorsement, recommendation, or favoring by the United States Government or any agency thereof. The views and opinions of authors expressed herein do not necessarily state or reflect those of the United States Government or any agency thereof.

Appendix A. MLP-GD and MLP-GA Pseudocode

This section provides a high-level overview of the main algorithmic steps for the genetic algorithm (GA) and gradient descent (GD) neural network training algorithms. In terms of computational effort, the main differences between the two algorithms are that the GA requires a population of different neural networks while the GD algorithm only requires a single network but instead computes the error gradients at each training iteration. The main computational burden of the GA stems from processing the training data through each network at every iteration of the algorithm. In contrast, the GD algorithm's main computational effort stems from the calculation of expensive partial derivatives to determine the error gradient at each layer of the network for every iteration. In practice, the computational cost of the backpropagation step in the GD algorithm outweighs the repeated processing of training data through each network in the GA. The results in Section 4.1 provide a good demonstration of this.

Appendix A.1. MLP-GA

```
# Network Structure contains weight/bias values
#   that define each layer of the neural network

struct network
    bias1
    Weight1
    bias2
    Weight2
    bias3
    Weight3
    bias4
    fitness
end

# initialization function
function init() do
    instantiate n random networks
    return list of networks
end

# main forward pass
function update_fitness(X, Y, Networks) do
    foreach network in Networks do
        foreach x in X do
            y_predicted = network output
            error = |y_predicted - Y|
        end
        network.fitness = sum(error)
    end
end

# mutation operator
function mutate_weights(weight_array) do
    foreach weight in weight_array do
        weight = weight + random_noise
    end
    return weight_array
end

# Genetic Algorithm
function GA(X, Y) do
    population = init()
    n_epochs = N

    for i = 1:N do
        update_fitness(X, Y, population)
        sort(population, on = fitness, order = ascending)

        # retain the best k networks as parents
        # mutate parent weights to create new children
        for j = k+1:length(population) do
            rand = rand(1:k)
            population[j] = mutate_weights(population[rand])
```

```
        end
    end

    update_fitness(population)
    sort(population, on = fitness, order = ascending)
    best_entity = population[1]

    return best_entity
end
```

Appendix A.2. MLP-GD

```
# Network Structure contains weight/bias values
#    that define each layer of the neural network

struct network
    bias1
    Weight1
    bias2
    Weight2
    bias3
    Weight3
    bias4
end

# initialization function
function init() do
    instantiate single random network
    return network
end

# main forward pass
function forward_pass(X, Y) do
    foreach x in X do
        y_predicted = network output
        error = |y_predicted - Y|
    end
    return error
end

# main backward pass
function backpropagation(network, error, eta)do
    foreach layer in network do
        gradient(error, layer) = partial derivative of error
        layer.weights = layer.weights - eta*gradient
    end
    return network
end

# Training Algorithm
function train(X, Y) do
    network = init()
    n_epochs = N
    eta = learning rate

    for i = 1:N do
```

```
            error = forward_pass(X, Y)
            network = backpropagation(network, error, eta)
        end

        return network
    end
```

References

1. Rosenblatt, F. The perceptron: A probabilistic model for information storage and organization in the brain. *Psychol. Rev.* **1958**, *65*, 386. [CrossRef] [PubMed]
2. Cybenko, G. Approximation by superpositions of a sigmoidal function. *Math. Control Signals Syst.* **1989**, *2*, 303–314. [CrossRef]
3. Hornik, K.; Stinchcombe, M.; White, H. Multilayer feedforward networks are universal approximators. *Neural Netw.* **1989**, *2*, 359–366. [CrossRef]
4. Géron, A. *Hands-On Machine Learning with Scikit-Learn, Keras, and TensorFlow: Concepts, Tools, and Techniques to Build Intelligent Systems*; O'Reilly Media: Sebastopol, CA, USA 2019.
5. Hosseini, S.H.; Samanipour, M. Prediction of final concentrate grade using artificial neural networks from Gol-E-Gohar iron ore plant. *Am. J. Min. Metall.* **2015**, *3*, 58–62.
6. Rumelhart, D.E.; Hinton, G.E.; Williams, R.J. *Learning Internal Representations by Error Propagation*; Technical Report; California University San Diego La Jolla Institute for Cognitive Science: San Diego, CA, USA, 1985.
7. Werbos, P. Beyond Regression: New Tools for Prediction and Analysis in the Behavioral Sciences. Ph.D. Thesis, Harvard University, Cambridge, MA, USA, 1974.
8. Karras, T.; Laine, S.; Aittala, M.; Hellsten, J.; Lehtinen, J.; Aila, T. Analyzing and Improving the Image Quality of StyleGAN. In Proceedings of the IEEE/CVF Conference on Computer Vision and Pattern Recognition, Seattle, WA, USA, 14–19 June 2020.
9. Deng, J.; Dong, W.; Socher, R.; Li, L.J.; Li, K.; Fei-Fei, L. ImageNet: A Large-Scale Hierarchical Image Database. In Proceedings of the 2009 IEEE Conference on Computer Vision and Pattern Recognition, Miami, FL, USA, 20–25 June 2009.
10. Oh, K.S.; Jung, K. GPU implementation of neural networks. *Pattern Recognit.* **2004**, *37*, 1311–1314. [CrossRef]
11. LeCun, Y.; Bottou, L.; Bengio, Y.; Haffner, P. Gradient-based learning applied to document recognition. *Proc. IEEE* **1998**, *86*, 2278–2324. [CrossRef]
12. Yin, Q.; Wang, J.; Luo, X.; Zhai, J.; Jha, S.K.; Shi, Y.Q. Quaternion convolutional neural network for color image classification and forensics. *IEEE Access* **2019**, *7*, 20293–20301. [CrossRef]
13. Matsui, N.; Isokawa, T.; Kusamichi, H.; Peper, F.; Nishimura, H. Quaternion neural network with geometrical operators. *J. Intell. Fuzzy Syst.* **2004**, *15*, 149–164.
14. Brown, J.W.; Churchill, R.V. *Complex Variables and Applications*; McGraw-Hill Higher Education: Boston, MA, USA, 2009.
15. Hamilton, W.R. LXXVIII. On quaternions; or on a new system of imaginaries in Algebra: To the editors of the Philosophical Magazine and Journal. *Lond. Edinb. Dublin Philos. Mag. J. Sci.* **1844**, *25*, 489–495. [CrossRef]
16. Kuipers, J.B. *Quaternions and Rotation Sequences: A Primer with Applications to Orbits, Aerospace, and Virtual Reality*; Princeton University Press: Princeton, NJ, USA, 1999.
17. Arena, P.; Fortuna, L.; Occhipinti, L.; Xibilia, M.G. Neural networks for quaternion-valued function approximation. In Proceedings of the IEEE International Symposium on Circuits and Systems-ISCAS'94, London, UK, 30 May–2 June 1994; Volume 6, pp. 307–310.
18. Ujang, B.C.; Jahanchahi, C.; Took, C.C.; Mandic, D. Quaternion valued neural networks and nonlinear adaptive filters. *IEEE Trans. Neural Netw.* **2010**, submission. [CrossRef] [PubMed]
19. Xu, D.; Zhang, L.; Zhang, H. Learning algorithms in quaternion neural networks using ghr calculus. *Neural Netw. World* **2017**, *27*, 271. [CrossRef]
20. Kim, T.; Adalı, T. Approximation by fully complex multilayer perceptrons. *Neural Comput.* **2003**, *15*, 1641–1666. [CrossRef] [PubMed]
21. Xu, D.; Jahanchahi, C.; Took, C.C.; Mandic, D.P. Enabling quaternion derivatives: The generalized HR calculus. *R. Soc. Open Sci.* **2015**, *2*, 150255. [CrossRef]
22. Arena, P.; Fortuna, L.; Re, R.; Xibilia, M.G. On the capability of neural networks with complex neurons in complex valued functions approximation. In Proceedings of the 1993 IEEE International Symposium on Circuits and Systems, Chicago, IL, USA, 3–6 May 1993; pp. 2168–2171.
23. Isokawa, T.; Kusakabe, T.; Matsui, N.; Peper, F. Quaternion neural network and its application. In Proceedings of the International Conference on Knowledge-Based and Intelligent Information and Engineering Systems, Oxford, UK, 3–5 September 2003; Springer: Berlin/Heidelberg, Germany, 2003; pp. 318–324.
24. Greenblatt, A.; Mosquera-Lopez, C.; Agaian, S. Quaternion neural networks applied to prostate cancer gleason grading. In Proceedings of the 2013 IEEE International Conference on Systems, Man, and Cybernetics, Manchester, UK, 13–16 October 2013; pp. 1144–1149.

25. Parcollet, T.; Morchid, M.; Bousquet, P.M.; Dufour, R.; Linarès, G.; De Mori, R. Quaternion neural networks for spoken language understanding. In Proceedings of the 2016 IEEE Spoken Language Technology Workshop (SLT), San Diego, CA, USA, 13–16 December 2016; pp. 362–368.
26. Buchholz, S.; Le Bihan, N. Optimal separation of polarized signals by quaternionic neural networks. In Proceedings of the 2006 14th European Signal Processing Conference, Florence, Italy, 4–8 September 2006; pp. 1–5.
27. Fortuna, L.; Muscato, G.; Xibilia, M.G. A comparison between HMLP and HRBF for attitude control. *IEEE Trans. Neural Netw.* **2001**, *12*, 318–328. [CrossRef] [PubMed]
28. Parcollet, T.; Morchid, M.; Linares, G. A survey of quaternion neural networks. *Artif. Intell. Rev.* **2020**, *53*, 2957–2982. [CrossRef]
29. Gaudet, C.J.; Maida, A.S. Deep quaternion networks. In Proceedings of the 2018 International Joint Conference on Neural Networks (IJCNN), Rio de Janeiro, Brazil, 8–13 July 2018; pp. 1–8.
30. Parcollet, T.; Ravanelli, M.; Morchid, M.; Linarès, G.; Trabelsi, C.; De Mori, R.; Bengio, Y. Quaternion recurrent neural networks. *arXiv* **2018**, arXiv:1806.04418.
31. Stanley, K.O.; Miikkulainen, R. Evolving Neural Networks through Augmenting Topologies. *Evol. Comput.* **2002**, *10*, 99–127. [CrossRef]
32. Stanley, K.O.; D'Ambrosio, D.B.; Gauci, J. A hypercube-based encoding for evolving large-scale neural networks. *Artif. Life* **2009**, *15*, 185–212. [CrossRef]
33. Such, F.P.; Madhavan, V.; Conti, E.; Lehman, J.; Stanley, K.O.; Clune, J. Deep neuroevolution: Genetic algorithms are a competitive alternative for training deep neural networks for reinforcement learning. *arXiv* **2017**, arXiv:1712.06567.
34. Ojha, V.K.; Abraham, A.; Snášel, V. Metaheuristic design of feedforward neural networks: A review of two decades of research. *Eng. Appl. Artif. Intell.* **2017**, *60*, 97–116. [CrossRef]
35. Fister, I.; Yang, X.S.; Brest, J.; Fister, I., Jr. Modified firefly algorithm using quaternion representation. *Expert Syst. Appl.* **2013**, *40*, 7220–7230. [CrossRef]
36. Papa, J.; Pereira, D.; Baldassin, A.; Yang, X.S. On the harmony search using quaternions. In Proceedings of the IAPR Workshop on Artificial Neural Networks in Pattern Recognition, Ulm, Germany, 31 August–2 September 2006; Springer: Berlin/Heidelberg, Germany, 2016; pp. 126–137.
37. Papa, J.P.; de Rosa, G.H.; Yang, X.S. On the Hypercomplex-Based Search Spaces for Optimization Purposes. In *Nature-Inspired Algorithms and Applied Optimization*; Springer: Berlin/Heidelberg, Germany, 2018; pp. 119–147.
38. Khuat, T.T.; Le, M.H. A genetic algorithm with multi-parent crossover using quaternion representation for numerical function optimization. *Appl. Intell.* **2017**, *46*, 810–826. [CrossRef]
39. Papa, J.P.; Rosa, G.H.; Pereira, D.R.; Yang, X.S. Quaternion-based deep belief networks fine-tuning. *Appl. Soft Comput.* **2017**, *60*, 328–335. [CrossRef]
40. Ackley, D. *A Connectionist Machine for Genetic Hillclimbing*; Springer Science & Business Media: Berlin/Heidelberg, Germany, 2012; Volume 28.
41. Lorenz, E.N. Deterministic nonperiodic flow. *J. Atmos. Sci.* **1963**, *20*, 130–141. [CrossRef]
42. Arena, P.; Fortuna, L.; Muscato, G.; Xibilia, M.G. *Neural Networks in Multidimensional Domains: Fundamentals and New Trends in Modelling and Control*; Springer: Berlin/Heidelberg, Germany, 1998; Volume 234.
43. Arena, P.; Caponetto, R.; Fortuna, L.; Muscato, G.; Xibilia, M.G. Quaternionic multilayer perceptrons for chaotic time series prediction. *IEICE Trans. Fundam. Electron. Commun. Comput. Sci.* **1996**, *79*, 1682–1688.
44. Arena, P.; Baglio, S.; Fortuna, L.; Xibilia, M. Chaotic time series prediction via quaternionic multilayer perceptrons. In Proceedings of the 1995 IEEE International Conference on Systems, Man and Cybernetics. Intelligent Systems for the 21st Century, Vancouver, BC, Canada, 22–25 October 1995; Volume 2, pp. 1790–1794.
45. Ujang, B.C.; Took, C.C.; Mandic, D.P. Split quaternion nonlinear adaptive filtering. *Neural Netw.* **2010**, *23*, 426–434. [CrossRef]
46. Isokawa, T.; Nishimura, H.; Matsui, N. Quaternionic Multilayer Perceptron with Local Analyticity. *Information* **2012**, *3*, 756–770. [CrossRef]
47. Morais, J.P.; Georgiev, S.; Sprößig, W. *Real Quaternionic Calculus Handbook*; Springer: Berlin/Heidelberg, Germany, 2014.
48. Innes, M. Flux: Elegant Machine Learning with Julia. *J. Open Source Softw.* **2018**. [CrossRef]

Article

Application of Artificial Intelligence and Gamma Attenuation Techniques for Predicting Gas–Oil–Water Volume Fraction in Annular Regime of Three-Phase Flow Independent of Oil Pipeline's Scale Layer

Abdulaziz S. Alkabaa [1], Ehsan Nazemi [2,*], Osman Taylan [1] and El Mostafa Kalmoun [3]

1. Department of Industrial Engineering, Faculty of Engineering, King Abdulaziz University, P.O. Box 80204, Jeddah 21589, Saudi Arabia; aalkabaa@kau.edu.sa (A.S.A.); otaylan@kau.edu.sa (O.T.)
2. Imec-Vision Lab, Department of Physics, University of Antwerp, 2610 Antwerp, Belgium
3. Department of Mathematics, Statistics and Physics, College of Arts and Sciences, Qatar University, Doha 2713, Qatar; ekalmoun@qu.edu.qa
* Correspondence: ehsan.nazemi@uantwerpen.be

Citation: Alkabaa, A.S.; Nazemi, E.; Taylan, O.; Kalmoun, E.M. Application of Artificial Intelligence and Gamma Attenuation Techniques for Predicting Gas–Oil–Water Volume Fraction in Annular Regime of Three-Phase Flow Independent of Oil Pipeline's Scale Layer. *Mathematics* **2021**, *9*, 1460. https://doi.org/10.3390/math9131460

Academic Editor: Aihua Wood

Received: 19 May 2021
Accepted: 19 June 2021
Published: 22 June 2021

Publisher's Note: MDPI stays neutral with regard to jurisdictional claims in published maps and institutional affiliations.

Copyright: © 2021 by the authors. Licensee MDPI, Basel, Switzerland. This article is an open access article distributed under the terms and conditions of the Creative Commons Attribution (CC BY) license (https://creativecommons.org/licenses/by/4.0/).

Abstract: To the best knowledge of the authors, in former studies in the field of measuring volume fraction of gas, oil, and water components in a three-phase flow using gamma radiation technique, the existence of a scale layer has not been considered. The formed scale layer usually has a higher density in comparison to the fluid flow inside the oil pipeline, which can lead to high photon attenuation and, consequently, reduce the measuring precision of three-phase flow meter. The purpose of this study is to present an intelligent gamma radiation-based, nondestructive technique with the ability to measure volume fraction of gas, oil, and water components in the annular regime of a three-phase flow independent of the scale layer. Since, in this problem, there are several unknown parameters, such as gas, oil, and water components with different amounts and densities and scale layers with different thicknesses, it is not possible to measure the volume fraction using a conventional gamma radiation system. In this study, a system including a ^{241}Am-^{133}Ba dual energy source and two transmission detectors was used. The first detector was located diametrically in front of the source. For the second detector, at first, a sensitivity investigation was conducted in order to find the optimum position. The four extracted signals in both detectors (counts under photo peaks of both detectors) were used as inputs of neural network, and volume fractions of gas and oil components were utilized as the outputs. Using the proposed intelligent technique, volume fraction of each component was predicted independent of the barium sulfate scale layer, with a maximum MAE error of 3.66%.

Keywords: annular regime; scale layer-independent; petroleum pipeline; volume fraction; dual energy technique

1. Introduction

Numerous applications for multiphase flow meters exist in the petrochemical and oil industries. For instance, there is a need to monitor multiphase flow continuously at some points. Some of these points include the gas–oil separator units and the wellhead collection lines. Monitoring at these points is of great value. The three-phase flow's volume fraction can be determined through several methodologies. One of the best methodologies is the gamma radiation-based technique, which is a nondestructive and reliable tool. Several studies have covered this area. One of the earliest studies was conducted in 1980, when Abouelwafa and Kendall introduced a method for metering three-phase flow. That is the dual-energy gamma-ray attenuation-based method [1]. In their study, Dong-hui et al. (2005) presented the dual-energy gamma-ray method. The method aimed to examine the volume fraction of various components on the multiphase pipe flow cross-section of gas–oil–water [2]. With the intent to measure the attenuation dose rate of the material, there was a need to design a data acquisition system and nuclear instruments. After designing the data acquisition system and the nuclear instruments, static tests followed. Oil–water–gas media has three phases. These phases were investigated to test the hypothesis that they effectively simulate different distributions of media volumetric fraction. The three phases were investigated when the oil–water–gas media was used in experimental vessels. During this investigation, the measurements of attenuation intensities were taken. There was also a study of the volumetric fraction's equations and the linear attenuation coefficients' arithmetic. When the attenuation equations were investigated for unexpected measurement error, the involvement of modified arithmetic was disclosed. Besides, the experimental research revealed that the system's accuracy was acceptable. In their study, Salgado et al. (2009) measured a three-phase flow volume fraction using the gamma radiation technique [3]. They used two radioactive sources whose energies were different. They also used three NaI detectors. In their estimation of the gas–oil–water's volume fraction, the researchers were aided by the artificial neural network (ANN). They considered several flow regimes in this process. The researchers replicated this methodology in 2010 to recognize the homogenous, stratified, and annular flow regime. They then estimated the water–gas–oil multiphase systems' volume fraction [4]. Hoffmann et al. (2011) measured phase fractions using a traversable gamma radiation-based instrument [5]. There was a need to ensure that the noisy measurements yielded relevant data. Therefore, the researchers had to be careful in their data analysis. In this regard, the researchers used the two-phase and single-phase flow calibration data and tested the three-phase flow data analysis technique against this data. The researchers found that the traversable gamma instrument's average density data was significantly related to the calibrated, stationary, single-energy gamma instrument's density measurements. However, more information was obtained from the traversable densitometer than from the single energy instrument. The rationale behind this is that it was possible to measure all the three phases' transient phase fraction over the pipeline's cross-section. By using this information, the flow pattern could be determined. Further studies in field of multiphase flow meters can be found in references [6–25].

According to the literature review, in the systems that used gamma radiation technique to determine the oil, water, and gas volume fraction, the existence of the scale layer's has not been considered. The scale layer forms gradually as mineral salts are deposited on the oil pipeline's inner surface. An example of such a layer is shown in Figure 1.

Figure 1. An oil pipeline before and after cleaning of the scale layer by chemical methods [26].

When the scale layer is compared to the fluid flow inside the oil pipeline, its density tends to be higher. In this regard, the three-phase flow meter's measuring precision can be reduced over time due to high photon attenuation that can occur due to the scale layer having a higher density. This study aims at presenting an intelligent gamma radiation-based system that can measure the oil, gas, and water components' volume fraction in a three-phase flow's annular regime independent of the scale layer. The article is structured as follows: Section 2 discusses the details of the proposed detection system and the implementation of the neural network. Section 3 reports the results of the neural network and calculates the accuracy of the designed neural networks. The last section provides both the summary and the conclusion.

2. Materials and Methods

2.1. Radiation Based System

Monte Carlo N Particle code (MCNP) [27] was implemented in the present investigation in order to model the radiation-based system. As pointed out in the abstract section, the aim of this investigation is to propose an intelligent, gamma radiation-based system with the ability of measuring volume fraction of gas, oil, and water components in annular regime of a three-phase flow independent of the scale layer. Since, in this problem, there are several unknown parameters, it is not possible to measure the volume fraction using a conventional gamma radiation system that includes one radiation source and one detector. To obtain more information from the fluid inside the pipe, a system including a dual energy source consisted of ^{241}Am and ^{133}Ba radioisotopes that emit photons with energies of 59 and 356 keV, respectively, and two NaI detectors for recording the transmitted photons, were used.

A steel pipe with internal radius and thickness of 10 cm and 0.5 cm, respectively, was considered in this study. In order to model the scale layer, a cylindrical shell of barium sulfate (BaSO$_4$) with density of 4.5 g·cm^{-3} and different thicknesses in the range of 0–3 cm, with a step of 0.5 cm, was considered on the internal wall of the steel pipe.

Annular regime of a three-phase flow was modeled inside the pipe. Air, gas, oil, and water with densities of 0.00125, 0.826, and 1 g·cm^{-3} were utilized as gas, oil, and water phases, respectively. For each scale thickness, various volume fractions were simulated for each component (seven different scale thickness×36 different volume fractions = totally 252 simulations were done).

As aforementioned, in this investigation, two 2.54 cm × 2.54 cm NaI detectors were applied to record the transmitted photons. Tally F8 was utilized in order to record photon spectra in both detectors. The first detector was positioned diametrically in front of the radioactive source (see Figure 2). For the second detector, at first, a sensitivity investigation was done in order to find the optimum position. In this regard, the center of the second detector was positioned in different orientation in the range of 5°–11°, with a step of 1°, and transmitted photons were recorded. Orientation of 5° was the minimum possible position for the second detector, because, at less than this orientation, the first and second detectors would interfere with each other. The reason for choosing orientation of 11° as the maximum position was that, at more than this orientation, there would be no more transmitted photons through the pipe that carries on useful information about the three-phase flow to reach the detector. At each position, sensitivity of the second detector relative to gas phase and oil phase volume fraction changes was investigated for both registered counts of gamma radiations emitted from ^{241}Am and ^{133}Ba radioisotopes. For instance, calculation of sensitivity of registered counts under ^{241}Am photo peak in the second detector relative to gas phase changes is indicated in Equation (1). It is worth mentioning that the sensitivity was calculated using the registered counts for gas phase fractions of 10% and 80%, which make the highest and lowest attenuation for photons.

$$Relative\ sensitivity\ (\%) = \left(\frac{C_{gvf80} - C_{gvf10}}{C_{gvf80}}\right) \times 100 \tag{1}$$

where, C_{gvf80} and C_{gvf10} refer to registered counts under ^{241}Am photo peak when the gas volume fraction is 80% and 10%, respectively. The results of sensitivity investigations are shown in Figure 3. For all four cases, by increasing orientation angle of the second detector, sensitivity starts to increase until it reaches a maximum value at the angle of 7° and then it gradually decreases. Based on the acquired results, it could be deduced that 7° is the optimal orientation angle for the second detector.

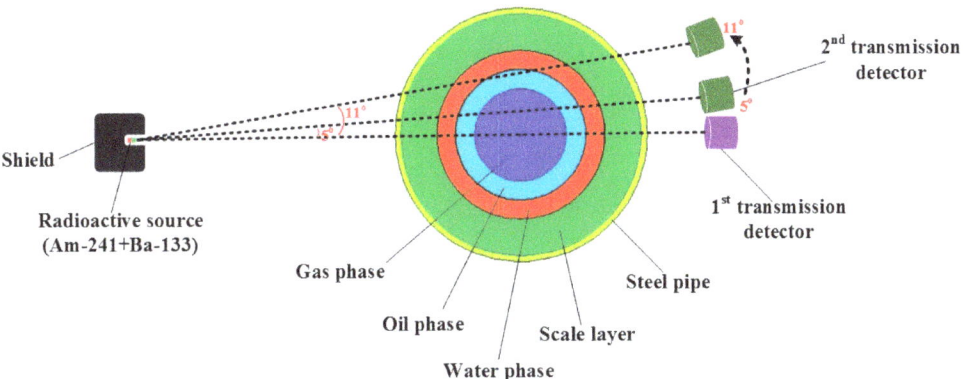

Figure 2. Investigation of optimum position for the second detector in the proposed system.

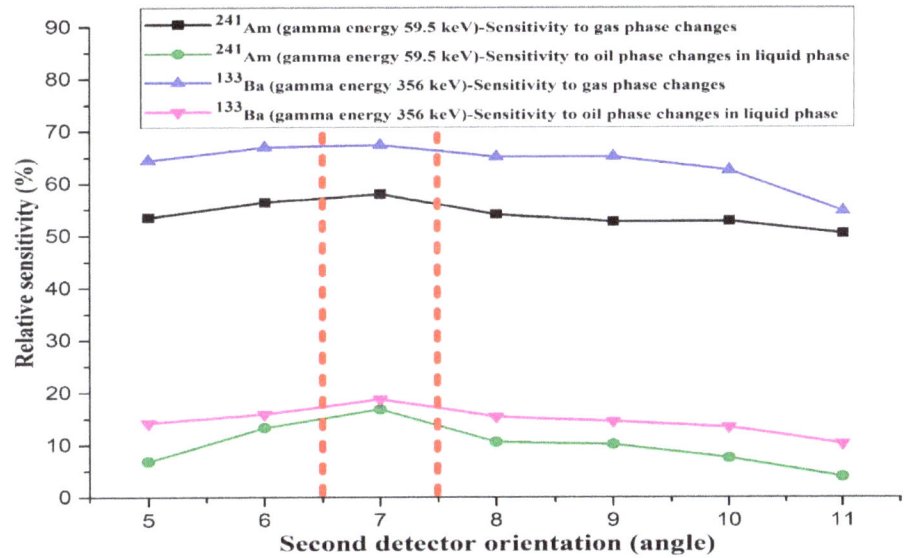

Figure 3. Sensitivity investigations results to find optimum position for the second detector.

It is worth mentioning that the simulated configuration in this work, especially the performance of the detectors, has been validated in our previous study using some experiments [28]. The corresponding experimental setup can be seen in Figure 4. A geometry identical to the experimental setup was simulated using MCNP code, and then the registered counts in both detectors were compared with the experimental ones. Calibrations of the gamma attenuation-based devices used for measuring the three-phase flow characteristics are usually done for three different extreme cases when the pipe is completely filled by gas, oil, and water. In the present study, a detection system the same as the validated one in our previous work was modeled. However, in the present study, a gas–oil–water three-phase flow was modeled instead of a two-phase flow.

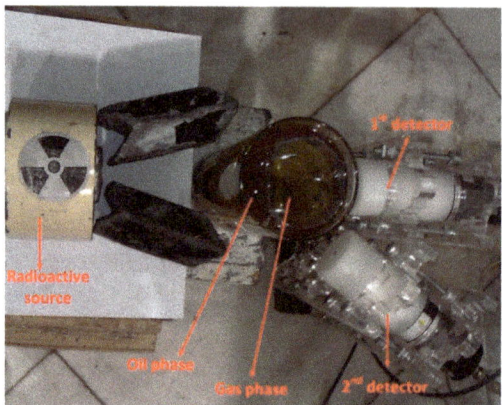

Figure 4. Experimental setup including 2 NaI detectors and one radioactive source [28].

2.2. Artificial Intelligence

In recent years, it has been proved that artificial intelligence can be implemented as a powerful tool for various engineering applications [29–65]. There are several kinds of ANNs, in which multilayer perceptron (MLP) is the most well-known kind of them. This kind of ANN has a good ability for regression and classification. This network is constructed from at least three layers of neurons: The input layer, the hidden layer (or hidden layers), and the output layer. There are different techniques to calculate the biases and weights of this mathematical network, of which Levenberg Marquardt (LM) is most well-known algorithm in this regard. In the present investigation, two different MLP–LM networks with four inputs and one output were considered. Four features were extracted from the recorded spectra in the detectors and were considered as MLP–LM inputs. The procedure of obtaining gas, oil, and water volume fraction percentages independent of scale layer thickness is indicated in Figure 5. The trained networks can estimate the percentages of gas and oil volume fraction based on input signals independent of scale layer thickness, correctly.

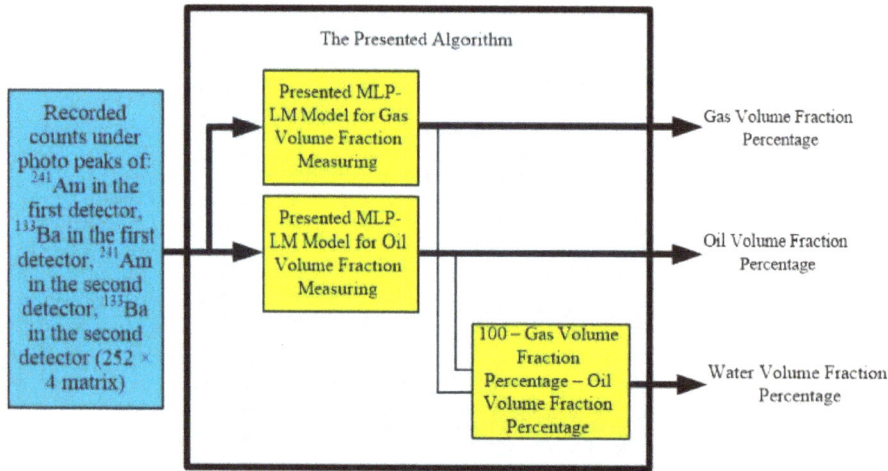

Figure 5. The procedure of obtaining gas, oil, and water volume fraction percentage in a three-phase flow independent of scale layer thickness.

A total of 252 different cases were simulated using MCNPX code; 177 cases were implemented for training the network, and 75 cases were used for testing the efficiency of presented MLP–LM. In order to obtain the optimum structure of proposed networks, different structures with various number of layers, neurons in each layer, epochs, and different activation functions were tested. For this purpose, different loops were defined, and, with trial and error, the optimum architecture was found. The mentioned algorithm is:

(1) The data set, counters, and error are defined.
(2) The data set is normalized.
(3) The parameters initial values are set.
(4) Several loops are created.
(5) Different number of layers, neurons in each layer, epochs, and different activation functions are tested.
(6) The efficiency of each network is checked.
(7) The best network with lowest error is saved.

The best structure of presented MLP–LM model for gas volume fraction measuring has one hidden layer consists of 9 neurons. The number of epochs was 685. The best structure of presented MLP–LM model for oil volume fraction measuring has one hidden layer consists of 10 neurons. The number of epochs was 750. Architectures of the ANN models were shown in Figure 6.

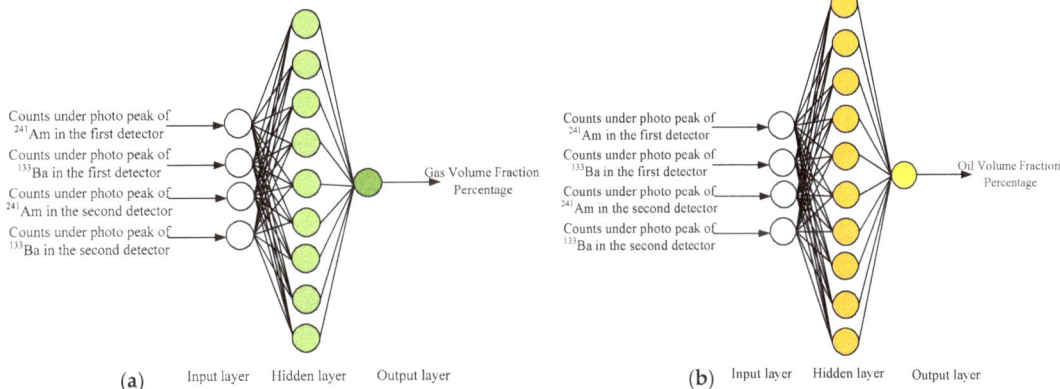

Figure 6. Architectures of presented MLP–LM models (**a**) for gas volume fraction measuring (**b**) for oil volume fraction measuring.

The mathematical equations for the first MLP–LM model are as follows. The input to the neuron **m** in the hidden layer is given by:

$$\eta_m = \sum_{u=1}^{4} (X_u W_{um}) + b_m \qquad m = 1, 2, \ldots, 9 \qquad (2)$$

The output from mth neuron of the hidden layer is given by:

$$U_m = f\left(\sum_{u=1}^{4} (X_u W_{um}) + b_m\right) \qquad m = 1, 2, \ldots, 9 \qquad (3)$$

The output of the neuron in the output layer is given by:

$$O = \sum_{u=1}^{9} (U_u W_u) + b \qquad (4)$$

where X is the input vector, b is the bias term, W is the weighting factor, and f is the activation function of the hidden layer.

3. Results and Discussions

After finding optimum positions for the detectors, orientation angle of 0° for the first detector and 7° for the second detector, counts under photo peaks of ^{241}Am and ^{133}Ba radioisotopes were recorded in both detectors for different scale layer thicknesses and volume fractions. Ternary contour plots of the recorded counts in both detectors for different volume fractions when the scale thickness is 0 and 3 cm, are shown in Figures 7–10. Comparing Figures 7 and 8 that correspond to the recorded counts in the first detector for ^{241}Am and ^{133}Ba radioisotopes, respectively, it can be said that dynamic range of registered counts relative to changes of gas volume fraction, or, better to say, sensitivity, for ^{133}Ba is more than ^{241}Am. A same response is also observed for the second detector. Comparing Figure 7a,b, it could be observed that, when scale layer is 0, sensitivity of detector relative to changes of gas, oil, and water components is much more than when the scale thickness is 3. In other words, by increasing thickness of scale layer, somehow information about the flow of inside the pipe starts fading. This manner can be also seen for both detectors and radioisotopes. Comparing Figures 7a and 9a, it can be observed that sensitivity of the second detector relative to changes of volume fractions is a little bit more than the first detector.

Regression diagrams of actual data and predicted data using presented MLP–LM models are shown in Figures 11 and 12. In Table 1, data number, scale layer thickness, actual outputs, and measured outputs for test data set are tabulated.

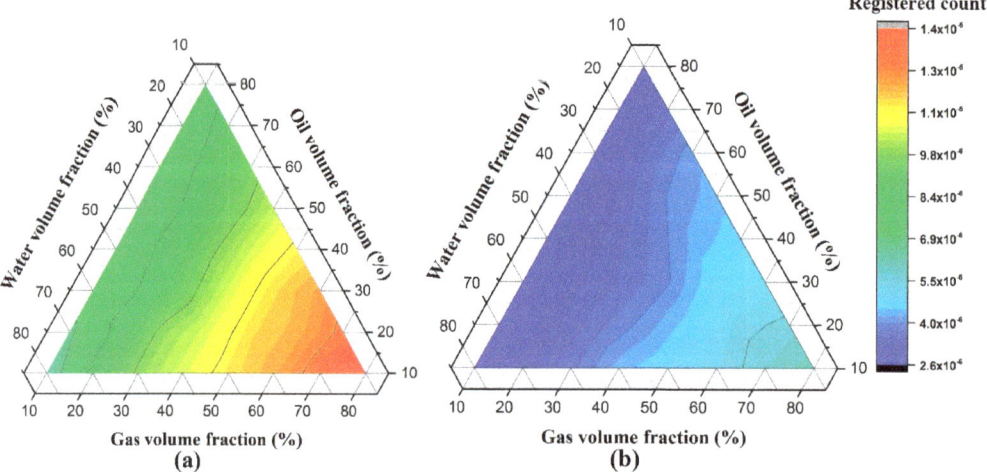

Figure 7. Recorded counts under photo peaks of ^{241}Am in the first detector versus gas, oil, and water volume fraction: (**a**) scale thickness is 0, (**b**) scale thickness is 3 cm.

Mean Absolute Error (MAE) and Root Mean Square Error (RMSE) of presented metering system were calculated using Equations (5) and (6).

$$\text{MAE} = \frac{1}{N} \sum_{i=1}^{Z} |X_i(\text{Actual}) - X_i(\text{Measured})| \quad (5)$$

$$\text{RMSE} = \left[\frac{\sum_{i=1}^{N}((X_i(\text{Actual}) - X_i(\text{Measured}))^2}{N} \right]^{0.5} \quad (6)$$

where N, X_i (Actual), and X_i (Measured) are the data number, real values, and estimated values, respectively. Performance criteria of the developed models were tabulated in Table 2 using Equations (5) and (6).

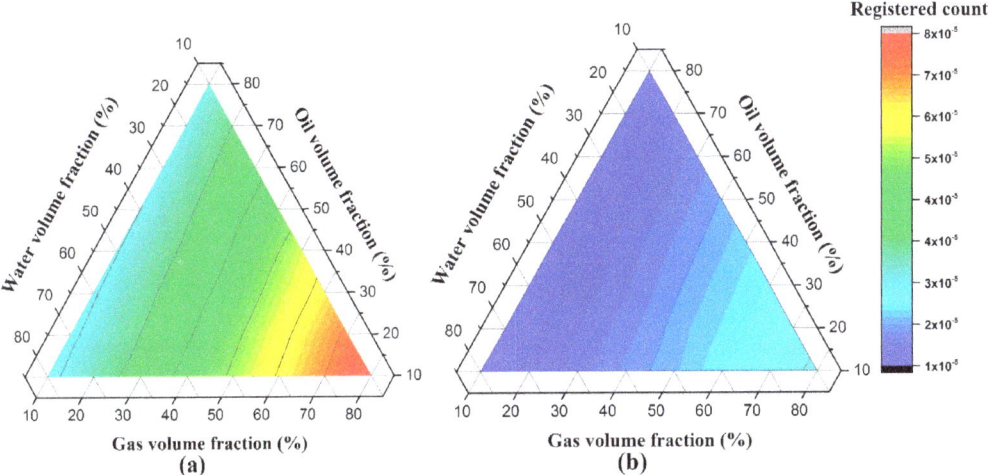

Figure 8. Recorded counts under photo peaks of ^{133}Ba in the first detector versus gas, oil, and water volume fraction: (**a**) scale thickness is 0, (**b**) scale thickness is 3 cm.

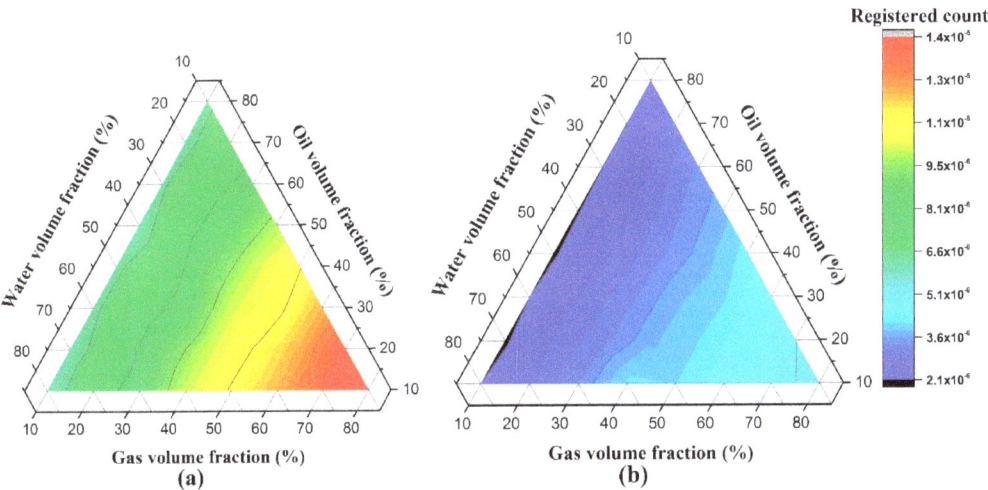

Figure 9. Recorded counts under photo peaks of ^{241}Am in the second detector versus gas, oil, and water volume fraction: (**a**) scale thickness is 0, (**b**) scale thickness is 3 cm.

It can be found from the obtained errors that the presented gauging system is reliable. This novel meter could be used in different industries for metering volume fraction of each phase independent of scale layer thickness. Radioisotope sources, detectors type, detectors position, extracted features from output signals, used data analysis algorithms and ANN architectures were selected appropriately in order to achieve the optimum performance for the proposed system.

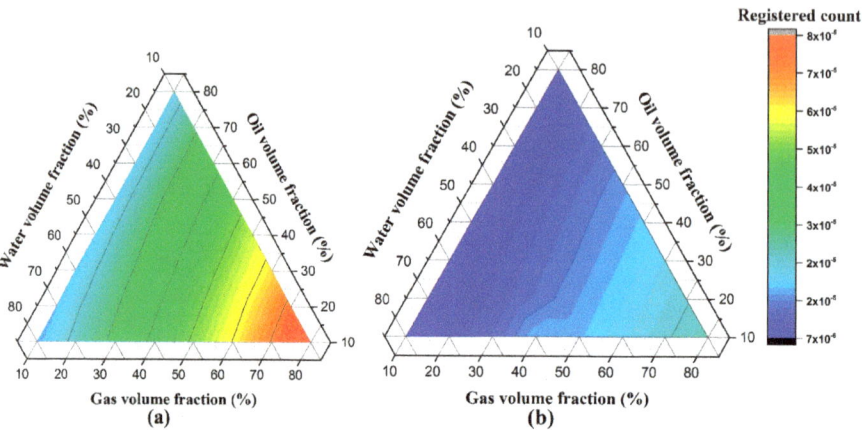

Figure 10. Recorded counts under photo peaks of ^{133}Ba in the second detector versus gas, oil, and water volume fraction: (**a**) scale thickness is 0, (**b**) scale thickness is 3 cm.

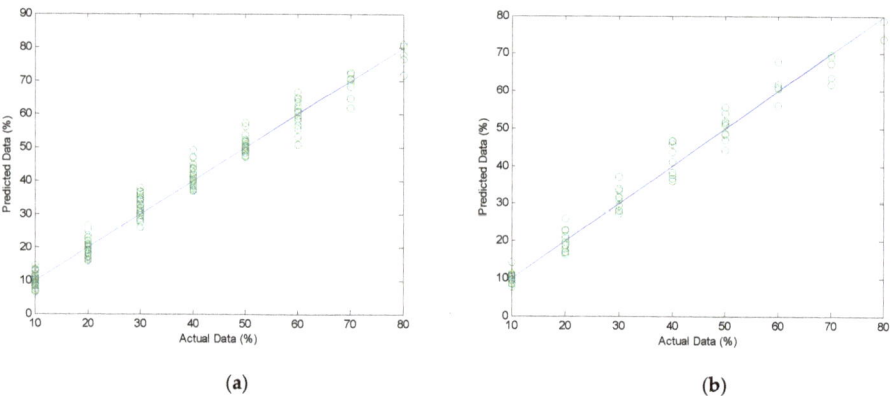

Figure 11. Regression diagrams of first model results (gas volume fraction) for (**a**) train data (**b**) test data.

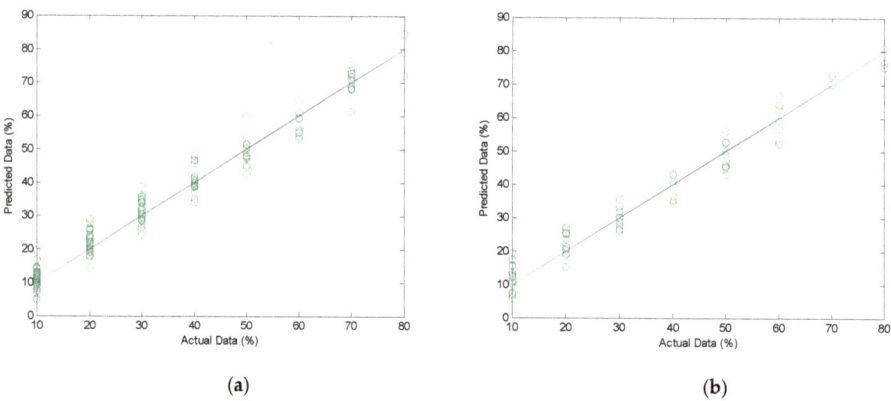

Figure 12. Regression diagrams of second model results (oil volume fraction) for (**a**) train data (**b**) test data.

Table 1. The test data with predicted values.

Data Number	Scale Layer Thickness	Actual Percentage of Gas Volume Fraction	Predicted Percentage of Gas Volume Fraction	Actual Percentage of Oil Volume Fraction	Predicted Percentage of Oil Volume Fraction	Data Number	Scale Layer Thickness	Actual Percentage of Gas Volume Fraction	Predicted Percentage of Gas Volume Fraction	Actual Percentage of Oil Volume Fraction	Predicted Percentage of Oil Volume Fraction
1	0	10	10.192	30	33.188	39	1.5	30	31.776	20	15.131
2	0	10	9.1929	70	72.778	40	1.5	30	33.395	50	42.774
3	0	20	18.589	20	23.656	41	1.5	40	38.145	20	24.954
4	0	20	22.769	50	45.377	42	1.5	50	44.139	10	19.463
5	0	30	27.957	20	21.208	43	1.5	60	61.639	10	10.063
6	0	30	27.197	50	44.794	44	1.5	70	69.562	10	17.143
7	0	40	40.997	20	26.426	45	2	10	9.568	10	13.678
8	0	40	46.476	50	46.974	46	2	10	9.3796	50	52.381
9	0	50	54.003	40	40.954	47	2	10	11.105	80	80.937
10	0	60	60.570	30	30.223	48	2	20	22.548	30	26.564
11	0	80	78.644	10	17.264	49	2	20	25.880	60	53.048
12	0.5	10	14.223	30	35.266	50	2	30	37.023	20	26.569
13	0.5	10	14.259	60	63.838	51	2	30	33.899	50	45.598
14	0.5	20	16.300	20	25.508	52	2	40	45.734	30	27.702
15	0.5	20	18.683	60	62.395	53	2	50	55.656	20	19.188
16	0.5	30	29.442	30	29.886	54	2	60	67.743	10	10.936
17	0.5	30	30.865	60	59.076	55	2	70	67.136	10	6.609
18	0.5	40	46.686	30	26.175	56	2.5	10	10.469	20	22.249
19	0.5	50	51.928	10	19.454	57	2.5	10	8.542	60	66.662
20	0.5	50	46.869	40	34.376	58	2.5	20	16.971	10	15.272
21	0.5	70	69.221	10	12.063	59	2.5	20	19.480	40	42.727
22	1	10	10.599	10	15.606	60	2.5	30	37.115	10	13.023
23	1	10	9.643	40	39.935	61	2.5	30	33.332	40	42.970
24	1	10	11.224	80	76.384	62	2.5	40	43.650	20	18.483
25	1	20	16.727	30	27.304	63	2.5	50	50.924	10	6.802
26	1	20	17.615	60	56.777	64	2.5	60	61.051	10	17.548
27	1	30	31.423	20	25.014	65	2.5	70	61.640	10	12.625
28	1	30	28.196	50	50.101	66	3	10	8.257	20	24.801
29	1	40	35.962	20	24.818	67	3	10	8.638	50	55.815
30	1	40	37.449	50	48.968	68	3	10	10.566	80	75.169
31	1	50	48.608	30	31.725	69	3	20	21.072	40	36.381
32	1	60	56.010	20	20.542	70	3	20	22.616	70	70.018
33	1	80	73.836	10	5.888	71	3	30	31.506	40	35.196
34	1.5	10	10.833	20	20.221	72	3	40	36.545	20	27.044
35	1.5	10	7.667	50	52.823	73	3	50	51.333	10	7.174
36	1.5	10	8.459	80	75.090	74	3	50	48.273	40	36.407
37	1.5	20	18.903	30	28.645	75	3	70	63.588	10	7.416
38	1.5	20	20.442	60	52.318						

Table 2. Performance criteria of the developed model.

Output	RMSE		MAE	
	Train	Test	Train	Test
Gas Volume Fraction Percentage	3.0956	3.3362	2.3266	2.6198
Oil Volume Fraction Percentage	3.5757	4.3268	2.7662	3.6579

4. Conclusions

In the present investigation, a novel and optimized radiation-based gauge, including two detectors and a dual energy source, was presented to measure volume fraction of gas, oil, and water components in annular regime of a three-phase flow independent of the scale layer. Position of the second detector was optimized. The percentages of gas, oil, and water volume fractions were measured independent of the barium sulfate scale layer. In fact, the presented measuring system can be used in different pipes with different thicknesses of scale layer. All the required data for modeling the presented system was achieved using MCNPX code. In order to model the metering system using MATLAB software, two different ANNs with four inputs and one output were considered. Recorded counts under photo peaks of ^{241}Am and ^{133}Ba were applied to both ANN models, as 252×4 input matrix and gas and oil volume fraction percentages were considered as the first and second ANN model outputs, respectively. The architectures of both ANNs were optimized using a presented algorithm. The dataset was divided to train set and test set. The accuracy of models was confirmed by good agreement of actual data and measured data in both sets. Finally, the volume fraction percentages were predicted with the RMSE of less than 4.33 and independent of scale layer.

Author Contributions: Conceptualization, E.N. and O.T.; Software, A.S.A., E.N., O.T. and E.M.K.; Writing—Review and Editing, A.S.A., E.N., O.T. and E.M.K.; Funding acquisition, A.S.A. and O.T. All authors have read and agreed to the published version of the manuscript.

Funding: This work was funded by the Deanship of Scientific Research (DSR), King Abdulaziz University, Jeddah, under grant No. (D-464-135-1441). The authors, therefore, gratefully acknowledge the DSR technical and financial support.

Institutional Review Board Statement: Not applicable.

Informed Consent Statement: Not applicable.

Data Availability Statement: Data is contained within the article.

Conflicts of Interest: The authors declare no conflict of interest.

References

1. Abouelwafa, M.; Kendall, E. The measurement of component ratios in multiphase systems using alpha-ray attenuation. *J. Phys. E Sci. Instrum.* **1980**, *13*, 341. [CrossRef]
2. Li, D.; Wu, Y.; Li, Z.; Zhong, X. Volumetric fraction measurement in oil–water–gas multiphase flow with dual energy gamma-ray system. *J. Zhejiang Univ. Sci. A* **2005**, *6*, 1405–1411.
3. Salgado, C.M.; Brandão, L.E.; Schirru, R.; Pereira, C.M.; da Silva, A.X.; Ramos, R. Prediction of volume fractions in three-phase flows using nuclear technique and artificial neural network. *Appl. Radiat. Isot.* **2009**, *67*, 1812–1818. [CrossRef]
4. Salgado, C.M.; Pereira, C.M.; Schirru, R.; Brandão, L.E. Flow regime identification and volume fraction prediction in multiphase flows by means of gamma-ray attenuation and artificial neural networks. *Prog. Nucl. Energy* **2010**, *52*, 555–562. [CrossRef]
5. Hoffmann, R.; Johnson, G.W. Measuring phase distribution in high pressure three-phase flow using gamma densitometry. *Flow Meas. Instrum.* **2011**, *22*, 351–359. [CrossRef]
6. Karami, A.; Roshani, G.H.; Khazaei, A.; Nazemi, E.; Fallahi, M. Investigation of different sources in order to optimize the nuclear metering system of gas–oil–water annular flows. *Neural Comput. Appl.* **2020**, *32*, 3619–3631. [CrossRef]
7. Meric, I.; Johansen, G.A.; Mattingly, J.; Gardner, R. On the ill-conditioning of the multiphase flow measurement by prompt gamma-ray neutron activation analysis. *Radiat. Phys. Chem.* **2014**, *95*, 401–404. [CrossRef]
8. Oliveira, D.F.; Nascimento, J.R.; Marinho, C.A.; Lopes, R.T. Gamma transmission system for detection of scale in oil exploration pipelines. *Nucl. Instrum. Methods Phys. Res. Sect. A Accel. Spectrometers Detect. Assoc. Equip.* **2015**, *784*, 616–620. [CrossRef]

9. Hanus, R.; Zych, M.; Mosorov, V.; Golijanek-Jędrzejczyk, A.; Jaszczur, M.; Andruszkiewicz, A. Evaluation of liquid-gas flow in pipeline using gamma-ray absorption technique and advanced signal processing. *Metrol. Meas. Syst.* **2021**, *28*, 145–159.
10. Roshani, M.; Phan, G.T.; Ali, P.J.M.; Roshani, G.H.; Hanus, R.; Duong, T.; Corniani, E.; Nazemi, E.; Kalmoun, E.M. Evaluation of flow pattern recognition and void fraction measurement in two phase flow independent of oil pipeline's scale layer thickness. *Alex. Eng. J.* **2021**, *60*, 1955–1966. [CrossRef]
11. Nazemi, E.; Feghhi, S.; Roshani, G.; Setayeshi, S.; Peyvandi, R.G. A radiation-based hydrocarbon two-phase flow meter for estimating of phase fraction independent of liquid phase density in stratified regime. *Flow Meas. Instrum.* **2015**, *46*, 25–32. [CrossRef]
12. Nazemi, E.; Feghhi, S.; Roshani, G.; Peyvandi, R.G.; Setayeshi, S. Precise Void Fraction Measurement in Two-phase Flows Independent of the Flow Regime Using Gamma-ray Attenuation. *Nucl. Eng. Technol.* **2016**, *48*, 64–71. [CrossRef]
13. Vlasák, P.; Chára, Z.; Matoušek, V.; Konfršt, J.; Kesely, M. Experimental investigation of fine-grained settling slurry flow behaviour in inclined pipe sections. *J. Hydrol. Hydromech.* **2019**, *67*, 113–120. [CrossRef]
14. Roshani, G.H.; Nazemi, E.; Roshani, M.M. Flow regime independent volume fraction estimation in three-phase flows using dual-energy broad beam technique and artificial neural network. *Neural Comput. Appl.* **2016**, *28*, 1265–1274. [CrossRef]
15. El Abd, A. Intercomparison of gamma ray scattering and transmission techniques for gas volume fraction measurements in two phase pipe flow. *Nucl. Instrum. Methods Phys. Res. Sect. A* **2014**, *735*, 260–266. [CrossRef]
16. Roshani, G.H.; Roshani, S.; Nazemi, E.; Roshani, S. Online measuring density of oil products in annular regime of gas-liquid two phase flows. *Measurement* **2018**, *129*, 296–301. [CrossRef]
17. Mosorov, V.; Rybak, G.; Sankowski, D. Plug Regime Flow Velocity Measurement Problem Based on Correlability Notion and Twin Plane Electrical Capacitance Tomography: Use Case. *Sensors* **2021**, *21*, 2189. [CrossRef]
18. Vlasák, P.; Matoušek, V.; Chára, Z.; Krupička, J.; Konfršt, J.; Kesely, M. Concentration distribution and deposition limit of medium-coarse sand-water slurry in inclined pipe. *J. Hydrol. Hydromech.* **2020**, *68*, 83–91. [CrossRef]
19. Zych, M.; Hanus, R.; Vlasák, P.; Jaszczur, M.; Petryka, L. Radiometric methods in the measurement of particle-laden flows. *Powder Technol.* **2017**, *318*, 491–500. [CrossRef]
20. Roshani, M.; Sattari, M.A.; Ali, P.J.M.; Roshani, G.H.; Nazemi, B.; Corniani, E.; Nazemi, E. Application of GMDH neural network technique to improve measuring precision of a simplified photon attenuation based two-phase flowmeter. *Flow Meas. Instrum.* **2020**, *75*, 101804. [CrossRef]
21. Mosorov, V.; Zych, M.; Hanus, R.; Sankowski, D.; Saoud, A. Improvement of Flow Velocity Measurement Algorithms Based on Correlation Function and Twin Plane Electrical Capacitance Tomography. *Sensors* **2020**, *20*, 306. [CrossRef]
22. Roshani, M.; Phan, G.; Faraj, R.H.; Phan, N.-H.; Roshani, G.H.; Nazemi, B.; Corniani, E.; Nazemi, E. Proposing a gamma radiation based intelligent system for simultaneous analyzing and detecting type and amount of petroleum by-products. *Nucl. Eng. Technol.* **2021**. [CrossRef]
23. Golijanek-Jędrzejczyk, A.; Mrowiec, A.; Hanus, R.; Zych, M.; Świsulski, D. Uncertainty of mass flow measurement using centric and eccentric orifice for Reynolds number in the range $10,000 \leq Re \leq 20,000$. *Measurement* **2020**, *160*, 107851. [CrossRef]
24. Zhang, F.; Chen, K.; Zhu, L.; Appiah, D.; Hu, B.; Yuan, S. Gas–Liquid Two-Phase Flow Investigation of Side Channel Pump: An Application of MUSIG Model. *Mathematics* **2020**, *8*, 624. [CrossRef]
25. Roshani, G.H.; Hanus, R.; Khazaei, A.; Zych, M.; Nazemi, E.; Mosorov, V. Density and velocity determination for single-phase flow based on radiotracer technique and neural networks. *Flow Meas. Instrum.* **2018**, *61*, 9–14. [CrossRef]
26. Chemical Cleaning. Available online: https://www.fourquest.com/services/chemical-cleaning (accessed on 15 February 2020).
27. Pelowitz, D.B. *MCNP-X TM User's Manual, Version 2.5.0*. LA-CP-05e0369; Los Alamos National Laboratory: New Mexico, NM, USA, 2005.
28. Nazemi, E.; Roshani, G.H.; Feghhi, S.A.H.; Setayeshi, S.; Zadeh, E.E.; Fatehi, A. Optimization of a method for iden-tifying the flow regime and measuring void fraction in a broad beam gamma-ray attenuation technique. *Int. J. Hydrog. Energy* **2016**, *41*, 7438–7444. [CrossRef]
29. Versaci, M.; Morabito, F.C. Image edge detection: A new approach based on fuzzy entropy and fuzzy divergence. *Int. J. Fuzzy Syst.* **2021**, 1–19. [CrossRef]
30. Burrascano, P.; Ciuffetti, M. Early Detection of Defects through the Identification of Distortion Characteristics in Ultrasonic Responses. *Mathematics* **2021**, *9*, 850. [CrossRef]
31. Roshani, M.; Phan, G.; Roshani, G.H.; Hanus, R.; Nazemi, B.; Corniani, E.; Nazemi, E. Combination of X-ray tube and GMDH neural network as a nondestructive and potential technique for measuring characteristics of gas–oil–water three phase flows. *Measurement* **2021**, *168*, 108427. [CrossRef]
32. Versaci, M.; Angiulli, G.; di Barba, P.; Morabito, F.C. Joint use of eddy current imaging and fuzzy similarities to assess the integrity of steel plates. *Open Phys.* **2020**, *18*, 230–240. [CrossRef]
33. Pourjabar, S.; Choi, G.S. A High-Throughput Multi-Mode LDPC Decoder for 5G NR. *arXiv* **2021**, arXiv:2102.13228.
34. Karami, A.; Yousefi, T.; Harsini, I.; Maleki, E.; Mahmoudinezhad, S. Neuro-Fuzzy Modeling of the Free Convection Heat Transfer from a Wavy Surface. *Heat Transf. Eng.* **2015**, *36*, 847–855. [CrossRef]
35. Darbandi, M.; Ramtin, A.R.; Sharafi, O.K. Tasks mapping in the network on a chip using an improved optimization algorithm. *Int. J. Pervasive Comput. Commun.* **2020**, *16*, 165–182. [CrossRef]

36. Moradi, M.J.; Roshani, M.M.; Shabani, A.; Kioumarsi, M. Prediction of the load-bearing behavior of spsw with rectangular opening by RBF net-work. *Appl. Sci.* **2020**, *10*, 1185. [CrossRef]
37. Abolhasani, M.; Karami, A.; Rahimi, M. Numerical Modeling and Optimization of the Enhancement of the Cooling Rate in Concentric Tubes Under Ultrasound Field. *Numer. Heat Transf. Part A Appl.* **2015**, *67*, 1282–1309. [CrossRef]
38. Jamshidi, M.B.; Lalbakhsh, A.; Talla, J.; Peroutka, Z.; Hadjilooei, F.; Lalbakhsh, P.; Jamshidi, M.; La Spada, L.; Mirmozafari, M.; Dehghani, M.; et al. Artificial intelligence and COVID-19: Deep learning approaches for diagnosis and treatment. *IEEE Access* **2020**, *8*, 109581–109595. [CrossRef]
39. Xue, H.; Yu, P.; Zhang, M.; Zhang, H.; Wang, E.; Wu, G.; Li, Y.; Zheng, X. A Wet Gas Metering System Based on the Extended-Throat Venturi Tube. *Sensors* **2021**, *21*, 2120. [CrossRef]
40. Moradi, M.; Daneshvar, K.; Ghazi-Nader, D.; Hajiloo, H. The prediction of fire performance of concrete-filled steel tubes (CFST) using artificial neural network. *Thin Walled Struct.* **2021**, *161*, 107499. [CrossRef]
41. Aghakhani, M.; Ghaderi, M.R.; Karami, A.; Derakhshan, A.A. Combined effect of TiO2 nanoparticles and input welding parameters on the weld bead penetration in submerged arc welding process using fuzzy logic. *Int. J. Adv. Manuf. Technol.* **2014**, *70*, 63–72. [CrossRef]
42. Jamshidi, M.B.; Roshani, S.; Talla, J.; Roshani, S.; Peroutka, Z. Size reduction and performance improvement of a microstrip Wilkinson power divider using a hybrid design technique. *Sci. Rep.* **2021**, *11*, 1–15. [CrossRef]
43. Roshani, G.H.; Nazemi, E.; Feghhi, S.A.H.; Setayeshi, S. Flow regime identification and void fraction prediction in two-phase flows based on gamma ray attenuation. *Measurement* **2015**, *62*, 25–32. [CrossRef]
44. Arabi, M.; Dehshiri, A.M.; Shokrgozar, M. Modeling transportation supply and demand forecasting using artificial intelligence parameters (Bayesian model). *J. Appl. Eng. Sci.* **2018**, *16*, 43–49. [CrossRef]
45. Salimi, J.; Ramezanianpour, A.M.; Moradi, M.J. Studying the effect of low reactivity metakaolin on free and restrained shrinkage of high performance concrete. *J. Build. Eng.* **2020**, *28*, 101053. [CrossRef]
46. Roshani, G.H.; Nazemi, E.; Feghhi, S.A.H. Investigation of using 60Co source and one detector for determining the flow regime and void fraction in gas–liquid two-phase flows. *Flow Meas. Instrum.* **2016**, *50*, 73–79. [CrossRef]
47. Lotfi, S.; Roshani, S.; Roshani, S. Design of a miniaturized planar microstrip Wilkinson power divider with harmonic cancellation. *Turk. J. Electr. Eng. Comput. Sci.* **2020**, *28*, 3126–3136. [CrossRef]
48. Karami, A.; Veysi, F.; Mohebbi, S.; Ghashghaei, D. Optimization of Laminar Free Convection in a Horizontal Cavity Consisting of Flow Diverters Using ICA. *Arab. J. Sci. Eng.* **2014**, *39*, 2295–2306. [CrossRef]
49. Khaleghi, M.; Salimi, J.; Farhangi, V.; Moradi, M.J.; Karakouzian, M. Application of Artificial Neural Network to Predict Load Bearing Capacity and Stiffness of Perfo-rated Masonry Walls. *CivilEng* **2021**, *2*, 48–67. [CrossRef]
50. Roshani, G.H.; Nazemi, E.; Roshani, M.M. Usage of two transmitted detectors with optimized orientation in order to three phase flow metering. *Measurement* **2017**, *100*, 122–130. [CrossRef]
51. Pirasteh, A.; Roshani, S.; Roshani, S. Compact microstrip lowpass filter with ultrasharp response using a square-loaded modified T-shaped resonator. *Turk. J. Electr. Eng. Comput. Sci.* **2018**, *26*, 1736–1746. [CrossRef]
52. Arief, H.A.; Wiktorski, T.; Thomas, P.J. A Survey on Distributed Fibre Optic Sensor Data Modelling Techniques and Machine Learning Algorithms for Multiphase Fluid Flow Estimation. *Sensors* **2021**, *21*, 2801. [CrossRef]
53. Roshani, S.; Roshani, S. Two-Section Impedance Transformer Design and Modeling for Power Amplifier Applications. *Appl. Comput. Electromagn. Soc. J.* **2017**, *32*, 1042–1047.
54. Roshani, G.H.; Nazemi, E.; Roshani, M.M. Intelligent recognition of gas–oil–water three-phase flow regime and determination of volume fraction using radial basis function. *Flow Meas. Instrum.* **2017**, *54*, 39–45. [CrossRef]
55. Jahanshahi, A.; Sabzi, H.Z.; Lau, C.; Wong, D. GPU-NEST: Characterizing Energy Efficiency of Multi-GPU Inference Servers. *IEEE Comput. Archit. Lett.* **2020**, *19*, 139–142. [CrossRef]
56. Roshani, G.H.; Nazemi, E.; Roshani, M.M. Identification of flow regime and estimation of volume fraction independent of liquid phase density in gas-liquid two-phase flow. *Prog. Nucl. Energy* **2017**, *98*, 29–37. [CrossRef]
57. Moradi, M.J.; Hariri-Ardebili, M.A. Developing a library of shear walls database and the neural network based predictive meta-model. *Appl. Sci.* **2019**, *9*, 2562. [CrossRef]
58. Roshani, S.; Jamshidi, M.B.; Mohebi, F.; Roshani, S. Design and Modeling of a Compact Power Divider with Squared Resonators Using Artificial Intelligence. *Wirel. Pers. Commun.* **2020**. [CrossRef]
59. Jahanshahi, A.; Taram, M.K.; Eskandari, N. Blokus Duo game on FPGA. In Proceedings of the 17th CSI International Symposium on Computer Architecture & Digital Systems (CADS 2013), Tehran, Iran, 30–31 October 2013; pp. 149–152.
60. Bavandpour, S.K.; Roshani, S.; Pirasteh, A.; Roshani, S.; Seyedi, H. A compact lowpass-dual bandpass diplexer with high output ports isolation. *AEU Int. J. Electron. Commun.* **2021**, *135*, 153748. [CrossRef]
61. Roshani, G.H.; Nazemi, E. Intelligent densitometry of petroleum products in stratified regime of two phase flows using gamma ray and neural network. *Flow Meas. Instrum.* **2017**, *58*, 6–11. [CrossRef]
62. Jahanshahi, A. TinyCNN: A Tiny Modular CNN Accelerator for Embedded FPGA. *arXiv* **2019**, arXiv:1911.06777.
63. Sattari, M.A.; Roshani, G.H.; Hanus, R.; Nazemi, E. Applicability of time-domain feature extraction methods and artificial intelligence in two-phase flow meters based on gamma-ray absorption technique. *Measurement* **2021**, *168*, 108474. [CrossRef]

64. Nabavi, M.; Elveny, M.; Danshina, S.D.; Behroyan, I.; Babanezhad, M. Velocity prediction of Cu/water nanofluid convective flow in a circular tube: Learning CFD data by differential evolution algorithm based fuzzy inference system (DEFIS). *Int. Commun. Heat Mass Transf.* **2021**, *126*, 105373. [CrossRef]
65. Karami, A.; Roshani, G.H.; Nazemi, E.; Roshani, S. Enhancing the performance of a dual-energy gamma ray based three-phase flow meter with the help of grey wolf optimization algorithm. *Flow Meas. Instrum.* **2018**, *64*, 164–172. [CrossRef]

Article

Application of Feature Extraction and Artificial Intelligence Techniques for Increasing the Accuracy of X-ray Radiation Based Two Phase Flow Meter

Abdulrahman Basahel [1], Mohammad Amir Sattari [2], Osman Taylan [1] and Ehsan Nazemi [3,*]

[1] Department of Industrial Engineering, Faculty of Engineering, King Abdulaziz University, P.O. Box 80204, Jeddah 21589, Saudi Arabia; ambasahel@kau.du.sa (A.B.); otaylan@kau.edu.sa (O.T.)
[2] Friedrich Schiller University Jena, Fürstengraben 1, 07743 Jena, Germany; mohamadamir.satari@gmail.com
[3] Imec-Vision Lab, Department of Physics, University of Antwerp, 2610 Antwerp, Belgium
* Correspondence: ehsan.nazemi@uantwerpen.be

Abstract: The increasing consumption of fossil fuel resources in the world has placed emphasis on flow measurements in the oil industry. This has generated a growing niche in the flowmeter industry. In this regard, in this study, an artificial neural network (ANN) and various feature extractions have been utilized to enhance the precision of X-ray radiation-based two-phase flowmeters. The detection system proposed in this article comprises an X-ray tube, a NaI detector to record the photons, and a Pyrex-glass pipe, which is placed between detector and source. To model the mentioned geometry, the Monte Carlo MCNP-X code was utilized. Five features in the time domain were derived from the collected data to be used as the neural network input. Multi-Layer Perceptron (MLP) was applied to approximate the function related to the input-output relationship. Finally, the introduced approach was able to correctly recognize the flow pattern and predict the volume fraction of two-phase flow's components with root mean square error (RMSE), mean absolute error (MAE), and mean absolute percentage error (MAPE) of less than 0.51, 0.4 and 1.16%, respectively. The obtained precision of the proposed system in this study is better than those reported in previous works.

Keywords: radiation-based flowmeter; two-phase flow; feature extraction; artificial intelligence; time domain

1. Introduction

Optimization of separation processes is not possible except with enough knowledge of the quantitative measurement of gas and oil components. The type of flow pattern impacts the efficiency of the separation process in such a way that the percentage of each component indicates whether the drilling needs to be stopped or not. The mixture of gamma radiation and ANNs have contributed in a lot of researches as a practical tool [1–7]. In [1], a calculation of volumetric percentages in three-phase flows was performed by using a dual-energy source and three detectors. Simulations were performed by MCNP-4C code. In addition, Abro and Johansen [2] researched the gas volume ratio by measuring two-phase flows. Their method consisted of a single ^{241}Am source and three detectors. The MRE% achieved was less than 3%. Adineh et al. presented a method to study the two-phase flow by a two-detector model of NaI and a single ^{137}Cs source [3]. The use of feature extraction methods can definitely lead to a qualitative improvement in the accuracy of flowmetry. In this regard, Sattari et al. [4] introduced a ^{137}Cs source and a single NaI detector to perform the flow measurement. In the research, the input ports of the GMDH neural network were time-domain features, which were extracted from the recorded spectrum. In similar studies, to establish the volume percentages and type of flow patterns with high accuracy, researchers evaluated many time- and frequency-domain characteristics, and they presented the best separator characteristics using an innovative method [5,6]. Some structures of MLP neural networks were investigated for the volume fraction calculation in 3-phase flows [7]. In that research, annular and homogeneous flow patterns were considered as the main flow regime. Complete diagnosis of the kind of flow regime and determining the volume fraction with an RMSE of 1.28 were the research findings. In addition to radioisotope sources, it has been proved that X-ray tubes can be potentially used in radiation-based multiphase flowmeters [8,9]. The X-ray tube has some advantages over radioisotope sources, for example, it has the capability of energy adjustment of emitted photons, it releases photons more intensely than fundamental radioisotope sources; it has the capability of turning on and off, etc.

Although X-ray radiation-based two-phase flowmeters have a lot of advantages over the radioisotope-based ones, they suffer from lower measurement accuracy. One reason might be that the registered X-ray spectrum has been analyzed in a simple way. The X-ray sources generate multi-energy photons despite radioisotopes that generate single energy photons. Therefore, data analysis of radioisotope sources would be easier than X-ray ones. As mentioned, one of the problems researchers have encountered is the lower measurement accuracy of the X-ray radiation-based two-phase flowmeters. The current project's goal is to resolve this problem by improving the precision of the X-ray radiation-based two-phase flowmeter using an artificial neural network (ANN) and feature extraction techniques.

In Section 2.1, the details of the modeled detection system, including one X-ray tube and one detector, will be discussed. Sections 2.2 and 2.3 are dedicated to processing and extracting features of the registered signals. In Section 2.4, the employment of ANN for flow pattern identification and volume ratio prediction will be explained. The results of all four sub-sections in part 2, will be discussed in Section 3. Figure 1 depicts the flowchart of the presented methodology used in the current investigation.

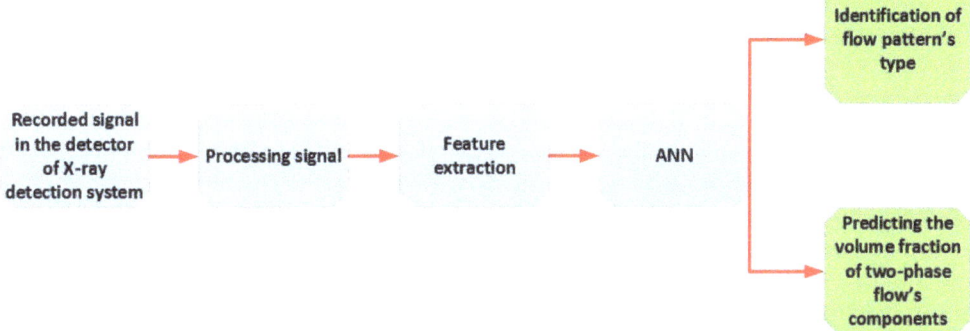

Figure 1. Flowchart of the presented methodology in this investigation.

2. Materials and Methods

2.1. X-ray System

In present article, the detection system that is consisted of an X-ray tube and a NaI detector which are located on both sides of the Pyrex-glass pipe, was modeled using the MCNP code. This code has been employed for modeling measuring instruments based on ionizing radiation [10–16].

In Figure 2, a geometric sketch of the designed system is shown. The emitted photons from the X-ray tube pass through the pipe, in which the two-phase flow components are being examined, and then, the portion of them that is not attenuated inside the pipe is detected by the detector. In fact, the attenuation of the radiation beam is based on the quantity of gas and liquid components inside the pipe.

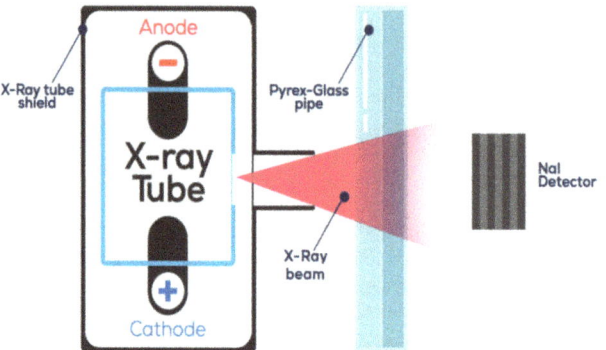

Figure 2. Modeled detection system using the MCNP code.

Since perfect modeling of an industrial X-ray tube including a cathode (electron source) and an anode (tungsten target) embedded in a cylindrical shield, using the MCNPX code is time consuming, in this work, a more efficient geometry including a photon source mounted inside a metal shield was defined. In other words, since photon tracking in the MCNPX code is much faster than electron tracking, a photon source mounted in a metal shield was just deemed in the present investigation instead of modeling the cathode-anode accumulation. To provide the X-ray energy spectrum for the photon source, the acquired spectrum by the TASMIC, a free software represented by Hernandez et al. [17], was employed. The employed X-ray spectrum including the X-ray characteristic peaks related to the tungsten anode is depicted in Figure 3. Fundamentally, the X-ray tube's cylinder-shaped shields are usually made of steel or lead to prevent leakage of radiation. On the shield surface, a section is left open, which is described as the output window to emit congenially produced X-ray photons. The output window's radius of the simulated X-ray in this study is 5 cm. To filter the low energy photons with the aim of reducing scattering, an aluminum filter having 2.5 mm thickness was embedded in front of the output window.

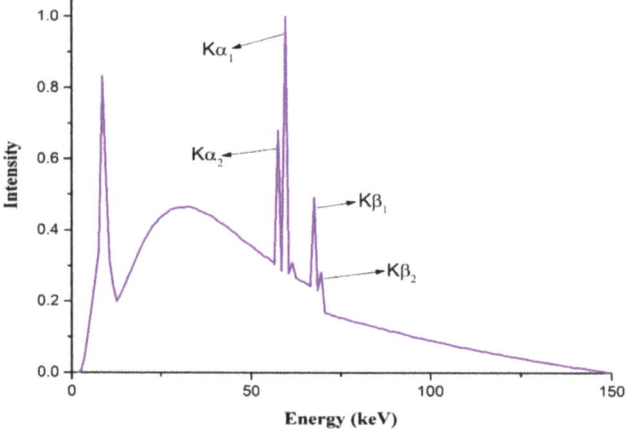

Figure 3. The applied X-ray energy spectrum in simulations obtained by the TASMIC package [17].

2.2. Signal Processing

In this investigation, 3 typical flow patterns (shown in Figure 4) and 19 different volume fractions from 5% void fraction to 95% with the step of 5%, are simulated (57 simulations were used in total). As an example, the recorded spectra in the detector for the 4 different void fractions of 25%, 45%, 65%, and 95% are shown in Figure 5.

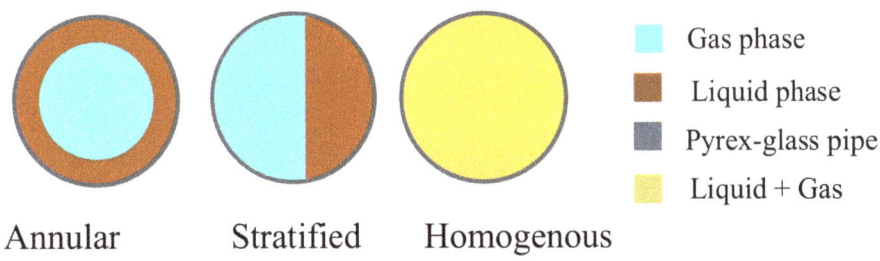

Figure 4. Simulated flow regimes.

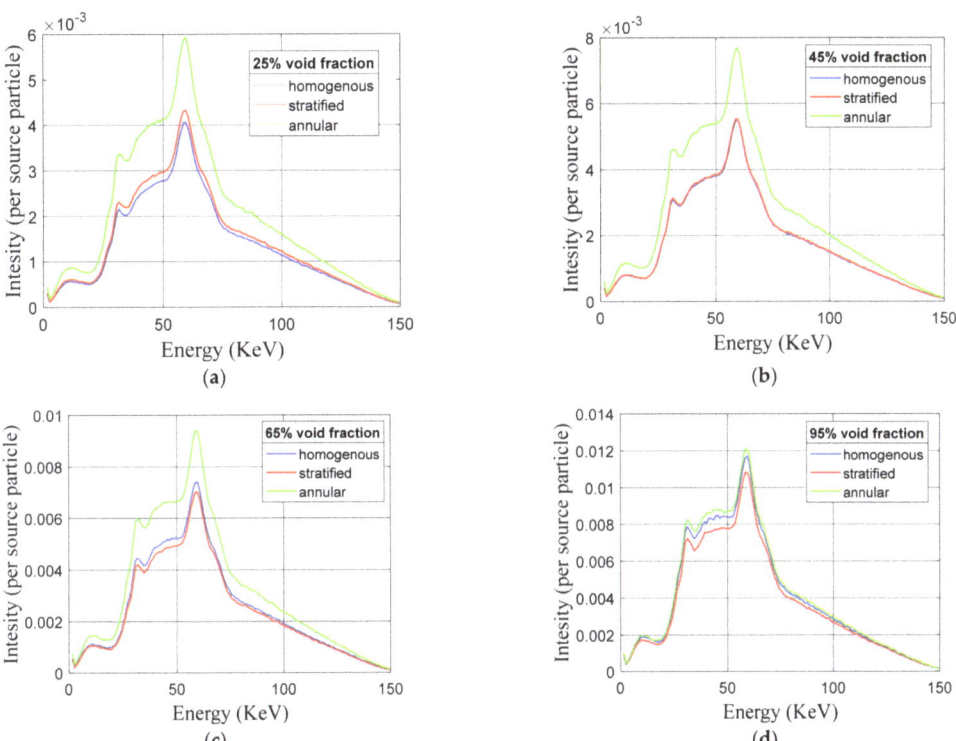

Figure 5. Recorded spectra in the detector for different void fractions of: (a) 25%, (b) 45%, (c) 65%, (d) 95%.

The tremendous data are collected for each simulation, as can be seen in Figure 5. In this context, to minimize the amount of data and preserve the data specifications simultaneously, the feature extraction techniques in the time domain have been employed. The extracted aspects are explained in detail in the following sections.

2.3. Feature Extraction

In this scrutiny, 12 time-domain characteristics (average value, variance, 4th-order moment, root mean square, skewness, kurtosis, median, waveform length (WL), SSR, MSR, SVER, and maximum value) were extracted from the recorded data. These characteristics were used as network inputs to determine the flow pattern and the volume percentages.

Therefore, an attempt was made to determine the efficient characteristics. To do this, various neural networks were configured with different combinations of the extracted features, and finally, it was observed that use of the Variance, Skewness, Kurtosis, SSR, and SVER can provide sufficient and high accuracy in detecting the mentioned parameters.

Five effective characteristics were extracted from the registered signals using the following equations. (x_n), $n = 1, \cdots, N$, where N is the number of datasets:

- Variance:

$$\sigma^2 = \frac{1}{N} \sum_{n=1}^{N} (x_n - m)^2, \; m = \frac{1}{N} \sum_{n=1}^{N} x_n \qquad (1)$$

- Skewness:

$$g_1 = \frac{m_3}{\sigma^3}, \; m_3 = \frac{1}{N} \sum_{n=1}^{N} [x_n - m]^3 \qquad (2)$$

- Kurtosis:

$$g_2 = \frac{m_4}{\sigma^4}, \; m_4 = \frac{1}{N} \sum_{n=1}^{N} [x_n - m]^4 \qquad (3)$$

- Summation of square roots (SSR):

$$\text{SSR} = \frac{1}{N} \sum_{n=1}^{N} (x_n)^{0.5} \qquad (4)$$

- Summation of variable exponent roots (SVER):

$$\text{SVER} = \frac{1}{N} \sum_{n=1}^{N} (x_n)^{exp}, \; exp = \begin{cases} 0.05 \text{ if } (n > 0.25 \cdot N \text{ and } n < 0.75 \cdot N) \\ 0.75 \text{ otherwise} \end{cases} \qquad (5)$$

The extracted features are indicated in Figure 6.

Two variables that are uncorrelated are not certainly independent, however, they may have a nonlinear relationship. In fact, two variables that have a little or insufficient correlation may have a strong nonlinear relationship. Since in this study there is not sufficient linear relation between the input and target, an MLP network with nonlinear parameters was used to find the relationship between input and output with high accuracy. Correlation analysis of each feature with respect to the target value is shown in Figure 7.

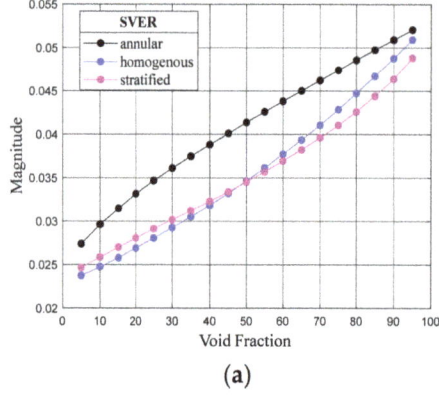

(a)

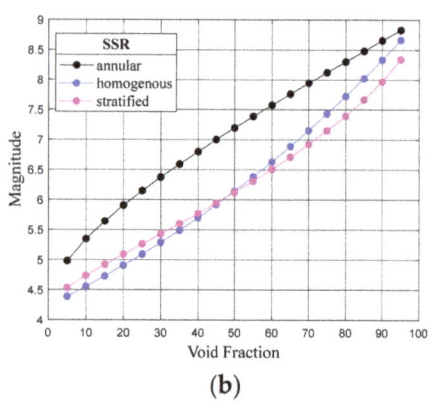

(b)

Figure 6. *Cont.*

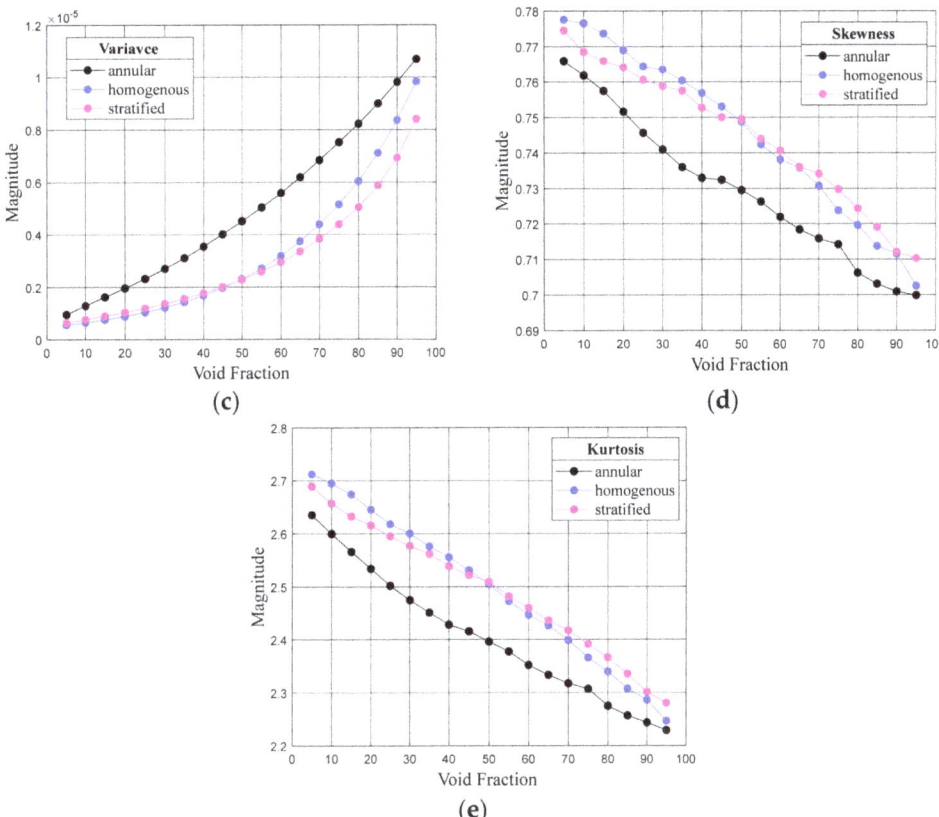

Figure 6. Extracted characteristics: (**a**) SVER, (**b**) SSR, (**c**) variance, (**d**) skewness, (**e**) kurtosis.

2.4. Artificial Intelligence

In the last few years, various computational techniques have been utilized for various applications in the engineering research area [18–32]. In this study, ANN has been implemented for flow pattern identification and volume ratio prediction. As a mathematical system, ANNs are described to be formed by the plain processing components called neurons acting in parallel and are produced as one or multiple layers [23,24]. MLP acquires nonlinear function mappings and could learn the abundant diversity of nonlinear decision surfaces as well. Figure 8 depicts the presented MLP model, in which the inputs are the extracted features described in the former section, and the outputs are the volume fraction ratios regardless of the flow pattern. The neuron output in the output layer is achieved by the following equations [33,34]:

$$x_l = \sum_{i=1}^{u} a_i w_{ij} + b \quad j = 1, 2, \cdots, m \tag{6}$$

$$y_j = f(\sum_{i=1}^{u} a_i w_{ij} + b) \quad j = 1, 2, \cdots, m \tag{7}$$

$$output = \sum_{n=1}^{j}(y_n w_n) + b \tag{8}$$

The equation related to the tansig activation function is given below:

$$tansig(x) = \frac{2}{(1+\exp(-2x))} - 1 \qquad (9)$$

where a, b, w, and f present the input parameters, the bias term, the weighting factor, and the activation function, respectively. The index i is the input number, and j is the neuron number in every hidden layer. The Levenberg–Marquardt algorithm was used for training of the presented MLP networks, where the 1st and 2nd derivatives (i.e., the gradient and Hessian) were utilized for network weight correction [35]. The available data are organized into three categories: training, validation, and testing data.

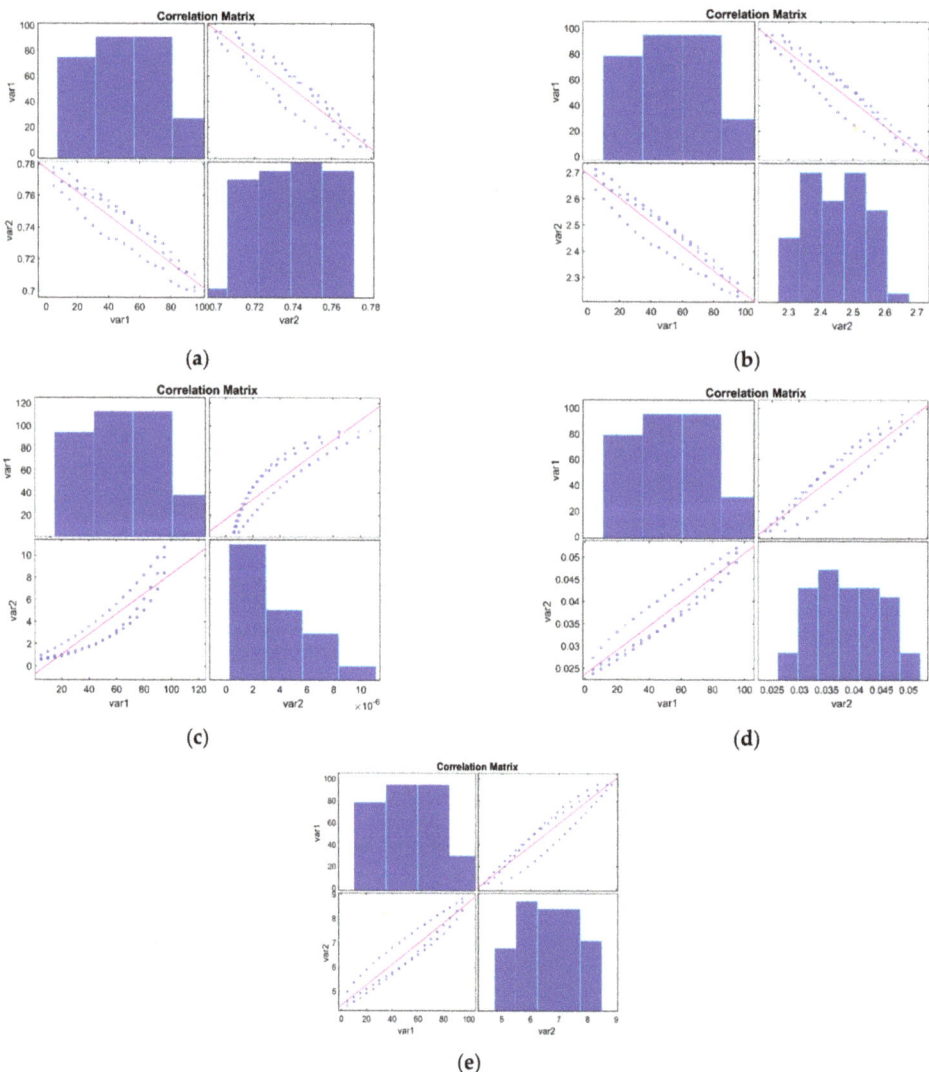

Figure 7. Correlation analysis of extracted features: (**a**) Skewness, (**b**) Kurtosis, (**c**) Variance, (**d**) SVER, and (**e**) SSR.

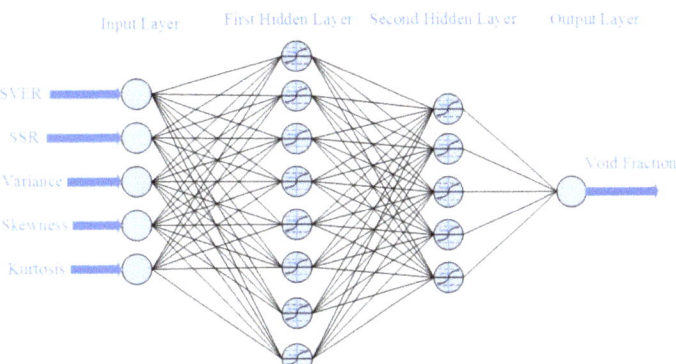

Figure 8. The configuration of the presented network for predicting void fraction.

Training dataset: The sample of data utilized to fit the model. The model sees and learns from these data.

Validation dataset: The validation set is utilized to assess the performance of a model, but this is for frequent evaluation. The model encounters these data on occasion, but never does it "Learn" from these data.

Testing dataset: The sample of data utilized to offer an unbiased assessment of a final model fit on the training dataset. The test dataset serves as the gold standard against which the model is assessed. It is only utilized once after completing the network training.

The use of validation data in the network training process as well as final network testing using test datasets will give us the reassurance to avoid under-fitting and over-fitting problems. The training, validation, and testing samples data are 39 (70% of data), 9 (15% of data), and 9 (about 15% of data), respectively. In the present article, two ANN models of MLP were trained to recognize the type of flow regime and to predict the volumetric fraction. Several ANN configurations were tested and enhanced to obtain the optimum ANN configuration with the least error. Several configurations with 1, 2, and 3 hidden layers owning different neuron numbers in every layer and diverse activation functions were examined. MATLAB-2018b was utilized for training the ANN model. The structure of neurons as predictors and clarifiers of ANNs are indicated in Figures 8 and 9, respectively. The specification of the implemented MLP ANNs is described in Tables 1 and 2. The outputs of the classifier network are the type of flow patterns: 1, 2 and 3 were deemed as the annular, homogenous and stratified flow pattern, respectively.

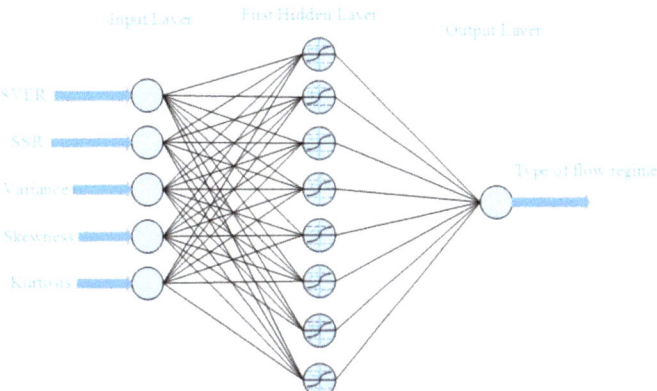

Figure 9. The configuration of presented ANN for classifying the flow regimes.

Table 1. Specifications of the presented network for void fraction prediction.

Neural Network	MLP
Nodes number in input layer	5
Nodes number in 1st latent layer	8
Nodes number in 2nd latent layer	5
Nodes number in output layer	1
Number of epochs	450
Activation function applied for hidden neurons	Tansig
Activation function applied for output neuron	purelin

Table 2. Characteristics of the presented flow regime classifier network.

Neural Network	MLP
Nodes number in input layer	5
Nodes number in 1st latent layer	8
Nodes number in output layer	1
Number of epochs	380
Activation function applied for hidden neurons	Tansig
Activation function applied for output neuron	purelin

Many different architectures with different configurations were tested based on the algorithm detailed below in order to find the optimized structure:

(1). The dataset is defined;
(2). The counter parameters with zero initial value are defined;
(3). The root mean square error is defined;
(4). The initial values of other parameters in order to break loops are set;
(5). Several nested loops are generated to test all of the structures;
(6). The ANN with various number of hidden layers, various number of neurons in each layer, various epochs and various activation functions are tested in created loops utilizing the specified counter parameters and other parameters' initial values;
(7). The network's effectiveness in each step is checked utilizing the specified error;
(8). The best network with lowest error is saved.

3. Results and Discussions

The function of the enacted network to the volumetric fraction project is displayed in Figures 10–12 using a fitting, regression, and histogram diagram. Both the given output and the network output are plotted in the fitting diagram. The blue star in the regression diagram depicts the network output, and the red line depicts the given output. Apparently, as the blue star is close to the red line, the planned network is more precise. The diagram of error histogram illustrates the error distribution. To show the precision of the flow regime classifier network, the confusion matrix is utilized and depicted in Figure 13 for training, validation, and testing of the dataset.

The MAPE, MAE, and RMSE of the network are computed by:

$$\text{MAPE\%} = 100 \times \frac{1}{N} \sum_{j=1}^{N} \left| \frac{X_j(Exp) - X_j(Pred)}{X_j(Exp)} \right| \qquad (10)$$

$$\text{RMSE} = \left[\frac{\sum_{j=1}^{N}(X_j(Exp) - X_j(Pred))^2}{N}\right]^{0.5} \quad (11)$$

$$\text{MAE} = \frac{1}{N} \times \sum_{j=1}^{N} |X_j(Exp) - X_j(Pred)| \quad (12)$$

where $X(Exp)$ and $X(Pred)$ denote the experimental and forecasted (ANN) void fractions, respectively. The errors of the given predictor network are listed in Table 3.

The results of the relevant investigations and the work are listed in Table 4. As it can be seen, the precision of the presented system in this paper is significantly higher than all of the previous meters in this category which demonstrates the superiority of the proposed method.

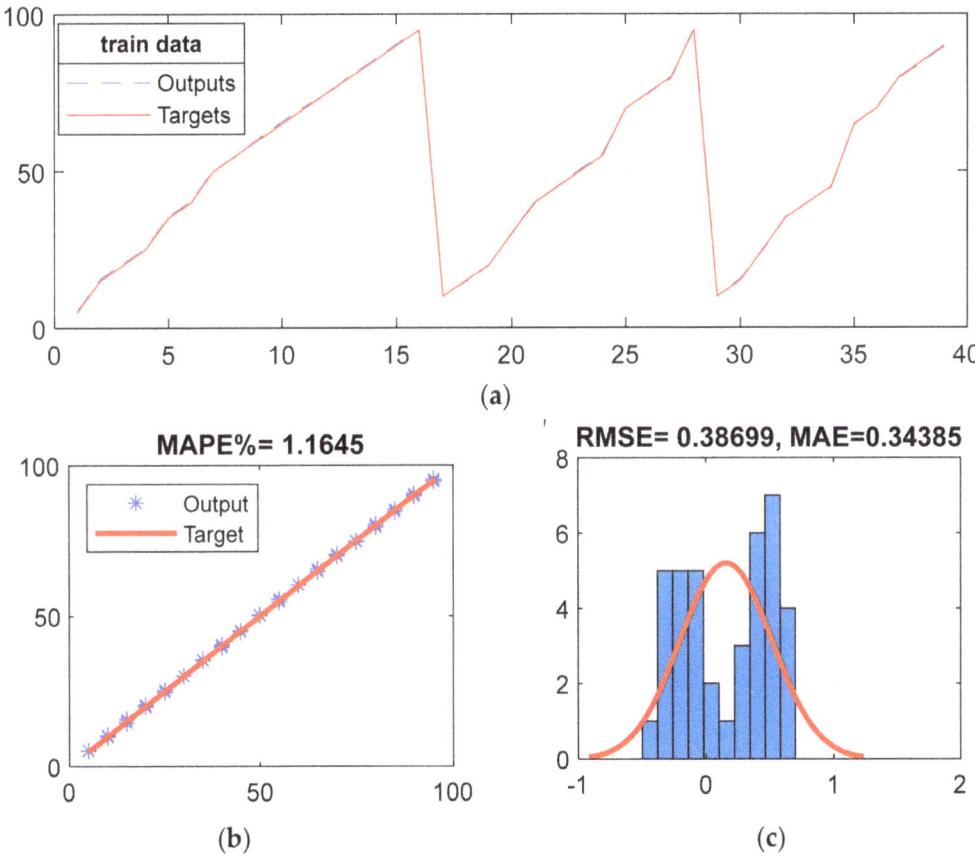

Figure 10. (a) Fitting, (b) regression, and (c) error histogram diagram for training the ANN to estimate the gas volume ratio percentage.

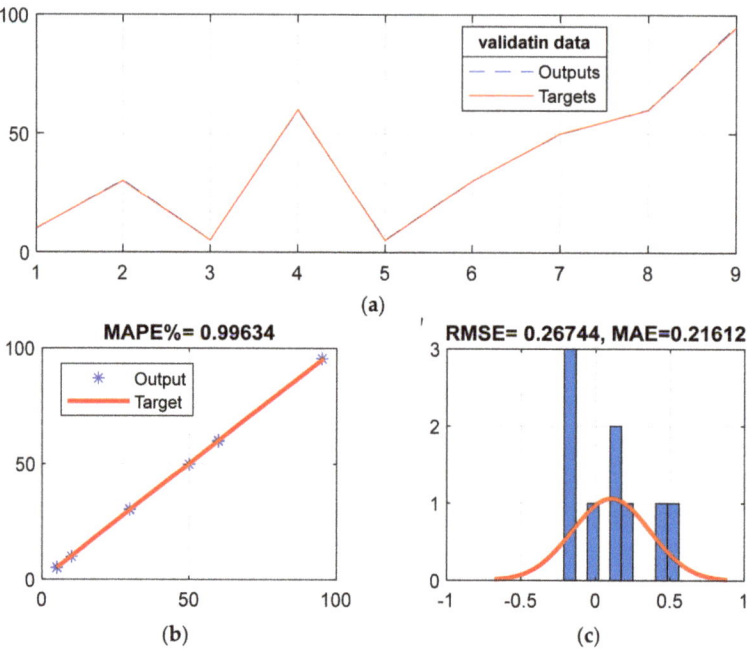

Figure 11. (**a**) Fitting, (**b**) regression, and (**c**) error histogram diagram for validation of the ANN to estimate the gas volume ratio percentage.

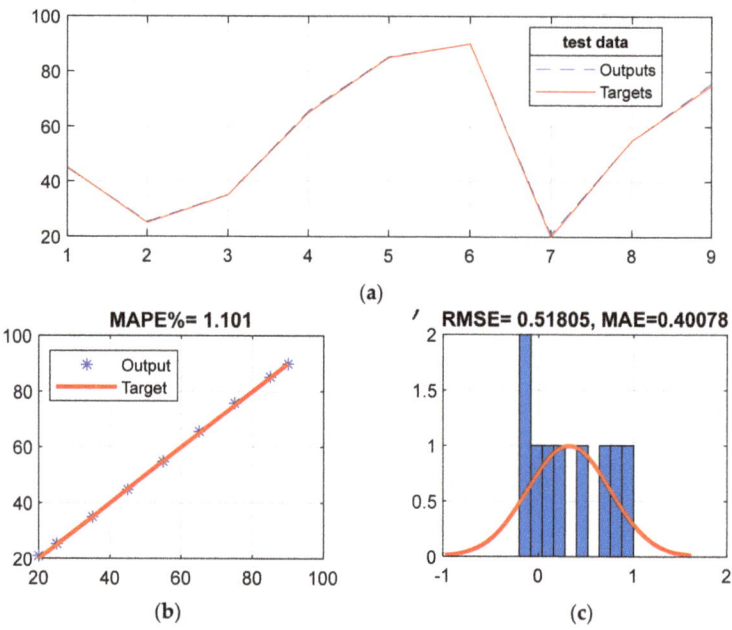

Figure 12. (**a**) Fitting, (**b**) regression, and (**c**) error histogram schematic for testing the ANN to estimate the gas volume ratio percentage.

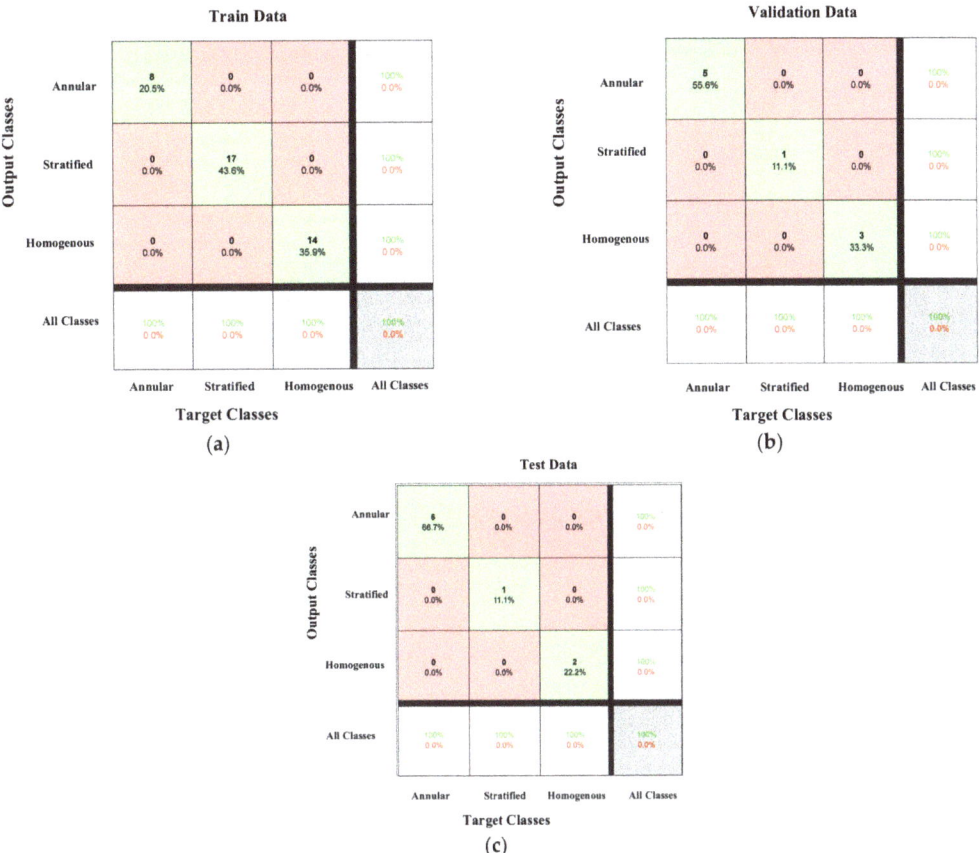

Figure 13. Precision of the flow regime classifier network: (**a**) train data, (**b**) validation data, (**c**) test data.

Table 3. Computed errors for training, validation, and testing dataset of the estimation network.

Dataset	MAPE	MAE	RMSE
Training dataset	1.16	0.34	0.38
Validation dataset	0.99	0.21	0.26
Testing dataset	1.1	0.4	0.51

Table 4. The results of the relevant investigations and the presented two-phase flow meter in this study.

Refs	Radiation Source	Number of Detectors/Type of Detector	Number of Considered Flow Regimes	Volume Fraction Prediction Accuracy (RMSE)	Volume Fraction Prediction Accuracy (MAPE)
[4]	Cs-137	1/NaI	3	1.11	5.32
[36]	Cs-137	2/NaI	3	1.29	1.48
[37]	Cs-137	1/NaI	2	6.12	1.17
[38]	Cs-137	2/NaI	3	2.12	1.32
[39]	Cs-137	1/NaI	1	3.57	-
[40]	Am-241	2/NaI	2	3.1	-
[13]	X-Ray Tube	1/NaI	2	5.54	4.49
[8]	X-Ray Tube	2/NaI	3	5.39	-
[Our study]	X-Ray Tube	1/NaI	3	0.51	1.16

4. Conclusions

The thirst of oil and gas companies for more and efficient access to fossil fuels has led the industry to find new development approaches for new production techniques. In this work, a system has been presented for accurate measurement of the volumetric percentages in two-phase flows independent of the flow pattern, in which an X-ray tube, a Pyrex-glass, and a sodium iodide detector, have been used. The different volume percentages of the three flow regimes have been simulated and the data obtained from each of them have been recorded. Five characteristics in the time-domain were acquired and deemed as the inputs of the multilayer perceptron. The capability for the proposed networks to classify the flow patterns with 100% accuracy and acquire the void percentages precisely with respect to the recorded values of 0.51 for RMSE and 1.16 for MAPE, represents the success of the approach presented in this work. The precision of the proposed X-ray-based system in this paper is significantly higher than all of the previous meters in this category. In addition, this meter has a safer and easier mechanism than other, radiation-based meters. The usage of appropriate soft computing methods and the suitable radiation source were the reason of this achievement.

Although the obtained results in this study are promising, the proposed methodology has been investigated for two-phase flow in static conditions. For future studies, it is planned to implement the proposed methodology for dynamic two-phase flows.

Author Contributions: Conceptualization, M.A.S., E.N. and O.T.; software, A.B., M.A.S., O.T.; writing—review and editing, M.A.S., E.N., O.T. and A.B.; funding acquisition, A.B., O.T. All authors have read and agreed to the published version of the manuscript.

Funding: This work was funded by the Deanship of Scientific Research (DSR), King Abdulaziz University, Jeddah, under grant No. (D1441-376-135). The authors, therefore, gratefully acknowledge the DSR technical and financial support.

Institutional Review Board Statement: Not applicable.

Informed Consent Statement: Not applicable.

Data Availability Statement: Data are contained within the article.

Conflicts of Interest: The authors declare no conflict of interest.

References

1. Salgado, C.M.; Brandão, L.E.; Schirru, R.; Pereira, C.M.; da Silva, A.X.; Ramos, R. Prediction of volume fractions in three-phase flows using nuclear technique and artificial neural network. *Appl. Radiat. Isot.* **2009**, *67*, 1812–1818. [CrossRef]
2. Åbro, E.; Johansen, G.A. Improved void fraction determination by means of multibeam gamma-ray attenuation measurements. *Flow Meas. Instrum.* **1999**, *10*, 99–108. [CrossRef]
3. Adineh, M.; Nematollahi, M.; Erfaninia, A. Experimental and numerical void fraction measurement for modeled two-phase flow inside a vertical pipe. *Ann. Nucl. Energy* **2015**, *83*, 188–192. [CrossRef]
4. Sattari, M.A.; Roshani, G.H.; Hanus, R. Improving the structure of two-phase flow meter using feature extraction and GMDH neural network. *Radiat. Phys. Chem.* **2020**, *171*, 108725. [CrossRef]
5. Sattari, M.A.; Roshani, G.H.; Hanus, R.; Nazemi, E. Applicability of time-domain feature extraction methods and artificial intelligence in two-phase flow meters based on gamma-ray absorption technique. *Measurement* **2021**, *168*, 108474. [CrossRef]
6. Hosseini, S.; Roshani, G.H.; Setayeshi, S. Precise gamma based two-phase flow meter using frequency feature extraction and only one detector. *Flow Meas. Instrum.* **2020**, *72*, 101693. [CrossRef]
7. Khayat, O.; Afarideh, H. Design and simulation of a multienergy gamma ray absorptiometry system for multiphase flow metering with accurate void fraction and water-liquid ratio approximation. *Nukleonika* **2019**, *64*, 19–29. [CrossRef]
8. Roshani, M.; Phan, G.; Roshani, G.H.; Hanus, R.; Nazemi, B.; Corniani, E.; Nazemi, E. Combination of X-ray tube and GMDH neural network as a nondestructive and potential technique for measuring characteristics of gas-oil–water three phase flows. *Measurement* **2021**, *168*, 108427. [CrossRef]
9. Song, K.; Liu, Y. A compact x-ray system for two-phase flow measurement. *Meas. Sci. Technol.* **2017**, *29*, 025305. [CrossRef]
10. Roshani, G.H.; Nazemi, E.; Roshani, M.M. Flow regime independent volume fraction estimation in three-phase flows using dual-energy broad beam technique and artificial neural network. *Neural Comput. Appl.* **2016**, *28*, 1265–1274. [CrossRef]

11. Karami, A.; Roshani, G.H.; Khazaei, A.; Nazemi, E.; Fallahi, M. Investigation of different sources in order to optimize the nuclear metering system of gas–oil–water annular flows. *Neural Comput. Appl.* **2020**, *32*, 3619–3631. [CrossRef]
12. Roshani, M.; Phan, G.; Faraj, R.H.; Phan, N.H.; Roshani, G.H.; Nazemi, B.; Corniani, E.; Nazemi, E. Proposing a gamma radi-ation based intelligent system for simultaneous analyzing and detecting type and amount of petroleum by-products. *Neural Eng. Technol.* **2021**, *53*, 1277–1283.
13. Amiri, S.; Ali, P.J.M.; Mohammed, S.; Hanus, R.; Abdulkareem, L.; Alanezi, A.A.; Eftekhari-Zadeh, E.; Roshani, G.H.; Nazemi, E.; Kalmoun, E.M. Proposing a Nondestructive and Intelligent System for Simultaneous Determining Flow Regime and Void Fraction Percentage of Gas–Liquid Two Phase Flows Using Polychromatic X-Ray Transmission Spectra. *J. Nondestruct. Eval.* **2021**, *40*, 1–12. [CrossRef]
14. Roshani, G.H.; Nazemi, E.; Feghhi, S.A.; Setayeshi, S. Flow regime identification and void fraction prediction in two-phase flows based on gamma ray attenuation. *Measurement* **2015**, *62*, 25–32. [CrossRef]
15. Roshani, G.; Hanus, R.; Khazaei, A.; Zych, M.; Nazemi, E.; Mosorov, V. Density and velocity determination for single-phase flow based on radiotracer technique and neural networks. *Flow Meas. Instrum.* **2018**, *61*, 9–14. [CrossRef]
16. Roshani, M.; Phan, G.T.; Ali, P.J.M.; Roshani, G.H.; Hanus, R.; Duong, T.; Corniani, E.; Nazemi, E.; Kalmoun, E.M. Evaluation of flow pattern recognition and void fraction measurement in two phase flow independent of oil pipeline's scale layer thickness. *Alex. Eng. J.* **2021**, *60*, 1955–1966. [CrossRef]
17. Hernandez, A.M.; Boone, J.M. Tungsten anode spectral model using interpolating cubic splines: Unfiltered x-ray spectra from 20 kV to 640 kV. *Med Phys.* **2014**, *41*, 042101. [CrossRef]
18. Zhang, F.; Chen, K.; Zhu, L.; Appiah, D.; Hu, B.; Yuan, S. Gas–Liquid Two-Phase Flow Investigation of Side Channel Pump: An Application of MUSIG Model. *Mathematics* **2020**, *8*, 624. [CrossRef]
19. Roshani, G.H.; Roshani, S.; Nazemi, E.; Roshani, S. Online measuring density of oil products in annular regime of gas-liquid two phase flows. *Measurement* **2018**, *129*, 296–301. [CrossRef]
20. Moradi, M.J.; Hariri-Ardebili, M.A. Developing a Library of Shear Walls Database and the Neural Network Based Predictive Meta-Model. *Appl. Sci.* **2019**, *9*, 2562. [CrossRef]
21. Ali, A.; Umar, M.; Abbas, Z.; Shahzadi, G.; Bukhari, Z.; Saleem, A. Numerical Investigation of MHD Pulsatile Flow of Micropolar Fluid in a Channel with Symmetrically Constricted Walls. *Mathematics* **2021**, *9*, 1000. [CrossRef]
22. Moradi, M.J.; Roshani, M.M.; Shabani, A.; Kioumarsi, M. Prediction of the load-bearing behavior of spsw with rectangular opening by RBF network. *Appl. Sci.* **2020**, *10*, 1185. [CrossRef]
23. Karami, A.; Roshani, G.H.; Nazemi, E.; Roshani, S. Enhancing the performance of a dual-energy gamma ray based three-phase flow meter with the help of grey wolf optimization algorithm. *Flow Meas. Instrum.* **2018**, *64*, 164–172. [CrossRef]
24. Khaleghi, M.; Salimi, J.; Farhangi, V.; Moradi, M.J.; Karakouzian, M. Application of Artificial Neural Network to Predict Load Bearing Capacity and Stiffness of Perforated Masonry Walls. *CivilEng* **2021**, *2*, 48–67. [CrossRef]
25. Roshani, G.; Nazemi, E.; Roshani, M. Identification of flow regime and estimation of volume fraction independent of liquid phase density in gas-liquid two-phase flow. *Prog. Nucl. Energy* **2017**, *98*, 29–37. [CrossRef]
26. Mosorov, V.; Zych, M.; Hanus, R.; Sankowski, D.; Saoud, A. Improvement of Flow Velocity Measurement Algorithms Based on Correlation Function and Twin Plane Electrical Capacitance Tomography. *Sensors* **2020**, *20*, 306. [CrossRef]
27. Moradi, M.; Daneshvar, K.; Ghazi-Nader, D.; Hajiloo, H. The prediction of fire performance of concrete-filled steel tubes (CFST) using artificial neural network. *Thin-Walled Struct.* **2021**, *161*, 107499. [CrossRef]
28. Roshani, G.; Nazemi, E. Intelligent densitometry of petroleum products in stratified regime of two phase flows using gamma ray and neural network. *Flow Meas. Instrum.* **2017**, *58*, 6–11. [CrossRef]
29. Waini, I.; Ishak, A.; Pop, I. Hybrid Nanofluid Flow over a Permeable Non-Isothermal Shrinking Surface. *Mathematics* **2021**, *9*, 538. [CrossRef]
30. Roshani, G.; Nazemi, E.; Roshani, M. Usage of two transmitted detectors with optimized orientation in order to three phase flow metering. *Measurement* **2017**, *100*, 122–130. [CrossRef]
31. Salimi, J.; Ramezanianpour, A.M.; Moradi, M.J. Studying the effect of low reactivity metakaolin on free and restrained shrinkage of high performance concrete. *J. Build. Eng.* **2020**, *28*, 101053. [CrossRef]
32. Roshani, G.H.; Nazemi, E.; Roshani, M.M. Intelligent recognition of gas-oil-water three-phase flow regime and determina-tion of volume fraction using radial basis function. *Flow Meas. Instrum.* **2017**, *54*, 39–45. [CrossRef]
33. Taylor, J.G. *Neural Networks and Their Applications*; John Wiley & Sons Ltd.: Brighton, UK, 1996.
34. Gallant, A.R.; White, H. On learning the derivatives of an unknown mapping with multilayer feedforward networks. *Neural Netw.* **1992**, *5*, e129–e138. [CrossRef]
35. Hagan, M.T.; Menhaj, M. Training feedforward networks with the Marquardt algorithm. *IEEE Trans. Neural Netw.* **1994**, *5*, e989–e993. [CrossRef]
36. Nazemi, E.; Roshani, G.; Feghhi, S.; Setayeshi, S.; Zadeh, E.E.; Fatehi, A. Optimization of a method for identifying the flow regime and measuring void fraction in a broad beam gamma-ray attenuation technique. *Int. J. Hydrogen Energy* **2016**, *41*, 7438–7444. [CrossRef]

37. Roshani, M.; Sattari, M.A.; Ali, P.J.M.; Roshani, G.H.; Nazemi, B.; Corniani, E.; Nazemi, E. Application of GMDH neural network technique to improve measuring precision of a simplified photon attenuation based two-phase flowmeter. *Flow Meas. Instrum.* **2020**, *75*, 101804. [CrossRef]
38. Nazemi, E.; Feghhi, S.; Roshani, G.; Peyvandi, R.G.; Setayeshi, S. Precise Void Fraction Measurement in Two-phase Flows Independent of the Flow Regime Using Gamma-ray Attenuation. *Nucl. Eng. Technol.* **2016**, *48*, 64–71. [CrossRef]
39. Falahati, M.; Vaziri, M.R.; Beigzadeh, A.; Afarideh, H. Design, modelling and construction of a continuous nuclear gauge for measuring the fluid levels. *J. Instrum.* **2018**, *13*, P02028. [CrossRef]
40. Hanus, R. Application of the Hilbert Transform to measurements of liquid–gas flow using gamma ray densitometry. *Int. J. Multiph. Flow* **2015**, *72*, 210–217. [CrossRef]

Article

Modeling and Simulation Techniques Used in High Strain Rate Projectile Impact

Derek G. Spear *, Anthony N. Palazotto and Ryan A. Kemnitz

Air Force Institute of Technology, Wright-Patterson AFB, Dayton, OH 45433, USA; anthony.palazotto@afit.edu (A.N.P.); ryan.kemnitz@afit.edu (R.A.K.)
* Correspondence: derek.spear@afit.edu

Abstract: A series of computational models and simulations were conducted for determining the dynamic responses of a solid metal projectile impacting a target under a prescribed high strain rate loading scenario in three-dimensional space. The focus of this study was placed on two different modeling techniques within finite element analysis available in the Abaqus software suite. The first analysis technique relied heavily on more traditional Lagrangian analysis methods utilizing a fixed mesh, while still taking advantage of the finite difference integration present under the explicit analysis approach. A symmetry reduced model using the Lagrangian coordinate system was also developed for comparison in physical and computational performance. The second analysis technique relied on a mixed model that still made use of some Lagrangian modeling, but included smoothed particle hydrodynamics techniques as well, which are mesh free. The inclusion of the smoothed particle hydrodynamics was intended to address some of the known issues in Lagrangian analysis under high displacement and deformation. A comparison of the models was first performed against experimental results as a validation of the models, then the models were compared against each other based on closeness to experimentation and computational performance.

Citation: Spear, D.G.; Palazotto, A.N.; Kemnitz, R.A. Modeling and Simulation Techniques Used in High Strain Rate Projectile Impact. *Mathematics* **2021**, *9*, 274. https://doi.org/10.3390/math9030274

Academic Editor: Simeon Reich and Clemente Cesarano
Received: 2 December 2020
Accepted: 18 January 2021
Published: 30 January 2021

Publisher's Note: MDPI stays neutral with regard to jurisdictional claims in published maps and institutional affiliations.

Copyright: © 2021 by the authors. Licensee MDPI, Basel, Switzerland. This article is an open access article distributed under the terms and conditions of the Creative Commons Attribution (CC BY) license (https://creativecommons.org/licenses/by/4.0/).

Keywords: high strain rate impact; modeling and simulation; smoothed particle hydrodynamics; finite element analysis

1. Introduction

High strain rate impact testing entails the high velocity collision of a projectile with a target, then observing the effects of the interaction. This is primarily done in terms of deformation and fragmentation, considering both the projectile and target, and evaluating the target failure pattern and projectile penetration depth [1]. Due to the expense of these impact experiments, especially considering the rising interest in hyper-velocity ballistic applications, much of this research must be done utilizing numerical and computational modeling and simulation methods [2]. Modeling and simulation of impact scenarios is complicated, involving the evaluation of several non-linear steps, such as contact between the projectile and target, high strain rates near the impact region, stress wave propagation through both projectile and target, and material deformation and separation. Due to these complexities, high strain rate impact has predominantly been modeled in two-dimensional space, only recently expanding into three dimensions with moderate success of correlating deformation with variation in impact velocity and projectile shape [3,4]. Progression into three-dimensional simulation is important to accurately characterize projectile impact, since in the real world impacts almost always occur across all three dimensions. Oblique impact, translation, and rotation all play a significant role in the outcome of an impact scenario, and these conditions can only truly be represented in three-dimensional space. Due to the overall complexity of the impact scenarios and expansion into three dimensions, being further combined with varied sources of non-linearity, it is essential to perform model validation. A thorough impact evaluation provides the validation of numerical failure models through the assessment of the energy absorbed during impact, penetration depth in a thick target or residual velocity of the projectile in the thin target, and the measured deformation of both projectile and target [5].

There have been several numerical methods devised to provide insight and evaluate an impact event, two of the more prevalent methods used to simulate and model impacts are the meshed finite elements and mesh-free discrete elements. The finite element method (FEM) has successfully been used to simulate and model impacts beyond a one-dimensional representation. Johnson first introduced two-dimensional and three-dimensional Lagrangian numerical models to simulate high velocity impact, noting that the elements were subjected to large distortions that required significant remeshing, which introduced error into the solution [6,7]. Utilizing two-dimensional FEM, Gailly and Espinosa were able to create an impact model that correlated well with experimental results at increased impact velocities, up to 1450 m per second [8]. Following the work of Gailly and Espinosa, Kurtaran et al. demonstrated that FEM is also capable of providing relatively accurate three-dimensional simulations at increased velocities, up to 1500 m per second [3]. As computing power increases, the use of three-dimensional FEM in modeling impact has also increased, with work expanding into impact modeling of fiber reinforced composites [9,10], concrete beams [11], and thin target perforation [12,13].

The second numerical method for analyzing impacts, which has been gaining renewed interest, is the Discrete Element Method (DEM). Although Cundall and Strack first suggested the use of the DEM in 1979 as a numerical method to evaluate the stresses and strains within a continuum of discontinuous materials [14], it was not until the early 1990s that it began to gain influence in impact modeling. Libersky et al. first adapted the Smoothed Particle Hydrodynamics (SPH) methodology within DEM to model the dynamic response of solids [15,16], later adapting their model to simulate high velocity impact [17]. Johnson et al. were some of the first to incorporate SPH into a computational model for impact simulation, which was used to address the severe distortions noted in traditional FEM [18]. The authors later introducing a normalized smoothing function into their algorithm which scoped the region of particle interaction within a coupled FEM-SPH model [19]. More recently, SPH modeling techniques have expanded into simulation of hypervelocity impact [20,21], penetration of ballistic gelatin [22,23], perforation of layered targets [24], and use of granular media [25,26]. Much of this work has remained in two-dimensional space as the computational requirements for modeling in three-dimensions are still high.

Within the Air Force Institute of Technology (AFIT), FEM has been used successfully in related impact analysis efforts preceding the current research effort. The previous work focused on two-dimensional stress field analysis for use in topology optimization of an additively manufactured projectile impacting a concrete target [27], expansion of the impact scenario to multiple target sets [28], inclusion of a multi-material projectile [29], and inclusion of a lattice substructure within the projectile wall [30]. The current line of research is expanding the high strain rate impact scenario to three dimensions with the inclusion of a lattice section into the projectile to aid in impact energy absorption and stress wave management [31,32]. Impact testing will also be used to determine the dynamic material properties of the lattice designs, along with characterizing plasticity and damage modeling parameters. With an emphasis on high strain rate impact, one of the first steps taken in this research was to develop and validate a three-dimensional physics based computational model capable of accurately representing the impact environment. As part of the modeling effort, an analysis between two particular techniques, traditional Lagrangian finite elements and SPH discrete methods, was performed to investigate the differences in methodology, modeling, and analysis capabilities.

2. Methodology

FEM is a numerical method used to find the solution to boundary value problems. FEM grew out of the aerospace industry in the 1960s, as a method for performing stress and thermal analysis on complex aircraft, rocket, and engine parts [33]. FEM uses the concept of discretization, by applying a fixed mesh, to divide a body into smaller units, finite elements, which are interconnected across shared nodes or boundaries. Applicable field quantities are approximated across the entire structure through piecewise element interpolation and summation [34].

FEM has been used with success to describe the failure modes of brittle materials during ballistic penetration, finding where transition from micro-cracking to pulverization occurs in ceramic armor [8]. Utilizing FEM techniques, two-dimensional impact models have been created that showed good correlation with experimental results even at higher impact velocities, up to 1450 m per second. More recently, FEM has proven capable of providing relatively accurate three-dimensional simulations at increased velocities, modeling the deformation patterns of projectile and target up to 1500 m per second, as well as showing the effects of thermal softening in target deformation at the higher velocities [3]. As demonstrated by the aforementioned studies, one benefit of FEM over other impact modeling techniques is the relative computational ease to perform simulations in both two and three dimensions while maintaining good correlation with experimental and real-world results.

Since Dassault Systèmes Abaqus finite element software was used as the primary means of computational analysis within this study, the specific applications and mathematical representations used within Abaqus are highlighted within the described theory and application sections.

2.1. Lagrangian Finite Element Analysis

Finite Element Analysis (FEA) is the numerical method for solving complex boundary value problems where analytical solutions may not be able to be obtained utilizing FEM. Within FEA, the explicit approach is most useful in solving problems that result in large deformations or are highly time dependent. Instead of solving the traditional finite element problem of global mass and stiffness matrices, the explicit technique solves Newton's Second Law of Motion, $F = ma$ for each element. This relies on a half-step time integration technique that evaluates a known nodal solution of displacement and acceleration component vectors at set time t with the velocity components at time $t - \frac{1}{2}$ [34]. This solution is then iterated with the next time increment, Δt. This analysis technique is formulated as an initial value problem that is capable of determining how a system evolves given an initial loading condition and position. Explicit analysis was required to be used in this study since it deals with the time dependent behavior of materials, the high strain rate effects of impact, and the progression of a wave through the system.

The explicit analysis approach, in order to account for the deformation rate and stress wave propagation of impact, must consider displacement, velocity, and acceleration on a node by node basis within the structure with respect to time. Abaqus incorporates a finite difference scheme incorporating central difference time integration (CDTI) that is used to calculate nodal field variables as the time step is incremented along with the use of lumped mass matrices. As this is a three-dimensional problem, the nodes create three-dimensional solid elements, which were represented by eight-noded stress bricks with reduced integration, or C3D8R elements, in this implementation. The C3D8R is considered a general purpose three-dimensional linear element with a single integration point. The reduced integration allows the element to overcome the shear locking issue prevalent in the full integration element, C3D8, under high values of plasticity. Shear locking is prevalent in all first-order, fully integrated solid elements, where the element cannot exhibit pure bending [35]. The failure to exhibit pure bending deformation leads to parasitic shear strain developed alongside the bending strain, where strain energy is absorbed by the shear strain causing the element to appear too stiff. Reduced integration of the element limits the development of parasitic shear strain. The first step in the CDTI process, using a half-step central differences scheme, is to create a force balance using the equations of motion at time t, then the acceleration values at time t are used to determine the nodal velocities at time $t + \frac{\Delta t}{2}$, which is then, in turn, used to determine the displacements at time $t + \Delta t$. This process is repeated throughout the time interval of interest, and evaluates deformation and stress wave propagation through the part. In this implementation of CDTI, the expressions for displacement (d), velocity ($\dot{d}$) and acceleration ($\ddot{d}$) are given by Equation (1), Equation (2) and Equation (3) respectively [36].

$$d_{(i+1)} = d_{(i)} + \Delta t_{(i+1)} \dot{d}_{(i+\frac{1}{2})} \tag{1}$$

$$\dot{d}_{(i+\frac{1}{2})} = \dot{d}_{(i-\frac{1}{2})} + \frac{\Delta t_{(i+1)} + \Delta t_{(i)}}{2} \ddot{d}_{(i)} \tag{2}$$

$$\ddot{d}_{(i)} = M^{-1}(F_{(i)} - R_{(i)}). \tag{3}$$

In these equations, the subscript i is the time increment number, and therefore $i - \frac{1}{2}$ and $i + \frac{1}{2}$ refer to the mid-increment values, Δt refers to the time step, M is the lumped mass matrix, and F is applied external load vector and R is the internal force vector [35]. The internal force vector is determined through an analysis of the element stress-strain relations.

In FEA, elemental strain, ε is determined from displacement through utilization of the strain-displacement matrix, shown in Equation (4).

$$\varepsilon = B u_{(i)} \tag{4}$$

The strain-displacement matrix, B, relates the displacements to the element strain component based on the derivatives of the element shape functions, or basis functions. The shape functions used by the C3D8R element are shown in their isoparametric form in Equation (5). In the isoparametric form, the reference coordinate system is changed to a natural coordinate system, where the element coordinates range from -1 to 1 regardless of the global element position. Here ξ, η, and ζ represent the element's x, y, and z isoparametric coordinates respectively.

$$\begin{aligned} N_1 &= \frac{1}{8}(1-\xi)(1-\eta)(1+\zeta) \\ N_2 &= \frac{1}{8}(1-\xi)(1-\eta)(1-\zeta) \\ N_3 &= \frac{1}{8}(1-\xi)(1+\eta)(1+\zeta) \\ N_4 &= \frac{1}{8}(1-\xi)(1+\eta)(1-\zeta) \\ N_5 &= \frac{1}{8}(1+\xi)(1-\eta)(1+\zeta) \\ N_6 &= \frac{1}{8}(1+\xi)(1-\eta)(1-\zeta) \\ N_7 &= \frac{1}{8}(1+\xi)(1+\eta)(1+\zeta) \\ N_8 &= \frac{1}{8}(1+\xi)(1+\eta)(1-\zeta) \end{aligned} \tag{5}$$

Thus, the strain-displacement matrix for this element can be found using the relationship shown in Equation (6), where the partial derivatives are taken within respect to the isoparametric coordinates found in the shape functions N_i presented in Equation (5).

$$[B] = [\partial]][N] \tag{6}$$

The strain determined from Equation (4) is the total strain, which is comprised of both the elastic and viscoplastic strain. Elastic strain is the deformation that is fully recoverable, and therefore is not reliant on deformation history. Viscoplastic strain on the other hand is not fully recoverable and results in permanent deformation, which means that is does rely on deformation history. Elastic strain is required to determine the internal force vector and is given by Equation (7).

$$\varepsilon_e = \varepsilon_{total} - \varepsilon_{vp(i)} \qquad (7)$$

where $\varepsilon_{vp(i)}$ is the viscoplastic strain at time increment i. The viscoplastic strain is determined through evaluation of a chosen plasticity model and determined in a later step of the explicit analysis. Having the elastic strain, the stress can now be determined through application of Hooke's Law, see Equation (8).

$$\sigma = E\varepsilon_e = E(\varepsilon_{total} - \varepsilon_{vp}) \qquad (8)$$

Here E is the stress-strain relationship matrix, or elasticity matrix, which relates stress and strain through the Elastic Modulus, Shear Modulus, and Poisson's Ratio. Figure 1 depicts a typical stress-strain curve with the total strain separated into viscoplastic and elastic strain components. As shown here, the viscoplastic strain is taken as the permanent set, or unrecoverable strain, when the curve is unloaded along the linear elastic slope. The recovered strain is the elastic strain portion, and that strain when multiplied by the elastic modulus provides the stress value at the total strain value as depicted. Therefore the stress is only reliant on the elastic strain and elastic matrix.

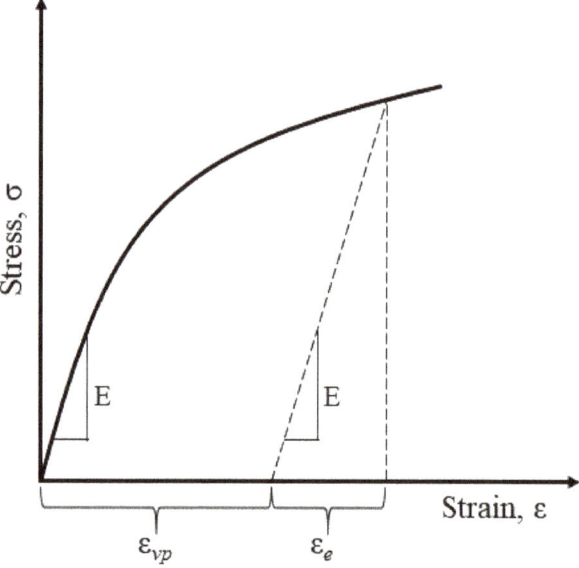

Figure 1. Typical Uniaxial Stress-Strain Response Curve.

Once the stress has been calculated, the internal force vector can be found using Equation (9), which is used to determine the acceleration in Equation (3) through a nodal force balance.

$$R = \int_V B^T \sigma dV \qquad (9)$$

In this integral, V represents the element volume at the current time step. The evaluation to determine R is based on the original shape elements, and as the elements deform the nodal integration across the volume is distorted when utilizing the Lagrangian coordinate systems for the B matrix. This allows for the development of errors within the solution. Abaqus uses the Eulerian coordinate system for certain functions, which is able to avoid significant distortion, to overcome this problem, although the coordinate transformations still introduce some error in this step. This error will ultimately be realized in the resultant nodal displacements determined through use of the CDTI Equations (1)–(3).

Plasticity is determined through an incremental approach, an associated flow rule, that is characterized through the use of an incremental stress function, the effective stress, and an incremental plastic strain function, the effective strain increment. For this impact study, the Johnson-Cook Flow Rule will be utilized, which allows for determination of the viscoplastic strain. The Johnson-Cook Flow Rule Condition evaluates the relationship between the Johnson-Cook Flow Stress and Static Stress, it is given by Equation (10). The Johnson-Cook equations are represented here as one-dimensional functions, when in reality they are multi-dimensional, but hold the same relationships as presented.

$$\bar{\sigma}_y \geq \bar{\sigma}_0. \tag{10}$$

$\bar{\sigma}_y$ is the Johnson-Cook Flow Stress is given by Equation (11), and $\bar{\sigma}_0$ is the Static Stress, given by Equation (12) [37].

$$\bar{\sigma}_y = [A + B(\bar{\varepsilon}_{vp})^n]\left[1 + C \ln\left(\frac{\dot{\bar{\varepsilon}}_{vp}}{\dot{\bar{\varepsilon}}_0}\right)\right][1 - (T^*)^m] \tag{11}$$

$$\bar{\sigma}_0 = [A + B(\bar{\varepsilon}_{vp})^n][1 - (T^*)^m]. \tag{12}$$

In Equations (11) and (12), A is the initial yield stress, B is the hardening modulus, C is a strain rate dependent coefficient, m is a thermal softening variable, and n is a work hardening variable. A, B, C, m, and n are material specific variables generally determined through experimentation. The first bracketed term accounts for the strain hardening, or plastic strain accumulation. The second bracketed term accounts for the effects of strain rate. The third, and final, bracketed term accounts for the effects of temperature. $\dot{\bar{\varepsilon}}_0$ is a reference strain rate, which is set to the strain rate for which A, B, and n were determined, and T^* is the homologous temperature, a ratio of the difference between the current temperature and room temperature over the difference between the material melting temperature and the current temperature, Equation (13). The temperature considered within the flow equation is a function of amount of plastic energy that is assumed present at the nodal locations. For impact problems, this is generally taken to be 90% or greater [38,39].

$$T^* = \frac{T - T_0}{T_m - T}. \tag{13}$$

Stress can be divided into two components, hydrostatic stress and deviatoric stress. Hydrostatic stress can be thought of as pressure stress, and is the average of the three principal stresses, see Equation (14) [40].

$$\sigma_{hyd} = \frac{1}{3}(\sigma_{11} + \sigma_{22} + \sigma_{33}). \tag{14}$$

Based on the assumption that metals are incompressible, it follows that hydrostatic stress cannot cause deformation. This means that deformation is caused by the deviatoric stress, which is the total stress minus the hydrostatic stress, presented in Equation (15) [40].

$$\sigma' = \sigma_{tot} - \sigma_{hyd}. \tag{15}$$

The flow stress equation, Equation (11), is a function of the von Mises stress equation, which is derived from the deviatoric stress, as shown in Equation (16).

$$\sigma_{vM} = \sqrt{\frac{3}{2}\sigma' : \sigma'}. \tag{16}$$

Here, : represents the double inner product, carried out across two second order stress tensor, which provides a scalar output. The von Mises stress is useful as an invariant effective stress, which can predict the onset of yield, which is why the von Mises stress can be utilized as the effective stress, σ_e, for determination of plasticity [34].

If the Johnson-Cook Flow Rule Condition evaluates as true, then yielding has occurred and the viscoplastic strain rate must be determined to calculate the equivalent viscoplastic strain. Otherwise, yielding has not occurred and the element is still in the elastic region, and therefore the viscoplastic strain rate is zero. The viscoplastic strain rate is determined through Equation (17).

$$\dot{\bar{\varepsilon}}_{vp_{(i+1/2)}} = \dot{\varepsilon}_0 e^{\left[\frac{1}{C}\left(\frac{\bar{\sigma}_y}{\bar{\sigma}_0}-1\right)\right]}. \tag{17}$$

The viscoplastic strain rate is then used to calculate an equivalent plastic strain, see Equation (18).

$$\bar{\varepsilon}_{vp_{(i+1)}} = \bar{\varepsilon}_{vp_{(i)}} + \sqrt{\frac{2}{3}\dot{\bar{\varepsilon}}_{vp_{(i+1/2)}} : \dot{\bar{\varepsilon}}_{vp_{(i+1/2)}}} \Delta t_{(i+1/2)}. \tag{18}$$

$\Delta t_{(i+1/2)}$ is the half time step increment. This equivalent plastic strain is carried forward to the next time step and used to determine the elastic strain. This time step iteration process is continued for all of the subject finite elements until the time interval is exhausted, or the elements are damaged, which is covered further in Section 2.4 [35].

An equation of state (EOS), see Section 2.3, is required to be used when modeling dynamic impacts in order to balance the physical properties of the projectile and target due to shock wave creation and propagation. The EOS is also used in the solution of the conservation of mass, momentum, and energy equations. This allows velocity to be a normalizing function, so that equilibrium conditions can be determined by relating the material's pressure and internal energy to its density and temperature.

2.2. Smoothed Particle Hydrodynamics

The large deformation and high strain rates of the projectile impact problem necessitates that another numerical technique be utilized, SPH. SPH is a computational technique used in FEA for the simulation of fluids and solid mechanics, primarily utilized due to its ability to trace failure events in a more physically representative manner. It was developed in the late 1970s for use in analyzing complex three-dimensional astrophysics problems related to asymmetry in stars [41] and fission of a rotating star [42] utilizing a mesh free method to determine element forces. As mentioned, this work was adapted and applied to solid materials in the early 1990s for use in a strength of materials elasticity model [15], then further refined for use in determining dynamic material response [16]. When applied to solid mechanics, it was found that the mesh free SPH method was able to handle larger displacements and distortions more accurately than the traditional Lagrangian grid-based methods, making SPH a useful tool in shock, impact, fracture, and damage analysis [4,43–45].

In SPH, the part being analyzed is discretized into a set of particles which retain all of the relevant field variable information within an associated volume, as well as maintain a mass [46]. Figure 2 shows the difference between a traditional grid-based FEM and SPH.

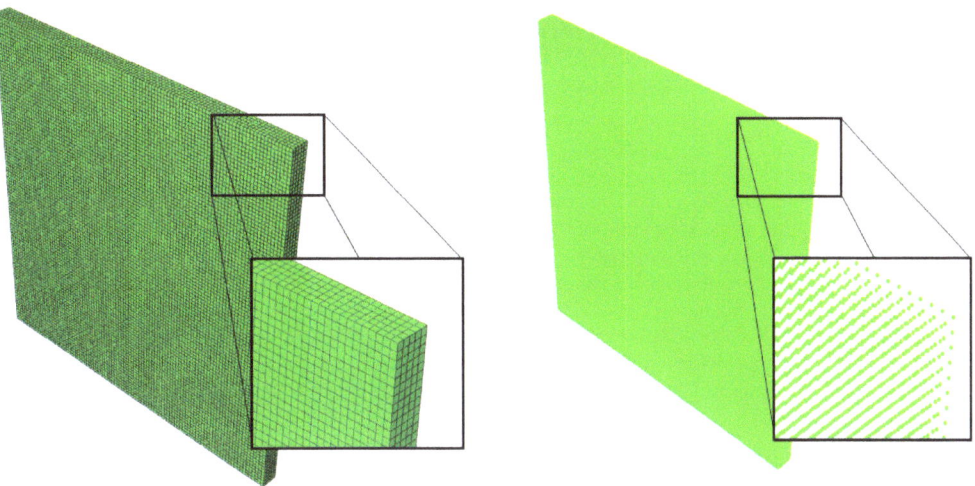

Figure 2. Traditional Grid Mesh versus Smoothed Particle Hydrodynamics Discretization.

As integrated within Abaqus, SPH is an extension of the explicit method, except instead of solving for residual forces through element volume integration to determine nodal displacements. SPH relates particles, via a weighting function, through satisfying the conservation equations about a point. This is accomplished through use of an interpolation method to express any field value function, f in terms of the particle set, defined by Equation (19).

$$f(x) = \int f(x')W(x - x', h)dx'. \tag{19}$$

In this equation, x represents the coordinates of the point of interest, x' denotes the particle positions, h is the smoothing length variable, and W is the interpolating kernel, which must satisfy the two properties presented in Equations (20) and (21). The smoothing length is the distance chosen to determine which particles within the model will influence the interpolation for the point of interest.

$$\int W(x - x', h)dx' = 1 \tag{20}$$

$$\lim_{h \to 0} W(x - x', h) = \delta(x - x'). \tag{21}$$

Here δ represents the Delta function [47]. As the kernel starts with the Delta function, which is point oriented, a smoothing function is required to make the integral numerically discrete, although the smoothing function also makes the integration in Equation (19) an approximation. An essential property of the smoothing function value for a particle is that it should be monotonically decreasing with increasing the distance away from the particle. This property is based on the physical consideration in that a nearer particle should have a larger influence on the particle under consideration than one further away. Additionally, as mentioned, the smoothing function should satisfy the Dirac delta function condition as the smoothing function approaches zero. This property of the smoothing function makes sure as the smoothing length tends towards zero, the approximation value approaches the function value [48]. The result of applying Equations (19)–(21) is determining the field variable values through convolution across the domain created by the smoothing length utilizing the kernel function. In the original SPH theory, the calculations were performed using a Gaussian kernel; however, the cubic spline kernel is more widely used today than the Gaussian in SPH, and is the default kernel used within the Abaqus software package. The cubic spline kernel is shown in Equation (22) [35].

$$W(x-x',h) = \frac{1}{h^3 \pi} \begin{cases} 1 - \frac{3}{2}\xi^2 + \frac{3}{4}\xi^3, & \text{for } 0 \leq \xi \leq 1; \\ \frac{1}{4}(2-\xi)^3, & \text{for } 1 \leq \xi \leq 2; \\ 0, & \text{for } \xi > 2, \end{cases} \qquad (22)$$

where $\xi = \frac{x}{h}$. The cubic spline kernel reduces the number of particles included in the calculations to those that are within twice the smoothing length of the particle of interest, while still maintaining C^2 continuity. The kernel function is generally chosen based on the type of problem being addressed, and a reduced interaction function, such as the cubic spline kernel, is often better suited for highly time dependent problems that occur over a short duration, such as impact events [49]. Regardless of kernel chosen, for numerical operations the field variables can be approximated though a kernel summation, which is presented in Equation (23).

$$f(x) \cong \sum_{k=1}^{N} f_k W(x - x_k, h) \frac{m_k}{\rho_k}. \qquad (23)$$

Here, k represents the particle index, N is the total number of particles, f_k is the field variable at the kth particle, x_k is the position of the kth particle, W and k remain the kernel function and smoothing length variable, m_k is the mass of the kth particle, and ρ_k is the density of the kth particle [44]. A visual example of a kernel function is depicted in Figure 3, where the black particles would be included in the computations and white particles would not be included.

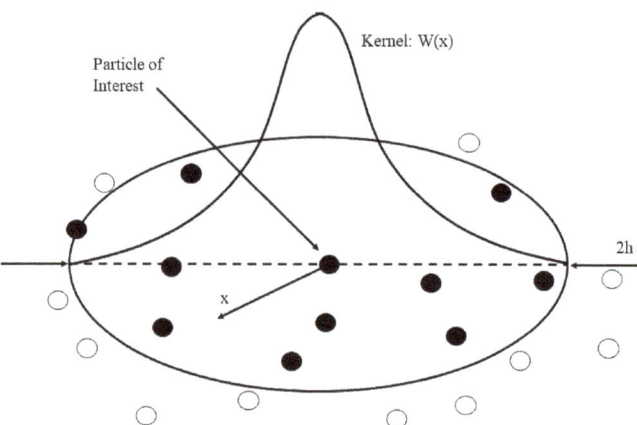

Figure 3. Visualization of the Cubic Spline Kernel Function.

As can be seen by viewing Equations (19) and (23), one of the primary differences between SPH and traditional grid based FEA is that SPH solves the field value differential functions through discretization of volume into particles versus a point-wise discretization of space-time [50]. Instead of solely using the traditional Lagrangian explicit technique based on Newton's Second Law of Motion, the conservation of mass, momentum, and energy equations are utilized to determine particle response [47]. SPH still uses a Lagrangian formulation; however in the SPH implementation, the conservation equations must be satisfied for each particle at each time increment and the updated field values will be carried forward to the next time step.

The first conservation equation that must be addressed is the conservation of mass, also known as the continuity equation. The particle density is calculated through the continuity equation, which is needed for the two remaining conservation equations. In evaluating the mass density interaction between a particular particle pair, notated as $a - b$, the continuity equation can be represented as the time rate of change of mass density shown in Equation (24). A subscript a denotes the properties at the particle of interest, and a subscript b denotes the properties at another particle within the field.

$$\frac{d\rho_a}{dt} = \sum_b m_b (v_a - v_b) \nabla_a W_{ab}. \tag{24}$$

In these equations, ρ is the particle density, m is the individual mass of the particle, v is the velocity of the particle, ∇ is the gradient function taken with respect to the coordinates of the particle of evaluation, and W_{ab} is the kernel function relating particles a and b.

Once the conservation of mass equation has been satisfied, the conservation of momentum, or momentum equation, will be evaluated. Again, in evaluating the interaction between a particle pair $a - b$, the momentum equation starts from a pressure gradient estimate shown in Equation (25) [47].

$$\rho_a \nabla P_a = \sum_b m_b (P_b - P_a) \nabla_a W_{ab}. \tag{25}$$

Here, P is the pressure at the particle under evaluation. In this form, the momentum equation is considered asymmetric, and can lead to unstable simulations by creating an inconsistent energy equation. To address this problem the pressure gradient can by symmetrized, then the equation can be rewritten using the relationship found in Equation (26) to arrive at Equation (27) [47].

$$\frac{\nabla P}{\rho} = \nabla \left(\frac{P}{\rho} \right) + \frac{P}{\rho^2} \nabla \rho \tag{26}$$

$$\frac{dv_a}{dt} = -\sum_b m_b \left(\frac{P_b}{\rho_b^2} + \frac{P_a}{\rho_a^2} \right) \nabla_a W_{ab}. \tag{27}$$

In Equation (27), dv/dt is the total time derivative, in the Lagrangian sense, of the velocity vector. By utilizing this form of the momentum equation both the linear and angular momenta are conserved, which may not be the case with an asymmetric pressure gradient term.

Finally, the conservation of energy, or energy equation, can be addressed. In the original SPH formulation, this was the rate of change of thermal energy per unit mass of the particles, or the hydrodynamic energy equation, absent heat sources or sinks as shown in Equation (28) [46].

$$\frac{de}{dt} = -\left(\frac{P}{\rho} \right) \nabla \cdot \mathbf{v}. \tag{28}$$

Here, de/dt is the time derivative of the specific internal energy, e. This equation can be rewritten to determine the conservation of energy at particle a, and simplified for adiabatic systems, as shown in Equation (29).

$$\frac{de_a}{dt} = \left(\frac{P_a}{\rho_a^2} \right) \sum_b m_b (\mathbf{v}_a - \mathbf{v}_b) \cdot \nabla_a W_{ab}. \tag{29}$$

There are several variations on the energy equation that can be used in SPH analysis, and it is important to note that several of these forms can present non-realistic physical solutions, such as negative internal energy. These issues are typically solved through use of a predictor-corrector approach for analysis. In this approach the governing conservation equations are used to predict the field variables utilizing the chosen kernel, the predictor phase. This may lead to an unbalanced energy solution, which is corrected for through a local restoration of the conservation of energy equation, the corrector phase. Following the corrector phase the field values are adjusted to meet the new particle state. This process is followed at every time step used in analysis, and with a sufficiently small time increment the numerical adjustments used in the predictor-corrector method do not affect the accuracy of the solution [44].

As with the traditional Lagrangian implementation in Abaqus, the SPH technique incorporates the Johnson-Cook Flow Stress to determine viscoplastic strain, Equation (11). Again, this assumes that the von Mises flow stress can be determined as the combination of plastic strain accumulation, strain rate effects, and thermal effects. The Johnson-Cook Flow Stress is also used in determining particle damage utilizing the Johnson-Cook Damage Model described in Section 2.4.

One of the concerns that arose from adapting SPH to solid mechanics is the issue of boundary effects. When the summation approximation, Equation (23), is applied near a boundary there is a truncation error, which results in an incomplete summation and C^0 continuity may not be maintained. This means that rigid body motion may not be determined correctly through the analysis process [50]. There have been many efforts to overcome this error to regain at a minimum C^1 continuity through the use of Lagrangian stabilization [51], symmetric formulation [52,53], Galerkin formulation [54], least squares methods [55,56], or the ghost particle method [57].

Abaqus incorporates the ghost particle method as a primary means to deal with SPH boundary surfaces. This method creates imaginary particles when an SPH body interacts with a solid Lagrangian boundary. Interaction is considered when a particle is within twice the smoothing length of a boundary surface. In this case, a virtual plane is formed along this boundary, with the ghost particles being formed across the plane from the physical particles. The number of ghost particles included within the simulation is based upon the smoothing length utilized and the boundary condition. The ghost particle's field properties are computed from those of the physical particles, as if the SPH part spanned the virtual plane, but they are assigned the opposite sign of the physical particles. The opposing field values are derived from a Lennard-Jones potential, which forms a repulsion force along the boundary surface, preventing SPH particle penetration of the solid [58–60]. Within the SPH methodology, the Lennard-Jones potential is used solely as a mathematical force generator for the required numerical boundary force. The use of ghost particles improves performance of SPH method integration along the boundaries, and it is worth noting that the ghost particles are not permitted to interact back across the virtual plane with the physical particles. As the ghost particles are still permitted to interact amongst themselves, care must be taken to avoid an excess of ghost particle mass [57]. An excess of ghost particle mass can start providing false inputs into the boundary surface integration, which in-turn can lead to errors in physical particle field values along the boundary. To limit the ghost particle mass the smoothing length can be decreased near a boundary, which is the application method used within Abaqus. See Figure 4 for a pictorial representation of the ghost particle method acting along a perpendicular boundary surface.

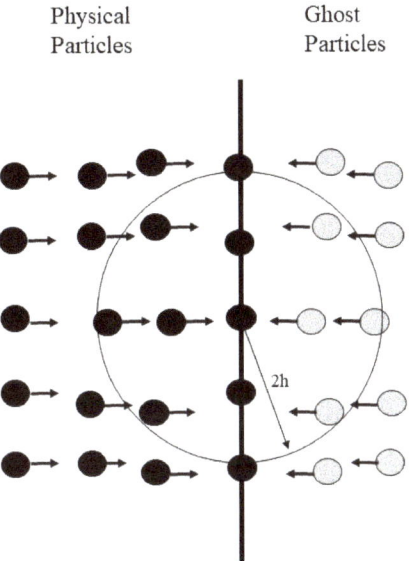

Figure 4. Ghost Particle Formulation along a Lagrangian Boundary Surface.

2.3. Equation of State

An EOS provides a hydrodynamic material model for use in FEA, in which the system's pressure, volume or density, temperature, and energy are related through application of a specific EOS [61]. EOS are used to relate the state variables together when the conservation equations are not enough, which is the case of high strain rate impact, along with all fluid dynamics cases. One of the most commonly used EOS in impact modeling is the Mie-Grüneisen equation, which has also proven effective at predicting the response of porous materials to compressive shock waves [62–65], and was the EOS used in the Abaqus impact simulations run as part of this research. In order to develop a relationship between pressure and volume there must be a link between statistical mechanics and thermodynamics, which starts with the analysis of the Helmholtz free energy equation, Equation (30) [62].

$$A_H = U - TS. \tag{30}$$

A_H represents the Helmholtz free energy, U is the internal energy of the system, T is the absolute temperature of the surroundings, and S is the entropy of the system. Within the EOS, entropy is linked to heat transfer, which ties the EOS to the same phenomena represented in the Johnson-Cook Flow Stress Equation, Equation (11), where temperature is a function of plastic energy concentrated at a nodal coordinate. The Helmholtz free energy is considered the useful energy within the system, as opposed to the energy related solely to the thermal environment. As such, it is defined as a measure of the work attainable within a closed thermodynamic system with constant temperature, or isothermal. Utilizing the fundamental thermodynamic relationship, see Equation (31), pressure is obtained through taking differential of the Helmholtz free energy with regards to volume, Equation (32) [66].

$$dA = TdS - PdV \tag{31}$$

$$P = \left| \frac{\partial A_H}{\partial V} \right|_T. \tag{32}$$

Through use of the virial theorem, mathematically represented in Equation (33) [67], Grüneisen obtained the pressure-energy relationship shown in Equation (34) [68]. For applications in mechanics, the virial theorem states that twice the total kinetic energy within a system is equal to the interaction of the particle forces, or virial, within the system [67]. Use of the total kinetic energy formulation provides valuable information on the system behavior, relating energy to force, which allows for the reformulation of the pressure-energy relationship from Equation (32) to Equation (34).

$$KE = -\frac{1}{2}\sum_{k=1}^{N}\langle F_k \cdot x_k \rangle. \tag{33}$$

Here KE represents the total kinetic energy, N is the total number of particles, F_k is the force applied to the kth particle, and x_k is the position vector of particle k in the local coordinates for a Lagrangian system [69].

$$P = \frac{\Gamma}{V}E. \tag{34}$$

In this equation, Γ is the Grüneisen coefficient, defined by Equation (35) [70], where V_s is the specific volume, e is the specific internal energy, and P is the pressure.

$$\Gamma = V_s(\partial_e P)_{V_s}. \tag{35}$$

From here the Grüneisen equations were modified, utilizing the work performed by Mie, by relating the current conditions to those of a known value, a point off of the Hugoniot curve. Additionally, a simplification to the Grüneisen equations was made by correlating the material volume to its density. These two associations led to the Mie-Grüneisen equation taking on a linear form with respect to energy, as shown in Equation (36) [35].

$$P - P_H = \Gamma \rho (e - e_H). \tag{36}$$

Here P is the current pressure, P_H is the Hugoniot pressure, Γ is a redefined Grüneisen ratio, ρ is the material density, e is the specific internal energy, and e_H is the Hugoniot specific energy. The Grüneisen ratio is presented in Equation (37), the Hugoniot pressure is presented in Equation (38), and the Hugoniot specific energy is presented in Equation (39).

$$\Gamma = \Gamma_0 \frac{\rho_0}{\rho}. \tag{37}$$

Γ_0 is the material specific Grüneisen coefficient, ρ_0 is a reference density, and ρ is the current density.

$$P_H = \frac{\rho_0 c_0^2 \eta}{(1 - s_H \eta)^2}. \tag{38}$$

In the Hugoniot pressure equation, c_0 is the material reference speed of sound, η is a nominal compressive strain, and s_H is the linear Hugoniot slope coefficient. The Hugoniot pressure is solely a function of density, and is determined through fit of experimental data.

$$e_H = P_H \frac{\eta}{2\rho_0}. \tag{39}$$

In Equations (38) and (39), the nominal strain is represented by Equation (40).

$$\eta = 1 - \frac{\rho_0}{\rho}. \tag{40}$$

Abaqus uses the form of the Mie-Grüneisen equation presented in Equation (41) for its analysis [35].

$$P = P_H\left(1 - \frac{\Gamma_0 \eta}{2}\right) + \Gamma_0 \rho_0 e. \tag{41}$$

This equation is solved simultaneously alongside the three conservation equations at each material point and time increment to ensure that the state variables are balanced throughout an impact event. In this implementation of the Mie-Grüneisen EOS, the reference material properties that will be used are those for the base material of the projectile and target.

2.4. Damage Modeling

As mentioned above, for both the Lagrangian and SPH finite analysis techniques, a damage model is required to determine when the elements or particles fail within a simulation so that they can be analyzed appropriately as part of the FEA process. The damage model used in this study, the Johnson-Cook failure model, was chosen due to its ability to model the failure of metals across a range of strain rates.

The Johnson-Cook failure model was developed in the 1980s, and It makes the assumption that the change in material properties between static and dynamic cases is due to strain rate effects, which can account for large strains, high temperatures, and high pressures [71]. It incorporates the Johnson-Cook material model and is based on plastic damage accumulation, that damage begins when plasticity begins. This model is commonly used for estimating the dynamics deformation of metals under high strain rates [72–74].

The Johnson-Cook failure model defines material damage as the sum of incremental equivalent plastic strain divided by the critical fracture strain, see Equation (42).

$$D = \sum \frac{\Delta \bar{\varepsilon}_p}{\bar{\varepsilon}_f}. \tag{42}$$

In this failure model, D is the Johnson-Cook damage coefficient, $\Delta \bar{\varepsilon}_p$ is the increment of equivalent plastic strain, and $\bar{\varepsilon}_f$ is the Johnson-Cook fracture strain, or strain at failure, see Equation (43). The right hand side of Equation (42) sums the incremental change in the element or particle plastic strain, and compares it as a ratio to the failure strain of the material, which is presented as the Johnson-Cook damage coefficient. The premise here is that as long as there is viscoplasticity, damage is accumulating. The damage coefficient has a range of zero to one. Where zero represents a pristine, or undamaged material, and a one represents the material being fully damaged and fracture will occur.

$$\bar{\varepsilon}_f = \left[D_1 + D_2 e^{D_3 \sigma^*}\right][1 + D_4 \varepsilon^*][1 + D_5 T^*]. \tag{43}$$

The Johnson-Cook relationship assumes that the damage effects can be decoupled. The first bracketed term contains the stress triaxiality effects. Where D_1, D_2, and D_3 are material specific model fit properties, and σ^* is triaxiality ratio, or the ratio of the average normal stress to von Mises equivalent stress. This term accounts for the static and quasi-static strain response of the finite element parts. The second bracketed term comprises the strain rate effects. Where D_4 is another material specific model fit property and ε^* is the dimensionless strain rate ratio of viscoplastic strain rate to reference strain rate. The reference strain rate here is the same as that used in Equation (11). The final bracketed term includes the effects of temperature on material failure. Where D_5 is a material specific model fit parameter, and T* is the material's homologous temperature, Equation (13). The D_i terms are traditionally found through experimentation, utilizing quasi-static compression testing and the Split Hopkinson Pressure Bar test, allowing for modeling of the material response across a broad range of strain rates.

The damage evolution presented by the Johnson-Cook failure model describes the degradation in the material stiffness once damage is initiated. This reduction in stiffness is formulated by Equation (44).

$$\sigma = (1 - D)\overline{\sigma}_y. \tag{44}$$

Here σ is the stress within the element or particle, $\overline{\sigma}_y$ is the Johnson-Cook Flow Stress given by Equation (11), and D is the damage variable as described above. When the element or particle is fully damaged, $D = 1$, it is removed from the analysis.

3. Results and Discussion

In this study, a three-dimensional physics based computational model was developed in Abaqus. This model was used to predict the damage and failure of both projectile and target under high strain rate impact using the Johnson-Cook plasticity and damage models native in the Abaqus software. The materials and dimensions for the initial model were chosen to match experimental validation conditions [4], in order to provide validation of the modeling techniques, configuration, and simulation execution.

The initial computational model used for this study was a single projectile-single target assembly constructed in Abaqus using the inherent explicit finite element solver. The models all started at the point of impact with calculated impact velocities from the experimental data. Contact between the two parts was modeling using the general contact algorithm native to Abaqus, which utilizes a penalty method to impose contact constraints through introduction of increased local stiffness. The general contact algorithm was used to enforce contact between two bodies, and model friction between parts. This algorithm allows for automatic contact definition based on surface inclusion. Within the Lagrangian systems, contact forces are generated based on node, face, and edge interactions. It is also capable of enforcing contact between Eulerian and Lagrangian systems, compensating for any discrepancies between the two constructs. Of the contact algorithms available within Abaqus, general contact is the only contact algorithm that can be used with three-dimensional models, and is capable of evaluating across a mixed model type simulation. The friction developed here follows the Coulomb friction model, which formulates the friction coefficient based on primarily on contact pressure for impact, but also includes surface slip and temperature at the contact point [35].

The projectile was configured with a cylindrical body 24.7 mm in length and 16.7 mm in diameter to develop an equivalent system to the experimental setup. It incorporated a blunt nose geometry, and was modeled using reference material properties of 6061-T6 Aluminum, shown in Table 1, along with the Johnson-Cook Parameters shown in Table 2. The impact velocity of 970 m per second was applied as a load to the rear face of the projectile, with no other boundary conditions enforced upon the projectile within the simulation.

Table 1. Material properties of 6061-T6 Aluminum [75].

Material Property	Value
Elastic Modulus, E	69 GPa
Poisson's Ratio, ν	0.33
Density, ρ	2700 g/m^3

Table 2. Johnson-Cook Model Parameters for 6061-T6 Aluminum [76].

Johnson-Cook Parameter	Value
Yield Stress, A	324.1 MPa
Strain Hardening Parameter, B	113.8 MPa
Strain Rate Parameter, C	0.002
Thermal Softening Exponent, m	1.34
Strain Hardening Exponent, n	0.42
Damage Constant, D_1	-0.77
Damage Constant, D_2	1.45
Damage Constant, D_3	-0.47
Damage Constant, D_4	0.0
Damage Constant, D_5	1.6

The targets were configured as square-faced plates that were 203 mm by 203 mm with a thickness of 12.7 mm, utilizing the same material set. The target had fixed boundary conditions applied at both the upper and lower surfaces, as if it were affixed in a mount, with the other edges left as free surfaces. A depiction of the simulation boundary and initial conditions is shown in Figure 5.

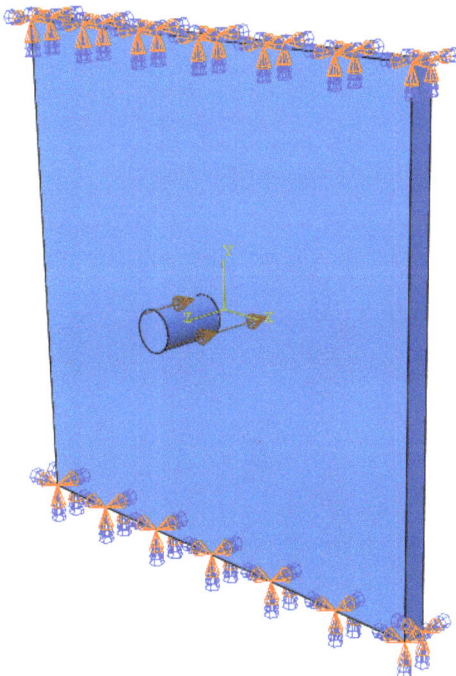

Figure 5. Depiction of Impact Simulation Loading and Boundary Conditions.

All of the simulations were evaluated for an impact time of 12 µs. Presented here are the validation cases, utilizing a variety of FEA techniques, compared to the previously acquired experimental results as means for evaluation of the modeling techniques [4]. Three variations of the base model were developed to evaluate different modeling techniques for use in this work. The first two models were based solely on Lagrangian FEA techniques, comprising a full scale model to match the real-world dimensionality of the experiment, and a symmetry reduced model that used dual-axis symmetry to achieve a quarter scale model. The third incorporated the use of SPH to model the target. The target was chosen for the use case of SPH as it would see larger deformations than the projectile.

Figure 6a presents the stress colormap that corresponds to traditional Lagrangian models, and Figure 6b presents the stress colormap scales that correspond to the mixed Lagrangian-SPH models.

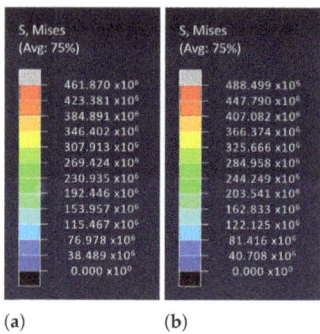

Figure 6. Stress Colormap Scale: (**a**) Traditional Lagrangian Model, (**b**) Mixed Traditional-SPH Model.

3.1. Lagrangian Models

The projectile and target were both symmetric across two axes, allowing the model to be cut along those axes and reduced in scale, to a quarter scale of the full model. This was only possible because of the shared planes of symmetry about the impact location of both the parts, forces, and boundary conditions. A visual comparison of the symmetry reduced and full scale model are provided in Figure 7.

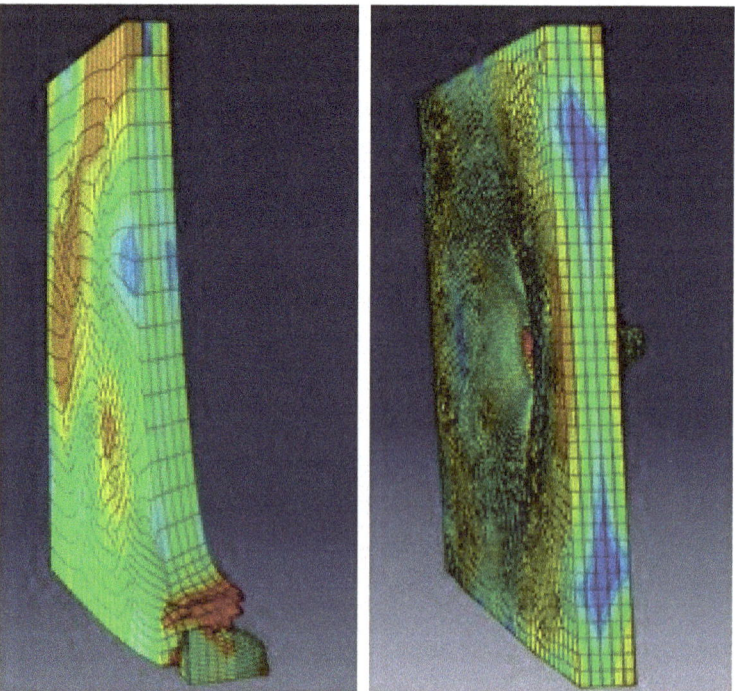

Figure 7. Symmetry Reduced versus Full Scale Model Impact Model.

As mentioned, in both the reduced and full scale models, the C3D8R eight-noded stress bricks with reduced integration were used as the elements for analysis of the projectile and target. In the full scale model, the projectile incorporated 7800 elements, and the plate target was comprised of 50,000 elements. The number of elements chosen for the parts was based on a convergence study performed by varying the element size from 1/50th down to 1/500th of the plate width using the symmetry reduced model. To save on computational run time the largest element size that still provided consistent results was chosen, which was 1/100th of the plate width. The Johnson-Cook plasticity and failure models were used to estimate the element failure modes of the materials, with element deletion occurring for cells with equivalent plastic strain values greater than 1.0. This value was found to most closely match the model data to previous experimental results. Figures 8 and 9 show a comparison between the full scale model deformation results against the experimental results of the projectile and target plate respectively [4].

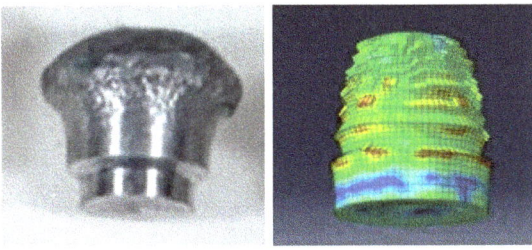

Figure 8. Visual Comparison of the Projectile with the Finite Element Results.

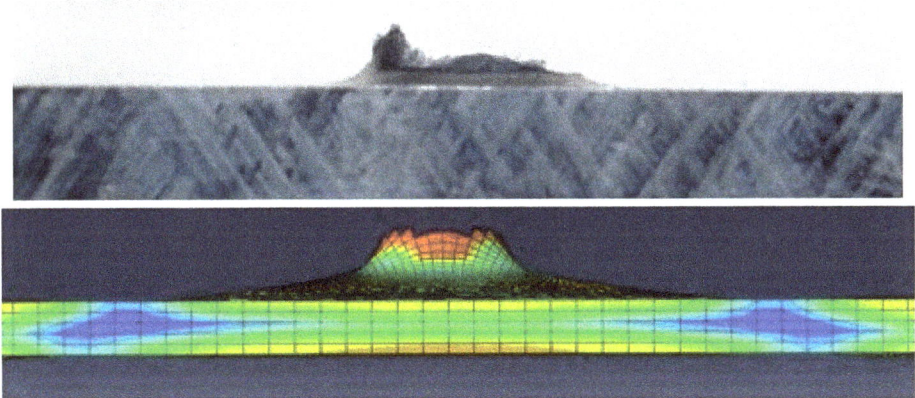

Figure 9. Visual Comparison of the Target with the Finite Element Results.

There is noticeable difference in the projectile results shown in Figure 8. The most significant aspect of this difference, showing the leading face of the projectile narrowing versus mushrooming, is due to inaccurate equivalent plastic strains for the impact face elements combined with the element deletion scheme used. At the projectile impact velocity, and subsequent strain rate, the traditional Lagrangian modeling technique is subject to mesh distortion causing errors which are compounded through the CDTI methodology. Ultimately this will lead to inaccuracies in the element strain and equivalent plastic strain, which would cause errant element deletion.

The symmetry reduced model provided nearly identical results as the full scale model, yet took approximately one third of the time to run, 0.3 processor hours versus 1.0 processor hours for the full model.

3.2. Mixed SPH-Lagrangian Model

While the deformation results of the traditional Lagrangian model appeared to match the experimental results adequately for the target, the deformation and residual velocity of the projectile were not well modeled. Residual velocity is the projectile's velocity upon exit of the target. Therefore, a mixed SPH-Lagrangian model was developed to evaluate the projectile and target dynamics and interactions more closely. As previously mentioned, it was decided that the target would be modeled utilizing SPH techniques, as it would be subject to larger deformations than the projectile based on the deformation seen in the traditional Lagrangian grid model and experimentation. The projectile was still modeled using the traditional C3D8R element with the same seeding as the full Lagrangian model, using 7800 elements. The plate was discretized into SPH particles to match the size of the elements used in the full Lagrangian model, resulting again in 50,000 elements used to model the target. However, due to the use of ghost particles in modeling boundary interaction within the SPH methodology, 100,000 particles were ultimately used in the computational analysis, which comprised the 50,000 particles used to represent the target plate and 50,000 particles used to model boundary interaction throughout the impact scenario. Figure 10 depicts the SPH target deformation following impact under the same parameters as the traditional model compared with the experimental results.

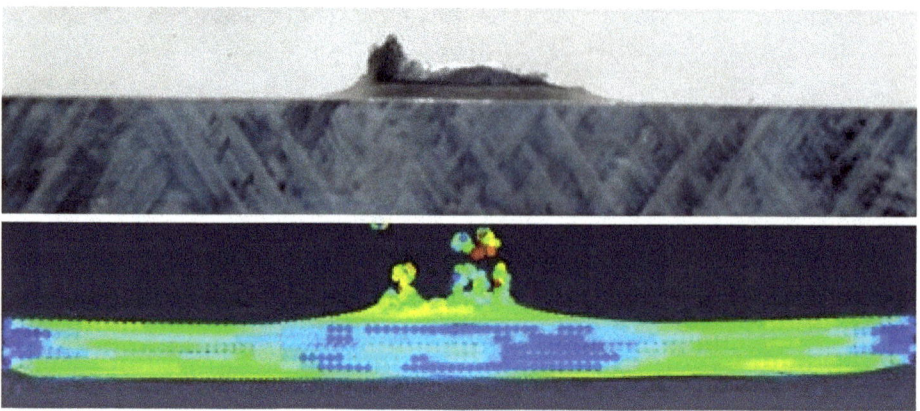

Figure 10. Visual Comparison of the Target with the Smoothed Particle Hydrodynamics Results.

In this figure, the particles that are no longer attached to the target plate would have been removed from the target during impact, and show as deleted under the traditional FEA method. As shown here, the SPH model more closely replicates the asymmetric shearing around the exit hole that was found in the experimental results. While the impact problem is described as a symmetrical problem, there are potential sources for the asymmetry in the simulation, such as asymmetric discretization of a part and numerical round off. The projectile was discretized the same between the two models, and some element asymmetry was noted in both the impact and rear faces. The asymmetry in the projectile discretization can still provide physically relevant data, as it can be seen as a similar effect to imperfections within the part, or non-homogeneity within the material or structure. As different mathematical methodologies are used in the two models, an asymmetric discretization could lead to asymmetry in the mixed model but not the traditional model. In a similar manner, numerical round offs could cause asymmetry in either model, but as different equations are utilized the round off would likely manifest differently between the models.

While in this model only the target was converted into an SPH model, the use of SPH in the assembly also gave a better appreciation of the projectile response throughout the interaction. Figure 11 shows the projectile following impact compared to the experimental projectile deformation.

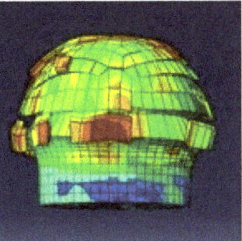

Figure 11. Visual Comparison of the Projectile with the Smoothed Particle Hydrodynamics Results.

Utilizing this model, the projectile shows a deformation pattern more similar to the experimental results than with the previous modeling technique. A primary contributor to the accuracy of the solution is the more precise displacement solution of the target through use of the SPH technique. With contact prescribed throughout much of the simulation, the displacement solution of the target has a direct impact on the forces imposed upon the projectile, which in-turn will dictate the plastic strain accumulation used in the damage model. An additional element that lead to the closeness in results is due to the equivalent plastic strain value used for element deletion being tuned more specifically for the projectile in the SPH-Lagrangian model than for the traditional model while maintaining a nearly identical target failure pattern. However, the SPH model took significantly longer to process than either the full scale or symmetry reduced models, with a run time of 5.2 processor hours.

3.3. Further Comparison of Models

As mentioned above, the traditional grid model did not provide an adequate result for the projectile's residual velocity, but the SPH model was able to very closely match the results seen through experimentation, see Table 3. Also shown here are the computer processing times required for each model.

Table 3. Comparison of Computational Models.

Model	Residual Velocity (m/s)	Residual Velocity Error (%)	Processor Time (h)
Experimental	336.194	-	-
Traditional, Full	30.186	91.021	1.0
Traditional, Reduced	30.189	91.020	0.3
SPH	335.406	0.234	5.2

Figure 12 shows the velocity plot of the projectile for both the traditional model and SPH model against the simulation time, with $t = 0$ being initial contact. The velocity values are taken from elements along the center-line of the projectile. As shown here, the SPH velocity follows a smooth and expected deceleration from the initial impact velocity to the projectile's residual velocity. On the other hand, the traditional grid velocity shows an initial acceleration within the first time step, then decelerates more quickly down to a velocity roughly one-tenth of that observed in experimentation.

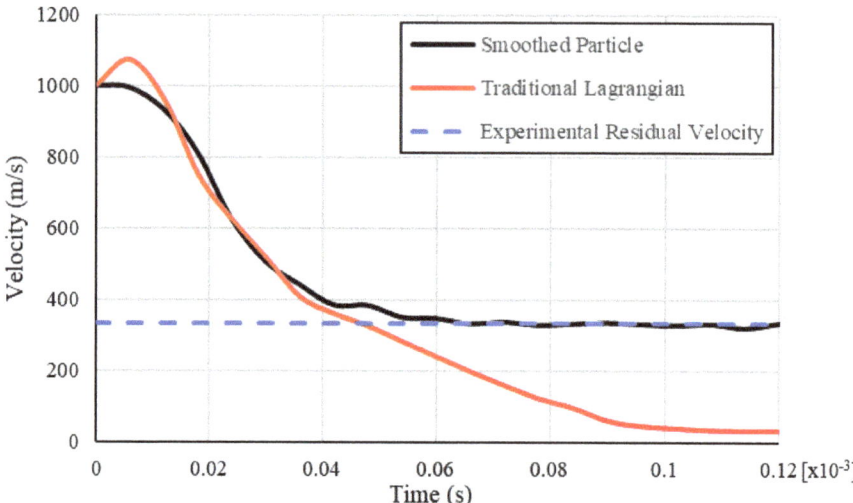

Figure 12. Comparison of Projectile Residual Velocity between Traditional Lagrangian Finite Element and Smoothed Particle Hydrodynamics Models.

The initial increase in velocity is likely due to the high strain rate of impact, which could not be accurately modeled by the traditional explicit methodologies. One of the most essential differences between the two methods is that SPH is meshless, and the problem domain is discretized with particles that do not have a fixed connectivity. Thus, large displacement problems are better evaluated since there is no need to evaluate the internal forces based on individual volume integration as required in the traditional approach. The traditional Lagrangian method requires a continuity of nodes, which requires the integration of the volume represented by the element geometry, and under large deformations may be so distorted that the evaluation will produce errors in balancing force distribution. At this impact velocity, and subsequent strain rate, the full Lagrangian model would have produced some error in the elemental volume integration required to determine the internal force vector that is used to determine the time step acceleration term. In the SPH method, there is no need to evaluate the integration of volume within the element as there are no element connecting nodes, rather the internal force vector is through a pre-established association with the neighboring particles by means of the kernel function. This association is predetermined and becomes part of the derivative included within the conservation equations. Another potential source of error that led to the lower residual velocity is the implementation of the Coulomb friction coefficient in Abaqus's general contact algorithm. If the friction coefficient is too high it can lead to binding in the model as it progresses. While the same friction coefficient was used in both models, the differences in relative motion of the projectile and target between the models would change the application of the friction and lead to errors in the velocity.

Figure 13, presents the impact axis, or z-axis, acceleration for a target particle and node on the edge of the initial contact. For reference, the model was oriented with initial impact velocity along the negative z-axis. The acceleration of the traditional Lagrangian element depicts a significantly larger rise in positive acceleration than the SPH particle in the beginning of the response. This difference is an important factor in the initial increase in velocity shown by the full Lagrangian model, and why the initial velocity of the SPH model stayed constant. The acceleration response of the full Lagrangian model also exhibits larger peaks and troughs, which is indicative of the errors manifesting in the internal energy volume integration.

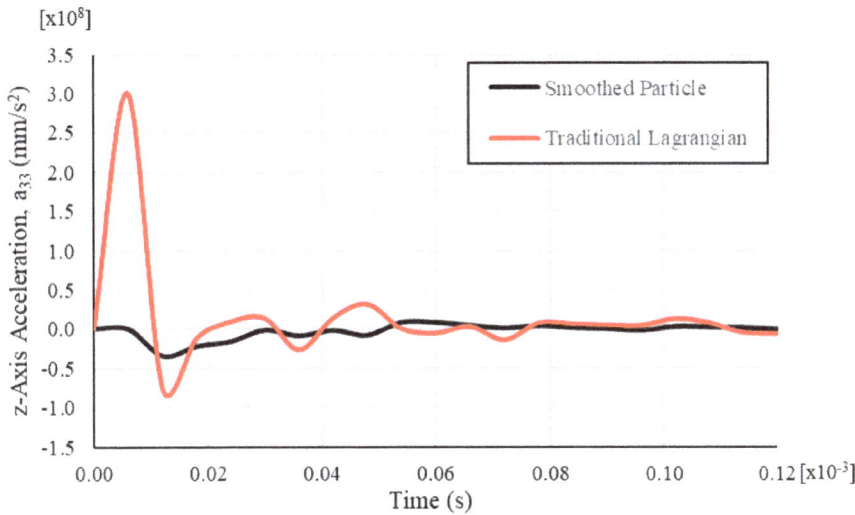

Figure 13. Comparison of Impact-Axis Acceleration between Traditional Lagrangian Finite Element and Smoothed Particle Hydrodynamics Models.

Furthermore, a comparison between the internal energy versus time for the two models is presented in Figure 14. Overall, the trend between the two curves is similar with the traditional model displaying higher internal energy values, see Figure 14a, but there is a unique artifact within the traditional model early within the simulation run, see Figure 14b. This variation within the internal energy curve is likely due to an error developed within the internal force calculations of the traditional method as mentioned above. Since the traditional Lagrangian model is based on Newton's Second Law, the error in the acceleration derived from the internal force calculation would have been carried forward to the velocity and displacement vectors through the CDTI methodology shown in Equations (1)–(3) and compounded throughout the time step integration process. This is likely the cause of the significant error in residual velocity realized by the two traditional models.

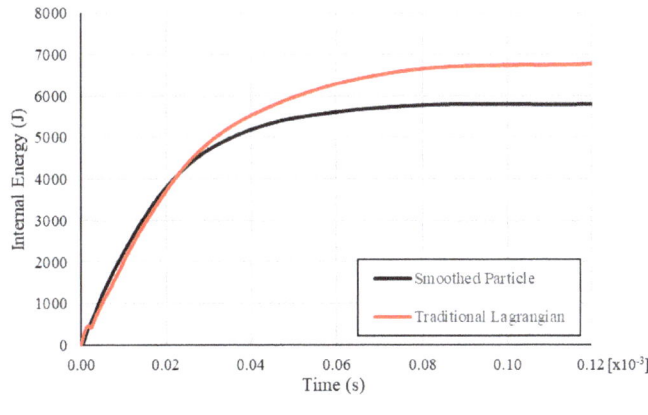

(a)

Figure 14. *Cont.*

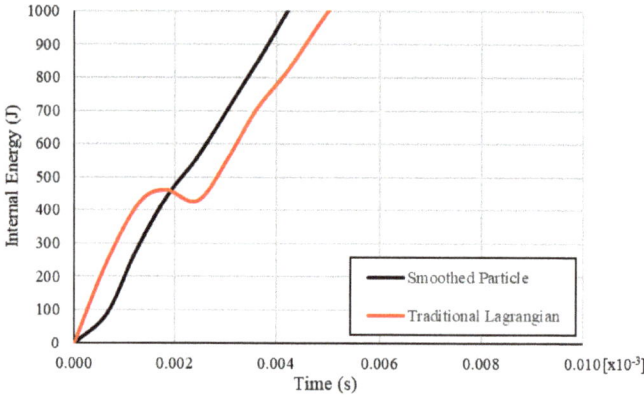

(b)

Figure 14. Model Internal Energy versus Time Comparison: (**a**) Full Run Duration, (**b**) Initial Reaction.

While both modeling techniques both rely upon a Lagrangian reference frame according to the internal interactions and external forces and thus evolve the system in time, within SPH the mathematical process of satisfying the three conservation equations alongside the equation of state reduces the likelihood of error. To highlight the differences in computed displacements between the two methods, a comparison of element strain over time is presented in Figure 15. Figure 15a compares the element strain of an element on the impact face of the projectile. As seen in the figure, both curve follow the same trend, although the SPH model strain is roughly twice that of the traditional model. Figure 15b shows a similar comparison for an element on the rear face of the projectile through the simulation duration. For this case, the response curves are not quite in alignment, although the general trend of the strain over time is comparable. The back face element shows the opposite case of the front face, in that here the strain values of the traditional model are higher than that of the SPH model, by roughly 70%. These figures emphasize the resultant difference in nodal displacements, and ultimately projectile strain, between the two models utilized.

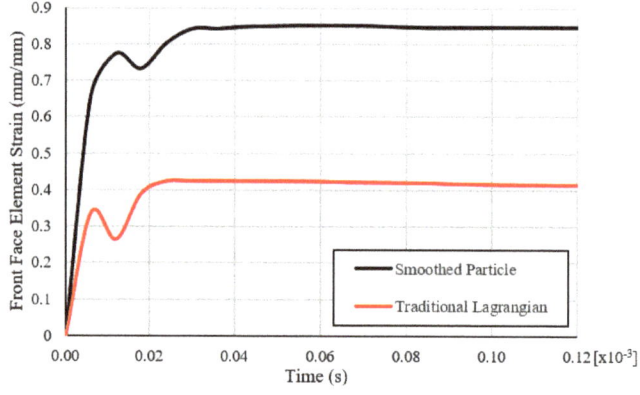

(a)

Figure 15. *Cont.*

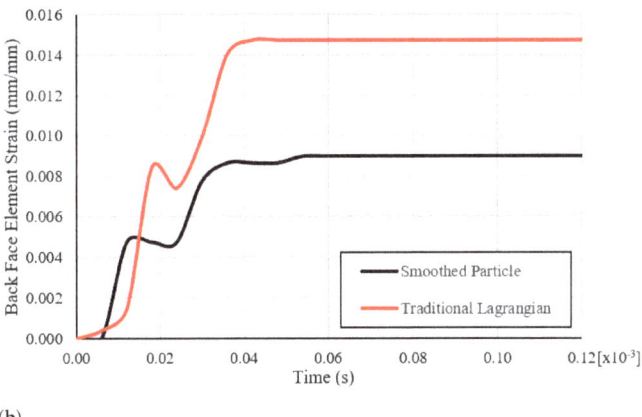

(b)

Figure 15. Element Strain versus Time: (**a**) Projectile Impact Face, (**b**) Projectile Back Face.

Figure 16 depicts the plastic strain accumulation of a single target element or particle on the outer edge of the initial contact between the particle and target. The Johnson-Cook Damage Model, which is used to determine element and particle failure, relies upon the plastic strain accumulation within the model. The SPH plastic strain accumulation was characterized by a smooth accumulation up to a maximum strain of 0.73 mm/mm. The traditional Lagrangian response was significantly more chaotic, displaying several discontinuities. The element reached a maximum plastic strain of 1.1 mm/mm, although the mean plastic strain in the plateau region was 0.85 mm/mm. The more stable response from the SPH target provides a reliable input source into the damage model for determining degradation of material properties throughout the impact. The erratic strain response of the full Lagrangian model could lead to either early or late element deletion, which would have a substantial impact on further evaluation within the model.

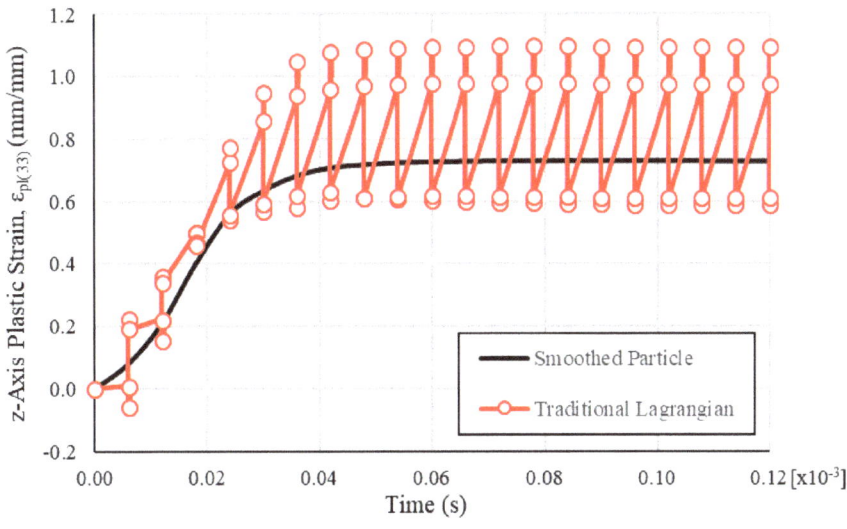

Figure 16. Comparison of Plastic Strain along the Impact Axis between Traditional Lagrangian Finite Element and Smoothed Particle Hydrodynamics Models.

During the impact period, the high rate force is applied and kinetic energy is partially transferred between the colliding bodies, the use of the conservation of momentum and energy specifically provide a more balanced solution than Newton's Second Law. Figure 17 shows a comparison of the model kinetic energy between the traditional model and mixed model. In Figure 17a, the model kinetic energy is presented over the entire simulation run, and it can be seen here that the SPH model retained a higher level of kinetic energy than the traditional model. This correlates with the higher residual velocity of the projectile in the SPH model; however, similar to the internal energy response, there is an interesting phenomenon that can be seen early in the simulation run, which is presented in Figure 17b. Early in the simulation, as shown in this figure, the traditional model appears to recover some kinetic energy, where the mixed model does not show this anomaly. This peculiar feature is a further indication of the errors produced in the Lagrangian impact model.

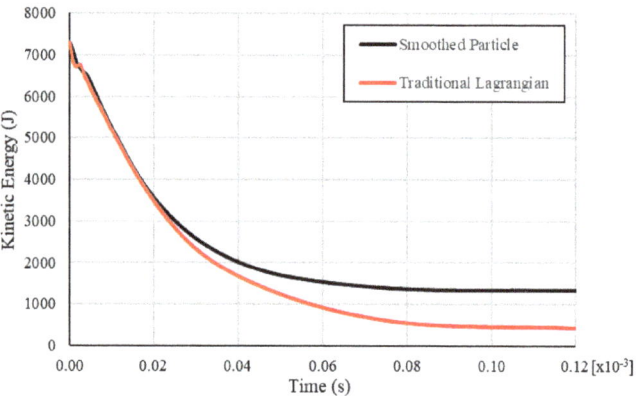

(a)

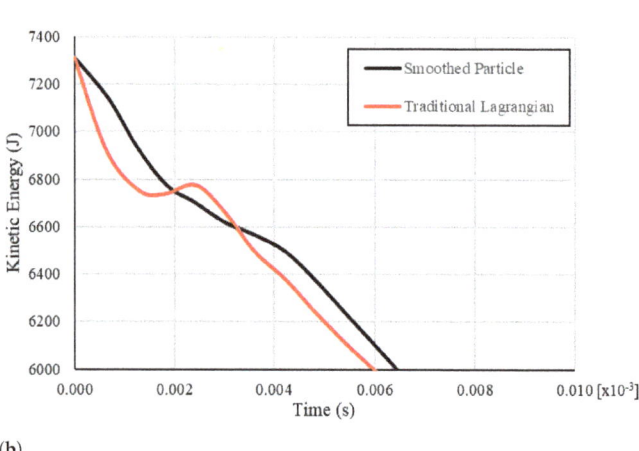

(b)

Figure 17. Model Kinetic Energy versus Time Comparison: (**a**) Full Run Duration, (**b**) Initial Reaction.

4. Conclusions

These two techniques show that there is a trade-off that must be made in the modeling of high-velocity projectile impact between computational cost and simulation performance. The first technique relied primarily on an explicit analysis of the impact using Lagrangian finite element methods and a finite difference time integration analysis to account for the dynamics. Lagrangian finite elements uses space-time discretization of the parts into a grid of elements connected by common nodes and boundaries. The time integration, CDTI, determines deformation through satisfying the equations of motions, which is used to find the strain and stress. Whereas the second technique, SPH, uses volume discretization into particles that retain volume, mass, and field properties. Then the three conservation equations are solved based on proximity to other particles through a kernel function. Considering the complexities of three-dimensional modeling, the Lagrangian model does an adequate job modeling physical deformation even at nearly 1000 m per second and at a reasonable computational cost. However, there are drawbacks to Lagrangian analysis. The most notable deficiency is that it does not model larger deformations or displacements well, which was plainly evident in the residual velocity results being one-tenth of that seen in experimentation. This deficiency is primarily caused by two factors. The first is in the volumetric integration used to determine the internal force vector, Equation (9), where the change in coordinate systems used between the Lagrangian and Eulerian systems introduces some error in the CDTI methodology, although less error than keeping the Lagrangian coordinate system through the integration. This miscalculation will propagate into the nodal displacement vector and can lead to significant mesh distortion, which will only further compound the error. A potential remedy to these issues within the full Lagrangian model is remeshing as the element distortion becomes too great to adequately evaluate the internal force integral. Remeshing adds significant complexity and time to the model, which would negate the benefit of a reduced run time. As the SPH technique does not rely on a mesh system, this error is not present in the SPH representation. The second source of error is in the mathematical methodology used, where the Lagrangian technique relies on solving Newton's Second Law of Motion at each time step and SPH satisfies the conservation equations at each time step. The balance of the conservation equations with the equation of state provides for a more accurate solution at each individual time step in the formulation used here, which prevents compounding error. For these reasons, the mixed model was capable of handling higher strain rates and larger deformations better, which was evidenced in the closeness of the simulation results to the experimental values. The mixed model was able to achieve a residual velocity within a quarter of a percent of the experimental results, and also displayed asymmetric results in the deformation pattern similar to experimentation. While the computational cost was five times that of the more traditional Lagrangian technique, the improved accuracy in the solution makes the mixed model preferable for simulating high strain rate projectile impact.

Author Contributions: Conceptualization, D.G.S. and A.N.P.; Methodology, D.G.S. and A.N.P.; Software, D.G.S.; Validation, D.G.S., A.N.P. and R.A.K.; Formal Analysis, D.G.S.; Investigation, D.G.S.; Resources, A.N.P.; Data Curation, D.G.S.; Writing–Original Draft Preparation, D.G.S.; Writing–Review and Editing, A.N.P. and R.A.K.; Visualization, D.G.S.; Supervision, A.N.P.; Project Administration, D.G.S.; Funding Acquisition, A.N.P. All authors have read and agreed to the published version of the manuscript.

Funding: This research is sponsored by the United States Air Force, Air Force Research Laboratory, Air Force Office of Scientific Research (AFOSR) program on Dynamic Materials and Interactions.

Institutional Review Board Statement: Not applicable.

Informed Consent Statement: Not applicable.

Data Availability Statement: Data will be made available upon reasonable request.

Acknowledgments: The authors would like to thank Martin Schmidt for his support of this research.

Conflicts of Interest: The authors declare no conflict of interest. The funders had no role in the design of the study; in the collection, analyses, or interpretation of data; in the writing of the manuscript, or in the decision to publish the results.

Abbreviations

The following abbreviations are used in this manuscript:

CDTI	Central Difference Time Integration
EOS	Equation of State
FEA	Finite Element Analysis
FEM	Finite Element Method
KE	Kinetic Energy
SPH	Smoothed Particle Hydrodynamics

References

1. Ben-Dor, G.; Dubinsky, A.; Elperin, T. Ballistic impact: Recent advances in analytical modeling of plate penetration dynamics-a review. *Appl. Mech. Rev.* **2005**, *58*, 355–370. [CrossRef]
2. Nair, R.P.; Rao, C.L. Simulation of depth of penetration during ballistic impact on thick targets using a one-dimensional discrete element model. *Sadhana Acad. Proc. Eng. Sci.* **2012**, *37*, 261–279. [CrossRef]
3. Kurtaran, H.; Buyuk, M.; Eskandarian, A. Ballistic impact simulation of GT model vehicle door using finite element method. *Theor. Appl. Fract. Mech.* **2003**, *40*, 113–121. [CrossRef]
4. Schwer, L.E. Aluminum Plate Perforation: A Comparative Case Study using Lagrange with Erosion, Multi-Material ALE, and Smooth Particle Hydrodynamics. In Proceedings of the 7th European LS-DYNA Conference, Salzburg, Austria, 14–15 May 2009; pp. 1–27.
5. Resnyansky, A.; Katselis, G. *Ballistic and Meterial Testing Procedures and Test Results for Composite Samples for the TIGER Helicopter Vulnerability Project*, dsto-tr-16 ed.; Australian Government, Department of Defence: Canberra, Australia, 2004. [CrossRef]
6. Johnson, G.R. High Velocity Impact Calculations in Three Dimensions. *J. Appl. Mech.* **1977**, *44*, 95–100. [CrossRef]
7. Johnson, G.R. Closure to "Discussion of 'High Velocity Impact Calculations in Three Dimensions'" (1977, ASME J. Appl. Mech., 44, pp. 793–794). *J. Appl. Mech.* **1977**, *44*, 794–795. [CrossRef]
8. Gailly, B.A.; Espinosa, H.D. Modelling of failure mode transition in ballistic penetration with a continuum model describing microcracking and flow of pulverized media. *Int. J. Numer. Methods Eng.* **2002**, *54*, 365–398. [CrossRef]
9. Grujicic, M.; Arakere, G.; He, T.; Bell, W.; Cheeseman, B.; Yen, C.F.; Scott, B. A ballistic material model for cross-plied unidirectional ultra-high molecular-weight polyethylene fiber-reinforced armor-grade composites. *Mater. Sci. Eng. A* **2008**, *498*, 231–241. [CrossRef]
10. Grujicic, M.; Glomski, P.; He, T.; Arakere, G.; Bell, W.C.; Cheeseman, B. Material Modeling and Ballistic-Resistance Analysis of Armor-Grade Composites Reinforced with High-Performance Fibers. *J. Mater. Eng. Perform.* **2009**, *18*, 1169–1182. [CrossRef]
11. Kantar, E.; Erdem, R.; Anıl, O. Nonlinear Finite Element Analysis of Impact Behavior of Concrete Beam. *Math. Comput. Appl.* **2011**, *16*, 183–193. [CrossRef]
12. Hazell, P. Numerical simulations and experimental observations of the 5.56-MM L2A2 bullet perforating steel targets of two hardness values. *J. Battlef. Technol.* **2003**, *6*, 1–4.
13. Borvik, T.; Dey, S.; Clausen, A. Perforation resistance of five different high-strength steel plates subjected to small-arms projectiles. *Int. J. Impact Eng.* **2009**, *36*, 948–964. [CrossRef]
14. Cundall, P.A.; Strack, O.D.L. A discrete numerical model for granular assemblies. *Géotechnique* **1979**, *29*, 47–65. [CrossRef]
15. Libersky, L.D.; Petschek, A.G. Smooth particle hydrodynamics with strength of materials. In *Advances in the Free-Lagrange Method Including Contributions on Adaptive Gridding and the Smooth Particle Hydrodynamics Method*; Trease, H.E., Fritts, M.F., Crowley, W.P., Eds.; Springer: Berlin/Heidelberg, Germany, 1991; pp. 248–257.
16. Libersky, L.D.; Petschek, A.G.; Carney, T.C.; Hipp, J.R.; Allahdadi, F.A. High strain lagrangian hydrodynamics a three-dimensional SPH code for dynamic material response. *J. Comput. Phys.* **1993**, *109*, 67–75. [CrossRef]
17. Randles, P.; Libersky, L. Smoothed Particle Hydrodynamics: Some recent improvements and applications. *Comput. Methods Appl. Mech. Eng.* **1996**, *139*, 375–408. [CrossRef]
18. Johnson, G.; Petersen, E.H.; Stryk, R.A. Incorporation of an SPH option into the EPIC code for a wide range of high velocity impact computations. *Int. J. Impact Eng.* **1993**, *14*, 385–394. [CrossRef]
19. Johnson, G.R.; Stryk, R.A.; Beissel, S.R. SPH for high velocity impact computations. *Comput. Methods Appl. Mech. Eng.* **1996**, *139*, 347–373. [CrossRef]
20. O'Toole, B.; Trabia, M.; Hixson, R.; Roy, S.K.; Pena, M.; Becker, S.; Daykin, E.; Machorro, E.; Jennings, R.; Matthes, M. Modeling Plastic Deformation of Steel Plates in Hypervelocity Impact Experiments. *Procedia Eng.* **2015**, *103*, 458–465. [CrossRef]
21. Poniaev, S.; Kurakin, R.; Sedov, A.; Bobashev, S.; Zhukov, B.; Nechunaev, A. Hypervelocity impact of mm-size plastic projectile on thin aluminum plate. *Acta Astronaut.* **2017**, *135*, 26–33. [CrossRef]
22. Taddei, L.; Awoukeng Goumtcha, A.; Roth, S. Smoothed particle hydrodynamics formulation for penetrating impacts on ballistic gelatine. *Mech. Res. Commun.* **2015**, *70*, 94–101. [CrossRef]
23. Frissane, H.; Taddei, L.; Lebaal, N.; Roth, S. SPH modeling of high velocity impact into ballistic gelatin. Development of an axis-symmetrical formulation. *Mech. Adv. Mater. Struct.* **2019**, *26*, 1881–1888. [CrossRef]
24. Xiao, Y.; Dong, H.; Zhou, J.; Wang, J. Studying normal perforation of monolithic and layered steel targets by conical projectiles with SPH simulation and analytical method. *Eng. Anal. Bound. Elem.* **2017**, *75*, 12–20. [CrossRef]

25. Chaussonnet, G.; Bravo, L.; Flatau, A.; Koch, R.; Bauer, H.J. Smoothed Particle Hydrodynamics Simulation of High Velocity Impact Dynamics of Molten Sand Particles. *Energies* **2020**, *13*, 5134. [CrossRef]
26. Soriano-Moranchel, F.A.; Sandoval-Pineda, J.M.; Gutiérrez-Paredes, G.J.; Silva-Rivera, U.S.; Flores-Herrera, L.A. Simulation of Bullet Fragmentation and Penetration in Granular Media. *Materials* **2020**, *13*, 5243. [CrossRef] [PubMed]
27. Graves, W.T.; Liu, D.; Palazotto, A.N. Impact of an Additively Manufactured Projectile. *J. Dyn. Behav. Mater.* **2017**, *3*, 362–376. [CrossRef]
28. Provchy, Z.A.; Palazotto, A.N.; Flater, P. Topology Optimization for Projectile Design. *J. Dyn. Behav. Mater.* **2018**, *4*, 129–137. [CrossRef]
29. Patel, A.; Palazotto, A.N. Investigation of Hybrid Material Projectile Impact Against Concrete Targets. In Proceedings of the 2018 AIAA/ASCE/AHS/ASC Structures, Structural Dynamics, and Materials Conference, Kissimmee, FL, USA, 8–12 January 2018. [CrossRef]
30. Beard, A.; Palazotto, A.N. Composite Material for High-Speed Projectile Outer Casing. In Proceedings of the 2020 AIAA Science and Technology Forum, Orlando, FL, USA, 6–10 January 2020. [CrossRef]
31. Spear, D.; Palazotto, A.N.; Kemnitz, R. Survivability and Damage Modeling of Advanced Materials. In Proceedings of the 2020 AIAA Science and Technology Forum, Orlando, FL, USA, 6–10 January 2020. [CrossRef]
32. Spear, D.; Palazotto, A.N.; Kemnitz, R. First Cell Failure of Lattice Structure under Combined Axial and Buckling Load. In Proceedings of the AIAA Scitech 2021 Forum, Nashville, TN, USA, 11–15 January 2021; pp. 1–16. [CrossRef]
33. Wilson, E.L. Structural analysis of axisymmetric solids. *AIAA J.* **1965**, *3*, 2269–2274. [CrossRef]
34. Cook, R.D.; Malkus, D.S.; Plesha, M.E.; Witt, R.J. *Concepts and Applications of Finite Element Analysis*; John Wiley & Sons, Inc.: Hoboken, NJ, USA, 2002.
35. Dassault Systèmes. *ABAQUS Version 6.12 Theory Manual*; Dassault Systemes Simulia Corp: Johnston, RI, USA, 2016.
36. KJ Bath. *Finite Element Procedures*, 2nd ed.; KJ Bathe: Watertown, MA, USA, 2014; p. 1037.
37. Johnson, G.; Cook, W. A Constitutive Model and Data for Metals Subjected to Large Strains, High Strain Rates, and High Temperatures. In Proceedings of the Seventh International Symposium on Ballistics, The Hague, The Netherlands, 19–21 April 1983; pp. 541–547.
38. Leseur, D. Experimental Investigations of Material Models for Ti-6Al-4V Titanium and 2024-T3 Aluminum. DOT/FAA/AR-00/25. 2000, Volume 9, pp. 1–29. Available online: https://www.osti.gov/biblio/11977-experimental-investigations-material-models-ti-t3 (accessed on 27 January 2021).
39. Kay, G. Failure Modeling of Titanium-6Al-4V and 2024-T3 Aluminum with the Johnson-Cook Material Model. DOT/FAA/AR-03/57. 2003, Volume 9, pp. 1–17. Available online: https://www.osti.gov/biblio/15006359-failure-modeling-titanium-t3-aluminum-johnson-cook-material-model (accessed on 27 January 2021). [CrossRef]
40. Shames, I.; Cozzarelli, F. *Elastic And Inelastic Stress Analysis*; Taylor & Francis: Abingdon, UK, 1997.
41. Gingold, R.A.; Monaghan, J.J. Smoothed particle hydrodynamics: Theory and application to non-spherical stars. *Mon. Not. R. Astron. Soc.* **1977**, *181*, 375–389. [CrossRef]
42. Lucy, L.B. A numerical approach to the testing of the fission hypothesis. *Astron. J.* **1977**, *82*, 1013. [CrossRef]
43. Wu, C.; Wu, Y.; Crawford, J.E.; Magallanes, J.M. Three-dimensional concrete impact and penetration simulations using the smoothed particle Galerkin method. *Int. J. Impact Eng.* **2017**, *106*, 1–17. [CrossRef]
44. Islam, M.R.I.; Chakraborty, S.; Shaw, A. On consistency and energy conservation in smoothed particle hydrodynamics. *Int. J. Numer. Methods Eng.* **2018**, *116*, 601–632. [CrossRef]
45. Chen, J.Y.; Peng, C.; Lien, F.S. Simulations for three-dimensional landmine detonation using the SPH method. *Int. J. Impact Eng.* **2019**, *126*, 40–49. [CrossRef]
46. Monaghan, J.J. An introduction to SPH. *Comput. Phys. Commun.* **1988**, *48*, 89–96. [CrossRef]
47. Monaghan, J.J. Smoothed Particle Hydrodynamics. *Annu. Rev. Astron. Astrophys.* **1992**, *30*, 543–574. [CrossRef]
48. Liu, M.; Liu, G. *Particle Methods for Multi-Scale and Multi-Physics*; World Scientific: Singapore, 2016. [CrossRef]
49. Zisis, I.; van der Linden, B.; Giannopapa, C. Towards a Smoothed Particle Hydrodynamics Algorithm for Shocks Through Layered Materials, Volume 4: Fluid-Structure Interaction. In Proceedings of the Pressure Vessels and Piping Conference, Paris, France, 14–18 July 2013; p. V004T04A005. [CrossRef]
50. Stranex, T.; Wheaton, S. A new corrective scheme for SPH. *Comput. Methods Appl. Mech. Eng.* **2011**, *200*, 392–402. [CrossRef]
51. Vignjevic, R.; Campbell, J.; Jaric, J.; Powell, S. Derivation of SPH equations in a moving referential coordinate system. *Comput. Methods Appl. Mech. Eng.* **2009**, *198*, 2403–2411. [CrossRef]
52. Batra, R.C.; Zhang, G.M. SSPH basis functions for meshless methods, and comparison of solutions with strong and weak formulations. *Comput. Mech.* **2008**, *41*, 527–545. [CrossRef]
53. Zhang, G.M.; Batra, R.C. Symmetric smoothed particle hydrodynamics (SSPH) method and its application to elastic problems. *Comput. Mech.* **2009**, *43*, 321–340. [CrossRef]
54. Cueto-Felgueroso, L.; Colominas, I.; Mosqueira, G.; Navarrina, F.; Casteleiro, M. On the Galerkin formulation of the smoothed particle hydrodynamics method. *Int. J. Numer. Methods Eng.* **2004**, *60*, 1475–1512. [CrossRef]
55. Dilts, G.A. Moving-Least-Squares-particle hydrodynamics—I. Consistency and stability. *Int. J. Numer. Methods Eng.* **1999**, *44*, 1115–1155. [CrossRef]

56. Dilts, G.A. Moving least-squares particle hydrodynamics II: Conservation and boundaries. *Int. J. Numer. Methods Eng.* **2000**, *48*, 1503–1524. [CrossRef]
57. Colagrossi, A.; Landrini, M. Numerical simulation of interfacial flows by smoothed particle hydrodynamics. *J. Comput. Phys.* **2003**, *191*, 448–475. [CrossRef]
58. Liu, G.; Liu, M. *Smoothed Particle Hydrodynamics: A Meshfree Particle Method*; World Scientific: Singapore, 2003. [CrossRef]
59. Danilewicz, A.; Sikora, Z. Numerical Simulation of Crater Creating Process in Dynamic Replacement Method by Smooth Particle Hydrodynamics. *Stud. Geotech. Mech.* **2015**, *36*, 3–8. [CrossRef]
60. Barbosa, D.; Piccoli, F. Comparing the force due to the Lennard-Jones potential and the Coulomb force in the SPH Method. *J. Ocean Eng. Sci.* **2018**, *3*, 310–315. [CrossRef]
61. Holzapfel, W.B. Equations of State and Thermophysical Properties of Solids Under Pressure. In *High-Pressure Crystallography*; Springer: Dordrecht, The Netherlands, 2004; pp. 217–236. [CrossRef]
62. Meyers, M.A. *Dynamic Behavior of Materials*; Wiley: Hoboken, NJ, USA, 1994. [CrossRef]
63. Roy, S.K.; Trabia, M.; O'Toole, B.; Hixson, R.; Becker, S.; Pena, M.; Jennings, R.; Somasoundaram, D.; Matthes, M.; Daykin, E.; Machorro, E. Study of Hypervelocity Projectile Impact on Thick Metal Plates. *Shock Vib.* **2016**, *2016*, 1–11. [CrossRef]
64. Zocher, M.; Maudlin, P.; Chen, S.; Flower-Maudlin, E. An Evaluation of Several Hardening Models using Taylor Cylinder Impact Data. In Proceedings of the European Congress on Computational Methods in Applied Sciences and Engineering (ECCOMAS 2000), Barcelona, Spain, 11–14 September 2000; Volume 53, pp. 1–20. [CrossRef]
65. Rice, M.H.; McQueen, R.G.; Walsh, J.M. Compression of Solids by Strong Shock Waves. *Solid State Phys. Adv. Res. Appl.* **1958**, *6*, 1–63. [CrossRef]
66. Mandl, F. *Statistical Physics*, 2nd ed.; Wiley: Hoboken, NJ, USA, 1991.
67. Marc, G.; McMillan, W.G. The Virial Theorem. In *Advances in Chemical Physics*; John Wiley & Sons, Inc.: Hoboken, NJ, USA, 2007; Volume 63, pp. 209–361. [CrossRef]
68. Heuzé, O. General form of the Mie-Grüneisen equation of state. *Comptes Rendus-Mec.* **2012**, *340*, 679–687. [CrossRef]
69. Goldstein, H.; Poole, C.; Safko, J. *Classical Mechanics*; Addison Wesley: Boston, MA, USA, 2002.
70. Grüneisen, E. Theorie des festen Zustandes einatomiger Elemente. *Ann. Phys.* **1912**, *344*, 257–306. [CrossRef]
71. Johnson, G.R.; Cook, W.H. Fracture characteristics of three metals subjected to various strains, strain rates, temperatures and pressures. *Eng. Fract. Mech.* **1985**, *21*, 31–48. [CrossRef]
72. Cowan, R.; Winer, W. Frictional Heating Calculations. *Frict. Lubr. Wear Technol. ASM Handb.* **1992**, *18*, 39–44.
73. Meyer, H.W.; Kleponis, D.S. Modeling the high strain rate behavior of titanium undergoing ballistic impact and penetration. *Int. J. Impact Eng.* **2001**, *26*, 509–521. [CrossRef]
74. Wang, X.; Shi, J. Validation of Johnson-Cook plasticity and damage model using impact experiment. *Int. J. Impact Eng.* **2013**, *60*, 67–75. [CrossRef]
75. Committee, A.H. *Properties and Selection: Nonferrous Alloys and Special-Purpose Materials*; ASM International: Almere, The Netherlands, 1990. [CrossRef]
76. Johnson, G.R.; Holmquist, T.J. *Test Data and Computational Strength and Fracture Model Constants for 23 Materials Subjected to Large Strains, High Strain Rates, and High Temperatures*; Los Alamos National Laboratory: Los Alamos, NM, USA, 1989.

Article

Numerical Investigation of MHD Pulsatile Flow of Micropolar Fluid in a Channel with Symmetrically Constricted Walls

Amjad Ali [1], Muhammad Umar [1], Zaheer Abbas [2], Gullnaz Shahzadi [3,*], Zainab Bukhari [1] and Arshad Saleem [1]

[1] Centre for Advanced Studies in Pure and Applied Mathematics, Bahauddin Zakariya University, Multan 60800, Pakistan; amjadali@bzu.edu.pk (A.A.); muhammadumar@bzu.edu.pk (M.U.); zainabbukhari398@gmail.com (Z.B.); arshad.saleem2100@gmail.com (A.S.)
[2] Department of Mathematics, The Islamia University of Bahawalpur, Bahawalpur 63100, Pakistan; zaheer.abbas@iub.edu.pk
[3] Department of Mechanical Engineering, École de Technologie Supérieure ÉTS, 1100 Notre-Dame W, Montreal, QC H3C 1K3, Canada
* Correspondence: gullnaz.shahzadi.1@ens.etsmtl.ca

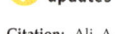

Citation: Ali, A.; Umar, M.; Abbas, Z.; Shahzadi, G.; Bukhari, Z.; Saleem, A. Numerical Investigation of MHD Pulsatile Flow of Micropolar Fluid in a Channel with Symmetrically Constricted Walls. *Mathematics* 2021, 9, 1000. https://doi.org/10.3390/math9091000

Academic Editor: Aihua Wood

Received: 23 March 2021
Accepted: 21 April 2021
Published: 28 April 2021

Publisher's Note: MDPI stays neutral with regard to jurisdictional claims in published maps and institutional affiliations.

Copyright: © 2021 by the authors. Licensee MDPI, Basel, Switzerland. This article is an open access article distributed under the terms and conditions of the Creative Commons Attribution (CC BY) license (https://creativecommons.org/licenses/by/4.0/).

Abstract: This article presented an analysis of the pulsatile flow of non-Newtonian micropolar (MP) fluid under Lorentz force's effect in a channel with symmetrical constrictions on the walls. The governing equations were first converted into the vorticity–stream function form, and a finite difference-based solver was used to solve it numerically on a Cartesian grid. The impacts of different flow controlling parameters, including the Hartman number, Strouhal number, Reynolds number, and MP parameter on the flow profiles, were studied. The wall shear stress (WSS), axial, and micro-rotation velocity profiles were depicted visually. The streamlines and vorticity patterns of the flow were also sketched. It is evident from the numerical results that the flow separation region near constriction as well as flattening of the axial velocity component is effectively controlled by the Hartmann number. At the maximum flow rate, the WSS attained its peak. The WSS increased in both the Hartmann number and Reynolds number, whereas it declined with the higher values of the MP parameter. The micro-rotation velocity increased in the Reynolds number, and it declined with increment in the MP parameter.

Keywords: micropolar fluid; constricted channel; MHD pulsatile flow; strouhal number; flow pulsation parameter

1. Introduction

MP fluids are non-Newtonian fluids consisting of the dilute suspension with an individual motion of thin, rigid cylindrical macromolecules. Incompressible MP fluids have significance in the study of various phenomena such as blood rheology in medical sciences and melted plastic mechanics in industries. MP fluid theory explains the micro-rotation effects. Eringen [1] first described micro-inertia effects. Several numerical studies have been conducted by researchers to study the behavior of internal and external MP fluid flows. Agarwal et al. [2] examined MP fluid flow on a porous stationary surface with heat transfer. The 2D stagnation point flow of MP fluid for the steady case over a stretching sheet was examined by Nazar et al. [3]. Lok et al. [4] researched the steady mixed convection boundary layer flow of MP fluid on a double-infinite, vertical flat plate near the stagnation point. The flow behavior and heat transfer effects of mixed convection in MP fluid flow over a vertical flat plate with conduction were analyzed by Chang et al. [5]. Magyari et al. [6] examined the flow of quiescent MP fluid over a doubly infinite plate accelerated from rest to a constant velocity. The impacts of radiation and viscous dissipation on MP fluid stagnation-point flow to a nonlinearly stretching surface with suction and injection were reported by Babu et al. [7]. The flow of MP fluid over a porous stretch surface with heat transfer was analyzed by Turkyilmazoglu [8]. Waqas et al. [9] provided a mixed convection flow of MP liquid in the occurrence of the magnetic field on a nonlinear stretched surface. Ramadevi et al. [10] carried out an analysis of the nonlinear MHD radiative flow of MP fluid on a stretching surface.

MHD MP fluid over an oscillating, infinite vertical plate embedded in a porous medium was analyzed by Sheik et al. [11]. Hussanan et al. [12] analytically examined MP fluid flow over a vertical plate with Newtonian heating in the presence of the magnetic field and the absence of thermal radiation. Kumar et al. [13] examined the heat transfer mechanism with variable heat sink/source, a nonlinear approximation of Rosseland and Biot number over a stretched field. Shamshuddin et al. [14] used the finite-element approach for solving MHD, incompressible, dissipative, and chemically reacting MP fluid flow with heat transfer as well as mass transfer on an inclined heat source/sink plate. Nadeem et al. [15] examined the flow of MP fluid over the Riga plate with exponential surface temperature and heating effects.

Si et al. [16] investigated the behavior of MP fluid flow in a porous channel with mutable walls. Lu et al. [17] considered the 2D creeping flow of MP fluid in a thin permeable channel with a variable absorption rate. Fakour et al. [18] studied heat and mass transfer of MP fluid flow inside a channel with permeable walls. Tutty [19] investigated the non-uniform channel which is used as a simple model of a constricted arterial vessel. There are several studies regarding fluid motion with pulsation in a constricted channel. Peristalsis is a mechanism in which progressive transverse waves produced by flexible channel/tube boundary walls transport the fluid. Peristaltic pumping is also very effective in the design of several biomedical devices for maintaining blood supply during critical operations. Mekheimer et al. [20] studied the effect of an induced magnetic field on the peristaltic transport in a symmetric channel of an incompressible MP conductive fluid. Hayat and Ali [21] examined the peristaltic wave motion for the endoscope impact via the distance among two concentric tubes, finding the inner tube to be rigid when moving outwards to allow the MP fluid to flow.

Under certain physical situations, the behavior of the pulsatile flow of Newtonian and non-Newtonian fluids has been examined, usually with assumptions of long wavelength and low Reynolds number to simplification. A numerical study of the MHD pulsatile flow of Newtonian fluid was carried out by Bandyopadhyay and Layek [22] in a single-constricted channel. Khair et al. [23] described the transition from laminar to the turbulent regime in a constricted channel for pulsatile flow. The steady and pulsatile flow of MHD Casson fluid in a constricted channel was studied by Ali et al. [24].

The present work's objective is to investigate the magnetohydrodynamic (MHD) pulsatile flow of non-Newtonian MP fluid in a channel having symmetrical constrictions on both the walls under the influence of the Lorentz force. The numerical method to solve the governing equations is based on the finite difference method on a Cartesian grid instead of the cylindrical one. The impacts of various parameters on the axial velocity, shear stress, and micro-rotation velocity are discussed. The streamlines and vorticity distributions of the pulsatile MP fluid flow are also shown. The flow separation region generated due to the constriction bumps is also discussed. The flow parameters under consideration for the study include the Hartmann number (M), Strouhal number (St), Reynolds number (Re), and MP parameter. The study finds applications in understarting the blood flow, modeled as non-Newtonian micropolar fluid, in stenotic arteries especially. The outcomes can be used in designing the biomedical devices and techniques for cardiovascular treatments, e.g., evaluating the thrombogenic potential of implantable cardiac devices [25]. The rest of the article is structured as follows. Section 2 explains the mathematical formulation of the problem and method. Section 3 presents the results and discussion. Section 4 displays the conclusions.

2. Materials and Methods

A two-dimensional pulsatile flow of MP fluid was analyzed by a uniform magnetic field applied perpendicular to the flow direction, as shown in Figure 1. The geometry under consideration was a constricted channel. The center of the constriction was placed at $x = 0$ with a total width of constriction as $2x_0$, as depicted in Equation (1). The constrictions on the walls of the channel are formulated and implemented as:

$$y_1(x) = \begin{cases} \frac{h_1}{2}\left[1 + \cos\left(\frac{\pi x}{x_0}\right)\right], & |x| \leq x_0 \\ 0, & |x| > x_0 \end{cases} \quad (1a)$$

$$y_2(x) = \begin{cases} 1 - \frac{h_2}{2}\left[1 + \cos\left(\frac{\pi x}{x_0}\right)\right], & |x| \leq x_0 \\ 1, & |x| > x_0 \end{cases} \quad (1b)$$

where y_1 and y_2 define the lower and upper walls with constriction heights h_1 and h_2, respectively.

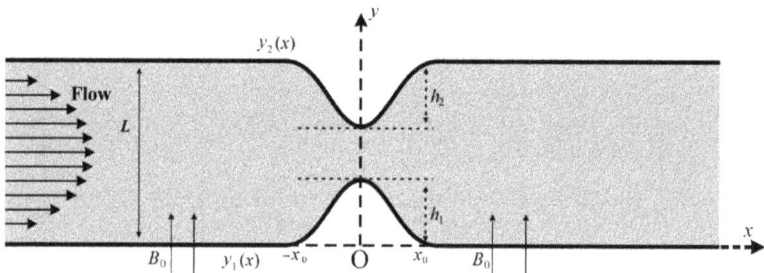

Figure 1. Constricted channel geometry.

The momentum equations of the unsteady flow are given by:

$$\frac{\partial \hat{u}}{\partial \hat{t}} + \hat{u}\frac{\partial \hat{u}}{\partial \hat{x}} + \hat{v}\frac{\partial \hat{u}}{\partial \hat{y}} = -\frac{1}{\rho}\frac{\partial \hat{p}}{\partial \hat{x}} + \left(\frac{\mu+k}{\rho}\right)\nabla^2 \hat{u} + \frac{1}{\rho}(\mathbf{J} \times \mathbf{B})_x + \frac{k}{\rho}\frac{\partial \hat{N}}{\partial \hat{y}} \quad (2)$$

$$\frac{\partial \hat{v}}{\partial \hat{t}} + \hat{u}\frac{\partial \hat{v}}{\partial \hat{x}} + \hat{v}\frac{\partial \hat{v}}{\partial \hat{y}} = -\frac{1}{\rho}\frac{\partial \hat{p}}{\partial \hat{y}} + \left(\frac{\mu+k}{\rho}\right)\nabla^2 \hat{v} - \frac{k}{\rho}\frac{\partial \hat{N}}{\partial \hat{x}} \quad (3)$$

$$\frac{\partial \hat{N}}{\partial \hat{t}} + \hat{u}\frac{\partial \hat{N}}{\partial \hat{x}} + \hat{v}\frac{\partial \hat{N}}{\partial \hat{y}} = -\frac{k}{\rho j}\left(2\hat{N} + \frac{\partial \hat{u}}{\partial \hat{y}} - \frac{\partial \hat{v}}{\partial \hat{x}}\right) + \frac{\gamma}{\rho j}\nabla^2 \hat{N} \quad (4)$$

The continuity equation is given by:

$$\frac{\partial \hat{u}}{\partial \hat{x}} + \frac{\partial \hat{v}}{\partial \hat{y}} = 0 \quad (5)$$

Here the velocity components along the $\hat{x}$- and $\hat{y}$-axis are $\hat{u}$ and $\hat{v}$, respectively. $\hat{p}$, ρ, and ν represent pressure, density, and kinematic viscosity, respectively. $\hat{N}$ represents the micro-rotation velocity, k represents vortex viscosity, $\mathbf{J} \equiv (J_x, J_y, J_z)$ the current density, $\mathbf{B} \equiv (0, B_0, 0)$ the magnetic field with uniform strength B_0, σ the electric conductivity, and μ the dynamic viscosity. $\gamma = j(\mu + k)/2$ represents the spin gradient viscosity, where j defines the micro-inertia density. If $\mathbf{E} \equiv (E_x, E_y, E_z)$ indicates the electric field directed along the normal to the flow plane, then $\mathbf{E} \equiv (0, 0, E_z)$. In addition, using Ohm's law:

$$J_x = 0, \quad J_y = 0, \quad J_z = \sigma(E_z + \hat{u}B_0) \quad (6)$$

Maxwell's equation $\nabla \times \mathbf{E} = 0$ for stationary flow implies that $E_z = a$, where a is a constant, assumed to be zero for simplicity. Then, $J_z = \sigma \hat{u} B_0$. Therefore, applying $\mathbf{J} \times \mathbf{B} = -\sigma \hat{u} B_0^2$, Equation (2) becomes:

$$\frac{\partial \hat{u}}{\partial \hat{t}} + \hat{u}\frac{\partial \hat{u}}{\partial \hat{x}} + \hat{v}\frac{\partial \hat{u}}{\partial \hat{y}} = -\frac{1}{\rho}\frac{\partial \hat{p}}{\partial \hat{x}} + \left(\frac{\mu+k}{\rho}\right)\nabla^2 \hat{u} + \frac{-\sigma \hat{u} B_0^2}{\rho} + \frac{k}{\rho}\frac{\partial \hat{N}}{\partial \hat{y}} \quad (7)$$

We define the dimensionless quantities:

$$x = \frac{\hat{x}}{L}, \quad y = \frac{\hat{y}}{L}, \quad u = \frac{\hat{u}}{U}, \quad v = \frac{\hat{v}}{U}, \quad t = \frac{\hat{t}}{T}, \quad St = \frac{L}{UT},$$

$$p = \frac{\hat{p}}{\rho U^2}, \quad Re = \frac{UL}{\nu}, \quad N = \frac{\hat{N}L}{U}, \quad K = \frac{k}{\mu}, \quad M = B_0 L \sqrt{\frac{\sigma}{\rho \nu}} \tag{8}$$

Here, T is the period of flow pulsation, L the maximum channel width, Re the Reynolds number, M the Hartmann Number, the micro-rotation velocity, and K the MP parameter.

Using the non-dimensional parameters from Equation (8), in Equations (7), (3), (4), and (5), we attain:

$$St\frac{\partial u}{\partial t} + u\frac{\partial u}{\partial x} + v\frac{\partial u}{\partial y} = -\frac{\partial p}{\partial x} + \left(\frac{1+K}{Re}\right)\nabla^2 u - \frac{M^2}{Re}u + \frac{K}{Re}\frac{\partial N}{\partial y} \tag{9}$$

$$St\frac{\partial v}{\partial t} + u\frac{\partial v}{\partial x} + v\frac{\partial v}{\partial y} = -\frac{\partial p}{\partial y} + \left(\frac{1+K}{Re}\right)\nabla^2 v - \frac{K}{Re}\frac{\partial N}{\partial x} \tag{10}$$

$$St\frac{\partial N}{\partial t} + u\frac{\partial N}{\partial x} + v\frac{\partial N}{\partial y} = \frac{-K}{Re}\left(2N + \frac{\partial u}{\partial y} - \frac{\partial v}{\partial x}\right) + \left(\frac{1}{Re} + \frac{K}{2Re}\right)\nabla^2 N \tag{11}$$

$$\frac{\partial u}{\partial x} + \frac{\partial u}{\partial y} = 0 \tag{12}$$

2.1. Vorticity–Stream Function Formulation

The dimensionless stream function (ψ) and vorticity function (ω) for the flow under consideration are as follows:

$$u = \frac{\partial \psi}{\partial y}, \quad v = -\frac{\partial \psi}{\partial x}, \quad \omega = \frac{\partial v}{\partial x} - \frac{\partial u}{\partial y} \tag{13}$$

Some manipulations with Equations (9) and (10) produce:

$$St\frac{\partial}{\partial t}\left(\frac{\partial v}{\partial x} - \frac{\partial u}{\partial y}\right) + u\frac{\partial}{\partial x}\left(\frac{\partial v}{\partial x} - \frac{\partial u}{\partial y}\right) + v\frac{\partial}{\partial y}\left(\frac{\partial v}{\partial x} - \frac{\partial u}{\partial y}\right)$$

$$= \left(\frac{1+K}{Re}\right)\left[\frac{\partial^2}{\partial x^2}\left(\frac{\partial v}{\partial x} - \frac{\partial u}{\partial y}\right) + \frac{\partial^2}{\partial y^2}\left(\frac{\partial v}{\partial x} - \frac{\partial u}{\partial y}\right)\right] + \frac{M^2}{Re}u - \frac{K}{Re}\left(\frac{\partial^2 N}{\partial x^2} - \frac{\partial^2 N}{\partial y^2}\right) \tag{14}$$

Using the quantities in Equation (13), we obtained the following vorticity transport equation as:

$$St\frac{\partial \omega}{\partial t} + \frac{\partial \psi}{\partial y}\frac{\partial \omega}{\partial x} - \frac{\partial \psi}{\partial x}\frac{\partial \omega}{\partial y} = \left(\frac{1+K}{Re}\right)\left[\frac{\partial^2 \omega}{\partial x^2} + \frac{\partial^2 \omega}{\partial y^2}\right] + \frac{M^2}{Re}\frac{\partial^2 \psi}{\partial y^2} - \frac{K}{Re}\left(\frac{\partial^2 N}{\partial x^2} - \frac{\partial^2 N}{\partial y^2}\right) \tag{15}$$

Again, using the quantities in Equation (13), Equation (11) becomes:

$$St\frac{\partial N}{\partial t} + \frac{\partial \psi}{\partial y}\frac{\partial N}{\partial x} - \frac{\partial \psi}{\partial x}\frac{\partial N}{\partial y} = \frac{-K}{Re}\left(2N + \frac{\partial^2 \psi}{\partial x^2} + \frac{\partial^2 \psi}{\partial y^2}\right) + \left(\frac{1}{Re} + \frac{K}{2Re}\right)\left(\frac{\partial^2 N}{\partial x^2} - \frac{\partial^2 N}{\partial y^2}\right) \tag{16}$$

The Poisson equation for ψ is:

$$\frac{\partial^2 \psi}{\partial x^2} + \frac{\partial^2 \psi}{\partial y^2} = -\omega \tag{17}$$

Here, u, v, and N are primitive variables, and ω and ψ are non-primitive variables.

2.2. Boundary Conditions

The steady case of the fluid flow from Equation (7) is considered to obtain the boundary conditions for the current problem:

$$-\frac{1}{\rho}\frac{\partial \hat{p}}{\partial \hat{x}} + \left(\frac{\mu + k}{\rho}\right)\frac{\partial^2 \hat{u}}{\partial \hat{y}^2} - \frac{1}{\rho}(\mathbf{J} \times \mathbf{B}) = 0 \tag{18}$$

where $\mathbf{J} \times \mathbf{B} = -\sigma(E_z + \hat{u}B_0)B_0$. By substituting in Equation (18) and rearranging:

$$(\rho v + k)\frac{\partial^2 \hat{u}}{\partial y^2} - \sigma \hat{u} B_0^2 = \frac{\partial \hat{p}}{\partial \hat{x}} + \sigma E_z B_0 \tag{19}$$

Using the dimensionless variables from Equation (8) and some manipulations results in the following:

$$C\frac{d^2 u}{dy^2} - M^2 u = \frac{L^2}{\rho v U}\left(\frac{\partial \hat{p}}{\partial \hat{x}} + \sigma E_z B_0\right) \tag{20}$$

Here $C = (1+K)$. Approximating the term on the right-hand side of Equation (20):

$$\frac{M^2 \cosh\left(\frac{M}{2}\right)}{8\sinh^2\left(\frac{M}{4}\right)} = -\frac{L^2}{\rho v U}\left(\frac{\partial \hat{p}}{\partial \hat{x}} + \sigma E_z B_0\right) \tag{21}$$

Solving Equation (20) gives:

$$u(y) = \frac{1}{8}\left[\frac{\cosh\left(\frac{M}{2}\right)\left[\cosh\left(\frac{M}{2\sqrt{C}}\right) - \cosh\left(\frac{M}{\sqrt{C}}\left(y-\frac{1}{2}\right)\right)\right]}{\sinh^2\left(\frac{M}{4}\right)\cosh\left(\frac{M}{2\sqrt{C}}\right)}\right], \quad v = 0, \quad M \neq 0 \tag{22}$$

The inlet velocity profile for $M = 0$ is:

$$u(y) = \frac{1}{C}(y - y^2), \quad v = 0, \quad M = 0 \tag{23}$$

where $u(y)$ represents the steady velocity profile given by Equations (22) and (23). A sinusoidal time-dependent flow is considered for pulsatile flow:

$$u(y,t) = u(y)[1 + \epsilon \sin(2\pi t)], \quad v = 0 \tag{24}$$

Further, $u = 0$ and $v = 0$ (i.e., no-slip conditions) are considered on the walls. The proper boundary conditions for N on both the walls are:

$$N = -\left[s\frac{\partial u}{\partial y}\right]_{y=0}, \quad N = \left[s\frac{\partial u}{\partial y}\right]_{y=1} \tag{25}$$

where $0 \leq s \leq 1$. $s = 0$, $s = 1/2$, and $s = 1$ are for the flow with high concentration, weak concentration, and turbulence, respectively. $N = 0$ is considered for the inlet boundary condition of the micro-rotation velocity function. The outlet boundary conditions are set considering the flow fully developed.

2.3. Coordinates Transformation

Consider the following relation for transforming the coordinates:

$$\xi = x, \quad \eta = \frac{y - y_1(x)}{y_2(x) - y_1(x)} \tag{26}$$

For computation purposes, we mapped the constriction to a straight channel which resulted in mapping the domain $[y_1, y_2]$ to $[0, 1]$. Equations (15)–(17) on applying Equation (26) result as follows:

$$St\frac{\partial \omega}{\partial t} + u\left(\frac{\partial \omega}{\partial \xi} - Q\frac{\partial \omega}{\partial \eta}\right) + vD\frac{\partial \omega}{\partial \eta}$$
$$= \left(\frac{1+K}{Re}\right)\left[\frac{\partial^2 \omega}{\partial \xi^2} - (P - 2QR)\frac{\partial \omega}{\partial \eta} - 2Q\frac{\partial^2 \omega}{\partial \xi \partial \eta} + (Q^2 + D^2)\frac{\partial^2 \omega}{\partial \eta^2}\right] + \frac{D^2 M^2}{Re}\frac{\partial^2 \psi}{\partial \eta^2}\frac{\partial^2 \omega}{\partial \xi^2} \tag{27}$$
$$- \frac{K}{Re}\left[\frac{\partial^2 N}{\partial \xi^2} - (P - 2QR)\frac{\partial N}{\partial \eta} - 2Q\frac{\partial^2 N}{\partial \xi \partial \eta} + (Q^2 + D^2)\frac{\partial^2 N}{\partial \eta^2}\right]$$

$$St\frac{\partial N}{\partial t} + u\left(\frac{\partial N}{\partial \xi} - Q\frac{\partial N}{\partial \eta}\right) + vD\frac{\partial N}{\partial \eta}$$
$$= \frac{-K}{Re}\left[2N + \frac{\partial^2 \psi}{\partial \xi^2} - (P - 2QR)\frac{\partial \psi}{\partial \eta} - 2Q\frac{\partial^2 \psi}{\partial \xi \partial \eta} + (Q^2 + D^2)\frac{\partial^2 \psi}{\partial \eta^2}\right] \qquad (28)$$
$$+ \left(\frac{1}{Re} + \frac{K}{2Re}\right)\left[\frac{\partial^2 N}{\partial \xi^2} - (P - 2QR)\frac{\partial N}{\partial \eta} - 2Q\frac{\partial^2 N}{\partial \xi \partial \eta} + (Q^2 + D^2)\frac{\partial^2 N}{\partial \eta^2}\right]$$

$$\frac{\partial^2 \psi}{\partial \xi^2} - (P - 2QR)\frac{\partial \psi}{\partial \eta} - 2Q\frac{\partial^2 \psi}{\partial \xi \partial \eta} + (Q^2 + D^2)\frac{\partial^2 \psi}{\partial \eta^2} = -\omega \qquad (29)$$

where:
$$P = P(\xi, \eta) = \frac{\eta y_2''(\xi) + (1-\eta)y_1''(\xi)}{y_2(\xi) - y_1(\xi)}, \quad R = R(\xi) = \frac{y_2'(\xi) - y_1'(\xi)}{y_2(\xi) - y_1(\xi)},$$
$$Q = Q(\xi, \eta) = \frac{\eta y_2'(\xi) + (1-\eta)y_1'(\xi)}{y_2(\xi) - y_1(\xi)}, \quad D = D(\xi) = \frac{1}{y_2(\xi) - y_1(\xi)} \qquad (30)$$

The velocity components u and v becomes:
$$u = D(\xi)\frac{\partial \psi}{\partial \eta}, \quad v = Q(\xi, \eta)\frac{\partial \psi}{\partial \eta} - \frac{\partial \psi}{\partial \xi} \qquad (31)$$

The boundary conditions at the walls, in the (ξ, η) coordinate system for ψ, ω, and N are:

$$\psi(\eta, t) = \left[\frac{\sqrt{C}\cosh\left(\frac{M}{2}\right)\tanh\left(\frac{M}{2\sqrt{C}}\right)}{8M\sinh^2\left(\frac{M}{4}\right)}\right][1 + \epsilon \sin(2\pi t)], \quad \text{at } \eta = 0$$

$$\psi(\eta, t) = \frac{\cosh\left(\frac{M}{2}\right)}{8\sinh^2\left(\frac{M}{4}\right)}\left[1 - \frac{\sqrt{C}}{M}\tanh\left(\frac{M}{2\sqrt{C}}\right)\right][1 + \epsilon \sin(2\pi t)], \quad \text{at } \eta = 1$$

$$\omega = -\left[(Q^2 + D^2)\frac{\partial^2 \psi}{\partial \eta^2}\right]_{\eta=0,1}$$

$$N = -\left[sD^2 \frac{\partial^2 \psi}{\partial \eta^2}\right], \quad \text{at } \eta = 0$$

$$N = \left[sD^2 \frac{\partial^2 \psi}{\partial \eta^2}\right], \quad \text{at } \eta = 1 \qquad (32)$$

The value of ϵ determines the nature of the flow, where 0 and 1 represent the steady and pulsatile flows, respectively.

2.4. Numerical Method

The finite difference method was employed to acquire the numerical solution of Equations (27)–(29) over a uniform structured Cartesian grid (ξ_i, η_j). The solution at time level $l + 1 = l + \Delta t$, for $l = 0, 1, 2, \cdots$, was computed using the known solution at time level l. To obtain the solution at the time level $l + 1$, firstly, the space derivatives of Equation (29) were discretized using the central difference, and the resulting linear system was solved for $\psi = \psi(\xi, \eta)$ by the tri-diagonal matrix algorithm (TDMA) method. Then, Equations (27) and (28) were solved for the vorticity function $\omega = \omega(\xi, \eta)$ and micro-rotation function $N = N(\xi, \eta)$ by the alternating direction implicit (ADI) method. The over-relaxation parameter used for computations was $\lambda = 1.4$. The execution time of the calculations could be reduced by parallel implementation of the computer program, Ali and Syed [26]. However, developing a parallel solution on any shared, distributed, or hybrid memory programming paradigms is not a trivial task.

Equation (29) is discretized for the solution at advanced time level $l + 1$, and for $l = 0, 1, 2, \cdots$, is given by:

$$\frac{\psi_{i+1,j} - 2\psi_{i,j} + \psi_{i-1,j}}{(d\xi)^2} - \{P_{i,j} - 2Q_{i,j}R_i\}\frac{\psi_{i,j+1} - \psi_{i,j-1}}{2d\eta}$$
$$-2Q_{i,j}\frac{\psi_{i+1,j+1} - \psi_{i+1,j-1} - \psi_{i-1,j+1} + \psi_{i-1,j-1}}{4d\xi d\eta} \qquad (33)$$
$$+\left\{Q_{i,j}^2 + D_i^2\right\}\frac{\psi_{i,j+1} - 2\psi_{i,j} + \psi_{i,j-1}}{(d\eta)^2} = -\omega_{i,j}^l$$

For the sake of simplicity, the $l + 1$ superscript from ψ is removed. Rearranging, Equation (33) results as:

$$A(j)\psi_{i,j-1} + B(j)\psi_{i,j} + C(j)\psi_{i,j+1} = S(j) \quad (34)$$

where $A(j)$, $B(j)$, $C(j)$, and $S(j)$ are given as:

$$A(j) = \frac{P_{i,j} - 2Q_{i,j}R_i}{2d\eta} + \frac{Q_{i,j}^2 + D_i^2}{(d\eta)^2}$$

$$B(j) = -\frac{2}{(d\xi)^2} - \frac{2}{(d\eta)^2}\left\{Q_{i,j}^2 + D_i^2\right\}$$

$$C(j) = -\frac{P_{i,j} - 2Q_{i,j}R_i}{2d\eta} + \frac{Q_{i,j}^2 + D_i^2}{(d\eta)^2}$$

$$S(j) = -\omega_{i,j}^l - \left(\frac{\psi_{i+1,j} + \psi_{i-1,j}}{(d\xi)^2}\right) + 2Q_{i,j}\frac{\psi_{i+1,j+1} - \psi_{i-1,j+1} - \psi_{i+1,j-1} + \psi_{i-1,j-1}}{4d\eta d\xi} \quad (35)$$

The solution of Equation (27) was computed at $l + \frac{1}{2}$ time level by incorporating the solution of level l in the ADI method's first half. The explicit and implicit schemes at time levels l and $l + \frac{1}{2}$ in ξ-direction and η-direction, respectively, were used while discretizing the derivatives of ω.

$$\begin{aligned}
&St\left[\frac{\omega_{i,j}^{l+1/2} - \omega_{i,j}^l}{\Delta t/2}\right] + u_{i,j}\left[\frac{\omega_{i+1,j}^l - \omega_{i-1,j}^l}{2d\xi} - Q_{i,j}\frac{\omega_{i,j+1}^{l+1/2} - \omega_{i,j-1}^{l+1/2}}{2d\eta}\right] \\
&+ v_{i,j}D_i\frac{\omega_{i,j+1}^{l+1/2} - \omega_{i,j-1}^{l+1/2}}{2d\eta} = \left(\frac{1+K}{Re}\right)\left[\frac{\omega_{i+1,j}^l - 2\omega_{i,j}^l + \omega_{i-1,j}^l}{(d\xi)^2}\right. \\
&- \{P_{i,j} - 2Q_{i,j}R_i\}\frac{\omega_{i,j+1}^{l+1/2} - \omega_{i,j-1}^{l+1/2}}{2d\eta} - 2Q_{i,j}\frac{\omega_{i+1,j+1}^l - \omega_{i+1,j-1}^l - \omega_{i-1,j+1}^l + \omega_{i-1,j-1}^l}{4d\eta d\xi} \\
&+ \left.\left(Q_{i,j}^2 + D_i^2\right)\frac{\omega_{i,j+1}^{l+1/2} - 2\omega_{i,j}^{l+1/2} + \omega_{i,j-1}^{l+1/2}}{(d\eta)^2}\right] + \frac{M^2}{Re}D_i^2\frac{\psi_{i,j+1} - 2\psi_{i,j} + \psi_{i,j-1}}{(d\eta)^2} \\
&- \frac{K}{Re}\left[\frac{N_{i+1,j} - 2N_{i,j} + N_{i-1,j}}{(d\xi)^2} - \{P_{i,j} - 2Q_{i,j}R_i\}\frac{N_{i,j+1} - N_{i,j-1}}{2d\eta}\right. \\
&- \left.2Q_{i,j}\frac{N_{i+1,j+1} - N_{i+1,j-1} - N_{i-1,j+1} + N_{i-1,j-1}}{4d\eta d\xi} + \left(Q_{i,j}^2 + D_i^2\right)\frac{N_{i,j+1} - 2N_{i,j} + N_{i,j-1}}{(d\eta)^2}\right]
\end{aligned} \quad (36)$$

Equation (36) can be rearranged as:

$$A_1(j)\omega_{i,j-1}^{l+1/2} + B_1(j)\omega_{i,j}^{l+1/2} + C_1(j)\omega_{i,j+1}^{l+1/2} = S_1(j)$$

where $A_1(j)$, $B_1(j)$, $C_1(j)$, and $S_1(j)$ are given as:

$$A_1(j) = u_{i,j}\frac{Q_{i,j}}{2d\eta} - v_{i,j}\frac{D_i}{2d\eta} - \left(\frac{1+K}{Re}\right)\frac{P_{i,j} - 2Q_{i,j}R_i}{2d\eta} - \left(\frac{1+K}{Re}\right)\frac{Q_{i,j}^2 + D_i^2}{(d\eta)^2}$$

$$B_1(j) = \frac{St}{\Delta t/2} + \left(\frac{1+K}{Re}\right)\frac{2\left(Q_{i,j}^2 + D_i^2\right)}{(d\eta)^2} \quad (1)$$

$$C_1(j) = -u_{i,j}\frac{Q_{i,j}}{2d\eta} + v_{i,j}\frac{D_i}{2d\eta} + \left(\frac{1+K}{Re}\right)\frac{P_{i,j} - 2Q_{i,j}R_i}{2d\eta} - \left(\frac{1+K}{Re}\right)\frac{Q_{i,j}^2 + D_i^2}{(d\eta)^2} \quad (2)$$

$$S_1(j) = \left(\frac{u_{i,j}}{2d\xi} + \frac{1}{(d\xi)^2}\left(\frac{1+K}{Re}\right)\right)\omega^l_{i-1,j} + \left(-\frac{u_{i,j}}{2d\xi} + \frac{1}{(d\xi)^2}\left(\frac{1+K}{Re}\right)\right)\omega^l_{i+1,j}$$
$$+ \left(\frac{St}{\Delta t/2} + \frac{2}{(d\xi)^2}\left(\frac{1+K}{Re}\right)\right)\omega^l_{i,j} - \left(\frac{1+K}{Re}\right)2Q_{i,j}\frac{\omega^l_{i+1,j+1} - \omega^l_{i+1,j-1} - \omega^l_{i-1,j+1} + \omega^l_{i-1,j-1}}{4d\eta d\xi}$$
$$+ \frac{M^2 D_i^2}{Re}\frac{\psi_{i,j+1} - 2\psi_{i,j} + \psi_{i,j-1}}{(d\eta)^2} - \frac{K}{Re}\left[\frac{N_{i+1,j} - 2N_{i,j} + N_{i-1,j}}{(d\xi)^2} - \{P_{i,j} - 2Q_{i,j}R_i\}\frac{N_{i,j+1} - N_{i,j-1}}{2d\eta}\right.$$
$$\left. - 2Q_{i,j}\frac{N_{i+1,j+1} - N_{i+1,j-1} - N_{i-1,j+1} + N_{i-1,j-1}}{4d\eta d\xi} + \left(Q_{i,j}^2 + D_i^2\right)\frac{N_{i,j+1} - 2N_{i,j} + N_{i,j-1}}{(d\eta)^2}\right]$$

The ω at both the walls is given as:

$$\omega^l_{i,0} = -2\left[Q^2_{i,0} + D_i^2\right]\frac{\psi_{i,1} - \psi_{i,0}}{(d\eta)^2}$$
$$\omega^l_{i,m} = -2\left[Q^2_{i,m} + D_i^2\right]\frac{\psi_{i,m-1} - \psi_{i,m}}{(d\eta)^2} \tag{37}$$

In the second step of the ADI method, using the solution computed at $l + 1/2$ level, the solution was obtained at the $l + 1$ time level. The explicit and implicit schemes at time levels $l + 1/2$ and $l + 1$ in the η-direction and ξ-direction, respectively, were used while discretizing the derivatives of ω.

$$St\left[\frac{\omega^{l+1}_{i,j} - \omega^{l+1/2}_{i,j}}{\Delta t/2}\right] + u_{i,j}\left[\frac{\omega^{l+1}_{i+1,j} - \omega^{l+1}_{i-1,j}}{2d\xi} - Q_{i,j}\frac{\omega^{l+1/2}_{i,j+1} - \omega^{l+1/2}_{i,j-1}}{2d\eta}\right]$$
$$+ v_{i,j}D_i\frac{\omega^{l+1/2}_{i,j+1} - \omega^{l+1/2}_{i,j-1}}{2d\eta} = \left(\frac{1+K}{Re}\right)\left[\frac{\omega^{l+1}_{i+1,j} - 2\omega^{l+1}_{i,j} + \omega^{l+1}_{i-1,j}}{(d\xi)^2}\right.$$
$$- \{P_{i,j} - 2Q_{i,j}R_i\}\frac{\omega^{l+1/2}_{i,j+1} - \omega^{l+1/2}_{i,j-1}}{2d\eta} - 2Q_{i,j}\frac{\omega^{l+1/2}_{i+1,j+1} - \omega^{l+1/2}_{i+1,j-1} - \omega^{l+1/2}_{i-1,j+1} + \omega^{l+1/2}_{i-1,j-1}}{4d\eta d\xi}$$
$$\left. + \left(Q_{i,j}^2 + D_i^2\right)\frac{\omega^{l+1/2}_{i,j+1} - 2\omega^{l+1/2}_{i,j} + \omega^{l+1/2}_{i,j-1}}{(d\eta)^2}\right] + \frac{M^2 D_i^2}{Re}\frac{\psi_{i,j+1} - 2\psi_{i,j} + \psi_{i,j-1}}{(d\eta)^2}$$
$$- \frac{K}{Re}\left[\frac{N_{i+1,j} - 2N_{i,j} + N_{i-1,j}}{(d\xi)^2} - \{P_{i,j} - 2Q_{i,j}R_i\}\frac{N_{i,j+1} - N_{i,j-1}}{2d\eta}\right.$$
$$\left. - 2Q_{i,j}\frac{N_{i+1,j+1} - N_{i+1,j-1} - N_{i-1,j+1} + N_{i-1,j-1}}{4d\eta d\xi} + \left(Q_{i,j}^2 + D_i^2\right)\frac{N_{i,j+1} - 2N_{i,j} + N_{i,j-1}}{(d\eta)^2}\right] \tag{38}$$

Equation (38) can be written as:

$$A_2(i)\omega^{l+1}_{i-1,j} + B_2(i)\omega^{l+1}_{i,j} + C_2(i)\omega^{l+1}_{i+1,j} = S_2(i) \tag{39}$$

where $A_2(j), B_2(j), C_2(j)$, and $S_2(j)$ are given as:

$$A_2(i) = \frac{-u_{i,j}}{2d\xi} - \frac{1}{(d\xi)^2}\left(\frac{1+K}{Re}\right)$$

$$B_2(i) = \frac{St}{\Delta t/2} + \frac{2}{(d\xi)^2}\left(\frac{1+K}{Re}\right) \tag{3}$$

$$C_2(i) = \frac{u_{i,j}}{2d\xi} - \frac{1}{(d\xi)^2}\left(\frac{1+K}{Re}\right) \tag{4}$$

$$S_2(i) = \left(\frac{v_{i,j}D_i}{2d\eta} + \frac{P_{i,j}-2Q_{i,j}R_i}{2d\eta}\left(\frac{1+K}{Re}\right) + \frac{Q_{i,j}^2+D_i^2}{(d\eta)^2}\left(\frac{1+K}{Re}\right) - \frac{u_{i,j}Q_{i,j}}{2d\eta}\right)\omega_{i,j-1}^{l+1/2}$$
$$+ \left(-\frac{v_{i,j}D_i}{2d\eta} - \frac{P_{i,j}-2Q_{i,j}R_i}{2d\eta}\left(\frac{1+K}{Re}\right) + \frac{Q_{i,j}^2+D_i^2}{(d\eta)^2}\left(\frac{1+K}{Re}\right) + \frac{u_{i,j}Q_{i,j}}{2d\eta}\right)\omega_{i,j+1}^{l+1/2}$$
$$+ \left(\frac{St}{\Delta t/2} + \frac{2}{(d\xi)^2}\left(\frac{1+K}{Re}\right)\right)\omega_{i,j}^{l+1/2} - 2Q_{i,j}\frac{\omega_{i+1,j+1}^l - \omega_{i+1,j-1}^l - \omega_{i-1,j+1}^l + \omega_{i-1,j-1}^l}{4d\eta d\xi}$$
$$\left(\frac{1+K}{Re}\right) + \frac{M^2}{Re}D_i^2 \frac{\psi_{i,j+1} - 2\psi_{i,j} + \psi_{i,j-1}}{(d\eta)^2} - \frac{K}{Re}\left[\frac{N_{i+1,j} - 2N_{i,j} + N_{i-1,j}}{(d\xi)^2}\right.$$
$$- \{P_{i,j} - 2Q_{i,j}R_i\}\frac{N_{i,j+1} - N_{i,j-1}}{2d\eta} - 2Q_{i,j}\frac{N_{i+1,j-1} - N_{i+1,j-1} - N_{i-1,j+1} + N_{i-1,j-1}}{4d\eta d\xi}$$
$$\left. + \left(Q_{i,j}^2 + D_i^2\right)\frac{N_{i,j+1} - 2N_{i,j} + N_{i,j-1}}{(d\eta)^2}\right]$$

In a similar way, using the ADI method, the solution of Equation (28) was computed.

3. Results and Discussion

A grid of 400×50 was found to be suitable for the current work after a grid independence test was carried out for multiple grids with $-10 \leq \xi \leq 10$ and $0 \leq \eta \leq 1$. For ξ and η directions, we considered the step length 0.05 and 0.02, respectively. The constriction length, i.e., x_0, was 2. The height of the constriction on both walls was considered as 0.35. The time step, Δt, was taken as 0.0001. We considered $t = 0, 0.25, 0.50, 0.75$ to show the influence of the flow controlling parameters in a pulsatile cycle. These four time levels were corresponding to the specific states of the flow pulsation: $t = 0$ corresponded to the start of pulsation motion, $t = 0.25$ corresponded to the maximum flow rate, $t = 0.50$ corresponded to the minimum flow rate, and $t = 0.75$ corresponded to the instantaneous zero flow rate. The magnitude of the WSS was the same for the upper and lower walls. Therefore, the WSS distribution was depicted only on the upper wall in the study.

Figure 2 presented the axial velocity (u) profile and micro-rotation velocity (N) profile for the four pulsation cycles at different values of η and at the center of constriction ($x = 0$) with $M = 5$, $St = 0.02$, $K = 0.6$, and $Re = 700$. The phase-amplitude of u profile increased, whereas a decrease in the shifting phase was observed as the distance from the bottom wall increased. An opposite behavior for N was observed. For validity of the present scheme, Figure 3 compares the present study with Bandyopadhyay and Layek [22] for the WSS by varying magnetic field strength. The results were found to be promising in the comparison.

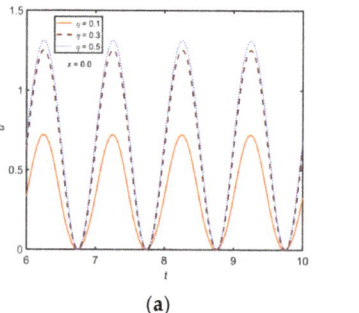
(a)

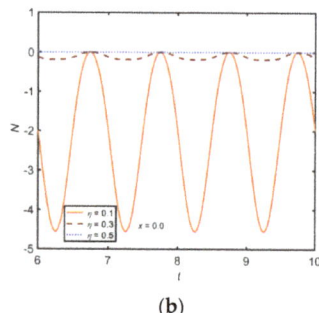
(b)

Figure 2. At $x = 0$ (a) the u profile and (b) the N profile is presented against η.

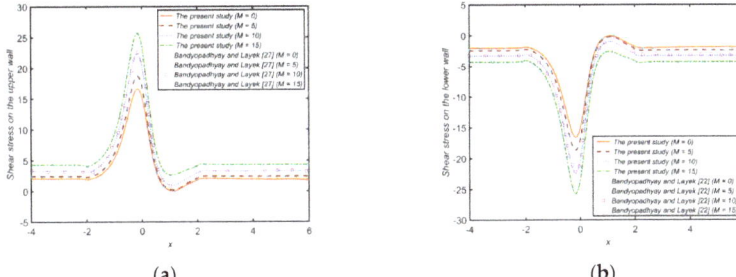

Figure 3. The WSS distribution on varying magnetic parameter values, (**a**) on the upper wall, and (**b**) on the lower wall.

The WSS on the upper wall in a pulsatile cycle at different times for $M = 0, 5, 10$, and 15 with $K = 0.6$, $St = 0.02$, and $Re = 700$ is shown in Figure 4. At $t = 0$, the WSS tended to increase with increasing M and attained its peak value at $x = 0$. The flow accelerated for $0 \leq t \leq 0.25$ during the pulse cycle. At $t = 0.25$, the WSS reached its extreme value at the maximum flow rate. During $0.25 < t < 0.75$, the flow started to decelerate, and the WSS decreased. The sign of the WSS changed at $t = 0.75$, when the net flow rate was zero. The flow separation region was maximum for $M = 0$, whereas it diminished for $M = 15$. The Hartmann number could be used to control the flow separation region.

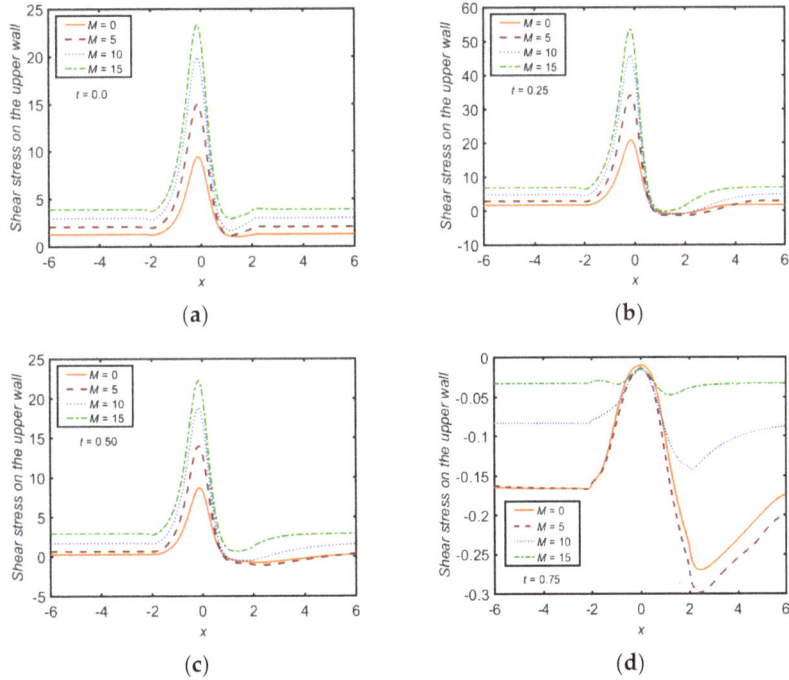

Figure 4. The WSS distribution on varying the values of M with $St = 0.02$, $Re = 700$, and $K = 0.6$ at (**a**) $t = 0$, (**b**) $t = 0.25$, (**c**) $t = 0.50$, and (**d**) $t = 0.75$.

The WSS on the upper wall in a pulsatile cycle at different times for $St = 0.02, 0.04$, 0.06, and 0.08 with $M = 5$, $K = 0.6$, and $Re = 700$ is shown in Figure 5. During $0 < t < 0.25$, the WSS increased with an increment in St. The flow separation region did not change significantly with an increase in the value of St. At $t = 0.25$, the WSS achieved the same peak value for all the values of St. During $0.25 < t < 0.75$, the flow started to decelerate, and the WSS decreased in the region with an increment in St and the flow separation region expanded. At $t = 0.75$, the sign of the WSS was changed.

The WSS at the four time levels for $K = 0.3, 0.6, 0.9$, and 1.2 with $St = 0.02$, $M = 5$, and $Re = 700$ is shown in Figure 6. The WSS fell with the increasing values of K during a complete cycle. During $0 < t < 0.25$, the flow separation region had an inverse relation with K. At $t = 0$, a decrease in the WSS was witnessed with increasing K. At $t = 0.25$, the WSS reached the maximum peak for all values of K. During $0.25 < t < 0.75$, the flow started to decelerate, and a decrease in the WSS for all the K's was observed. The flow separation region slightly expanded with the increasing values of K. The WSS altered its sign at $t = 0.75$.

The WSS at the four time levels for $Re = 500, 700, 900$, and 1100 with $St = 0.02$, $M = 5$, and $K = 0.6$ is shown in Figure 7. The WSS has an inciting trend towards Re. At $t = 0$, an increase in the WSS was witnessed with increasing Re. At $t = 0.25$, the WSS reached the maximum peak for all values of Re. The flow started to decelerate during $0.25 < t < 0.75$ and a decrease in the WSS for all the Re's was observed. The flow separation region expanded with the increasing values of Re. The WSS altered its sign at $t = 0.75$.

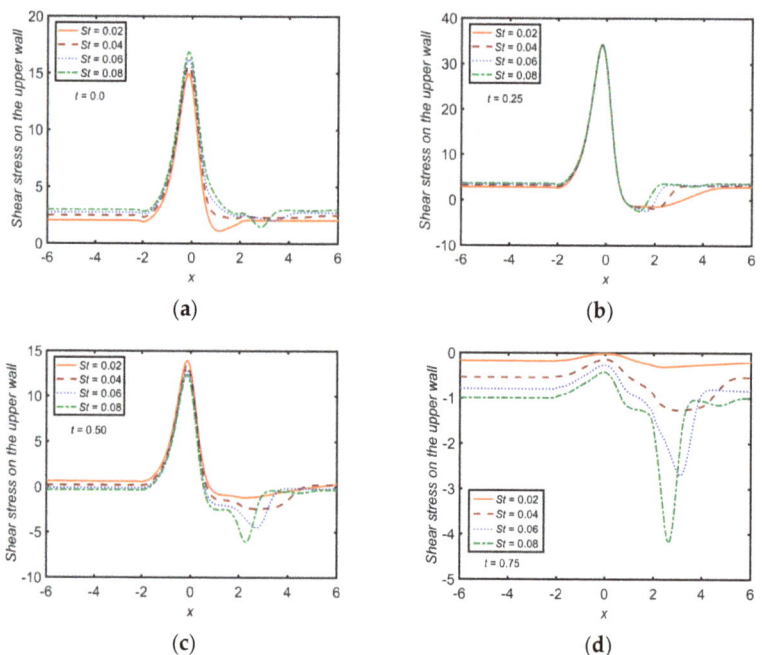

Figure 5. The WSS distribution on varying the values of St with $M = 5$, $K = 0.6$, and $Re = 700$ at (**a**) $t = 0$, (**b**) $t = 0.25$, (**c**) $t = 0.50$, and (**d**) $t = 0.75$.

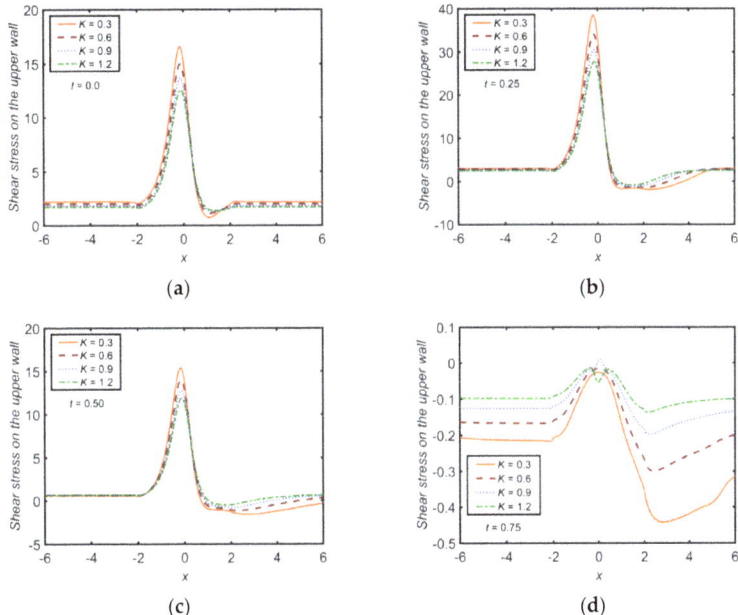

Figure 6. The WSS distribution on varying the values of K with $M = 5$, $St = 0.02$, and $Re = 700$ at (**a**) $t = 0$, (**b**) $t = 0.25$, (**c**) $t = 0.50$, and (**d**) $t = 0.75$.

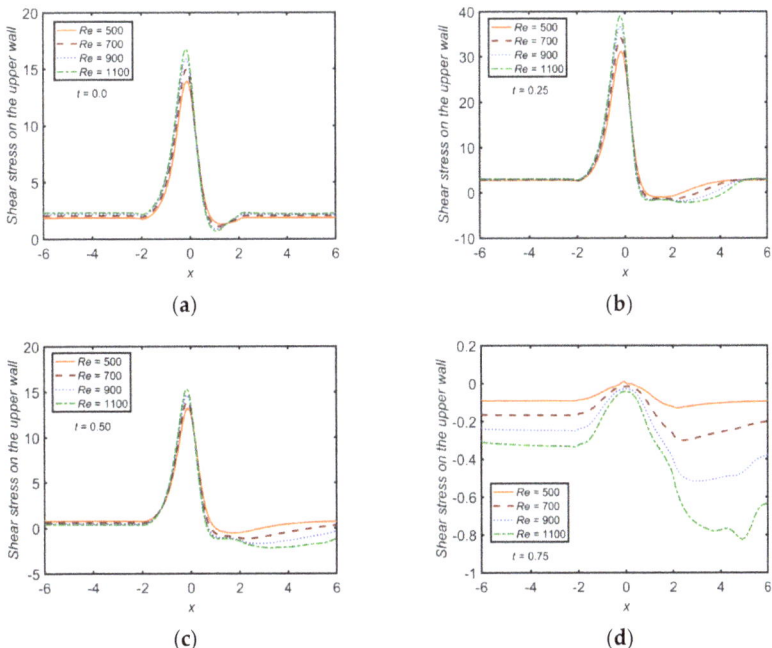

Figure 7. Impact on the WSS distribution on varying the values of Re with $M = 5$, $St = 0.02$, and $K = 0.6$ at (**a**) $t = 0$, (**b**) $t = 0.25$, (**c**) $t = 0.50$, and (**d**) $t = 0.75$.

Figure 8 displays the plots of u-velocity profiles at $x = -5, 2, 5$, and 7 for $M = 0, 5$, 10, and 15 with $K = 0.6$, $St = 0.02$, and $Re = 700$ at $t = 0.25, 0.5$, and 0.75. The u profiles became flattened as the value of M increased. There was flow symmetry at $x_0 = -2$ and $x_0 = 2$. The flow separation region appeared larger near the constriction for lower values of M. The flow separation region expanded at the end of the constriction. The backflow occurred near the walls when incoming flow tended to be zero at $t = 0.75$. Asymmetric behavior of the velocity contours can be seen in the case of $M = 0$. It was observed that in a complete pulsation cycle, the u profile decreased. This happened because a resistive force was produced due to the magnetic field, known as the Lorentz force, which opposed the fluid flow.

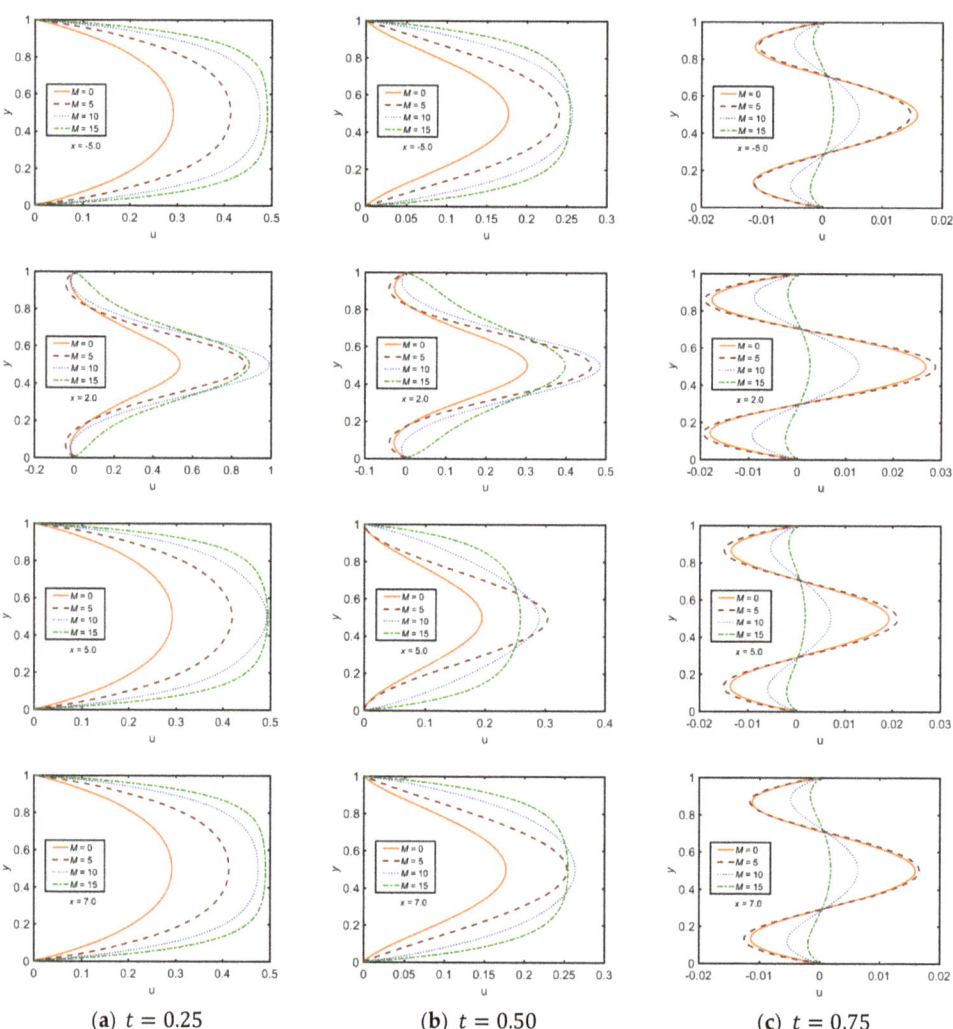

Figure 8. u profiles of the pulsatile flow at different instants of a pulsation cycle for different values of M at various x locations.

Figure 9a depicts the influence of K on N profile at $x = 0$ with fixed parameters $M = 5$, $St = 0.02$, $Re = 700$, and $K = 0.6$. The oscillations in the N profile were observed. The amplitude of N had a direct relation with K near the lower wall of the constricted channel, whereas it had an inverse relation with K near the upper wall. In Figure 9b,c, it can be seen that near the lower wall, the micro-rotation velocity boundary layer had a declining behavior towards Re and M. In contrast, it had inclining behavior towards Re and M near the upper wall.

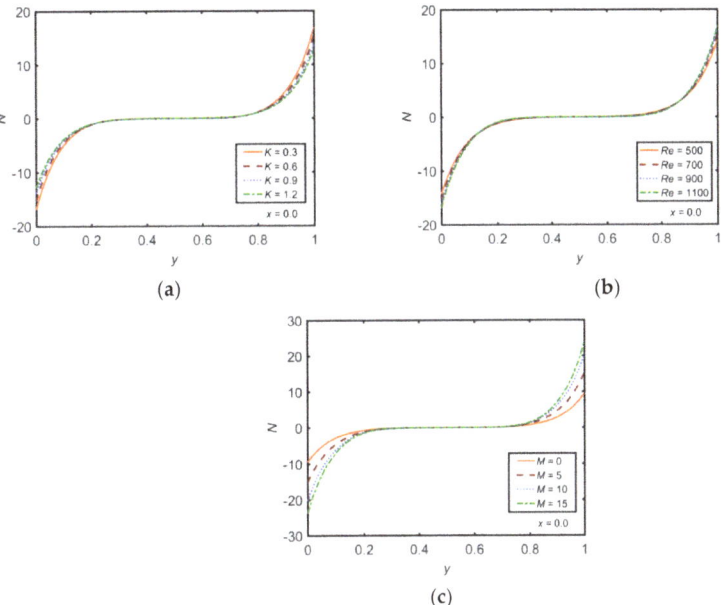

Figure 9. Effects of (**a**) micropolar parameter (**b**) Reynolds number, and (**c**) Hartmann number on micro-rotation velocity profile at $t = 0.25$.

Figure 10 presents the influence of M on the streamline at the four time levels. The flow separation region had a direct relation with M. At $t = 0$, the streamlines ran smoothly over the constriction. It can be seen that at $t = 0.25$ and $t = 0.50$, the flow separation region was decreased for larger values of M. At $t = 0.75$, vortices took up the largest portion of the channel. A symmetric behavior can be seen in the streamlines.

Figure 11 presents the effects of varying the Reynolds number on the streamline at the four time levels for $Re = 500, 700, 900$, and 1100 with $M = 5, K = 0.6$, and $St = 0.02$. The streamlines near the constriction were smooth. It is noted as well that the disturbance in the flow tended to grow, leading towards the turbulence, as the value of Re was increased. The flow separation region was maximum at $t = 0.25$ for $Re = 1100$, as can be seen in Figure 11d.

Figure 12 presents the vorticity of the flow for different values of Re at the four time levels. At $t = 0$, the vortex size had a direct relation with Re. Over time, the size of these vortices increased with increasing values of Re and eventually occupied a greater part of the channel downstream of the constrictions on increasing Re. The presence of the backflow was observed at $t = 0.75$. Symmetric behavior of contours could also be witnessed. Moreover, as the flow accelerated, a small vortex emerged near the constriction on the upper wall, which became larger with increasing values of St. The vortex that appeared near the constriction became smaller with increasing values of K.

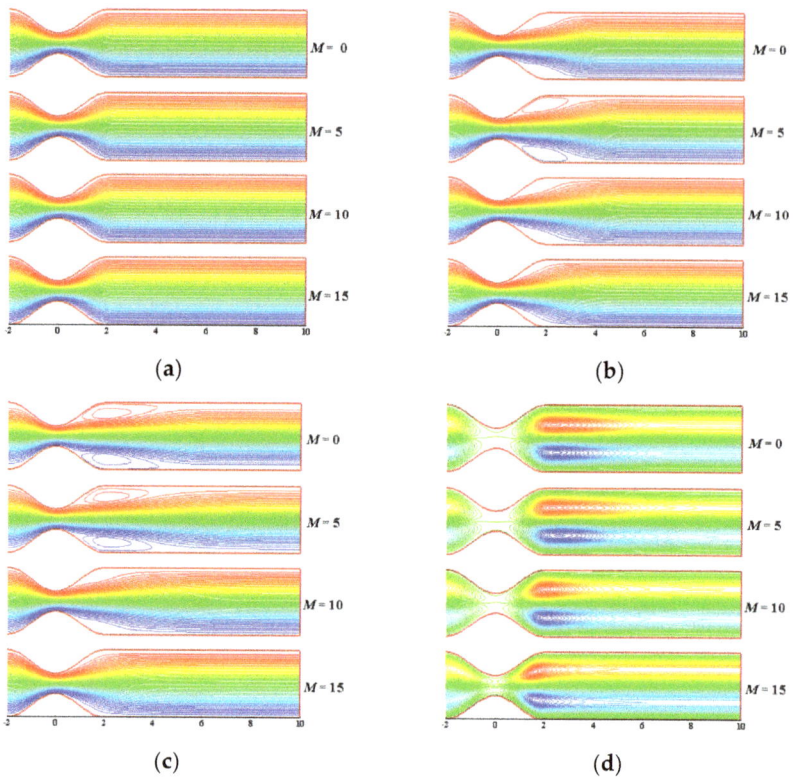

Figure 10. Streamlines of the pulsatile flow for different values of M with $St = 0.02$, $Re = 700$, and $K = 0.6$ at (**a**) $t = 0$, (**b**) $t = 0.25$, (**c**) $t = 0.50$, and (**d**) $t = 0.75$.

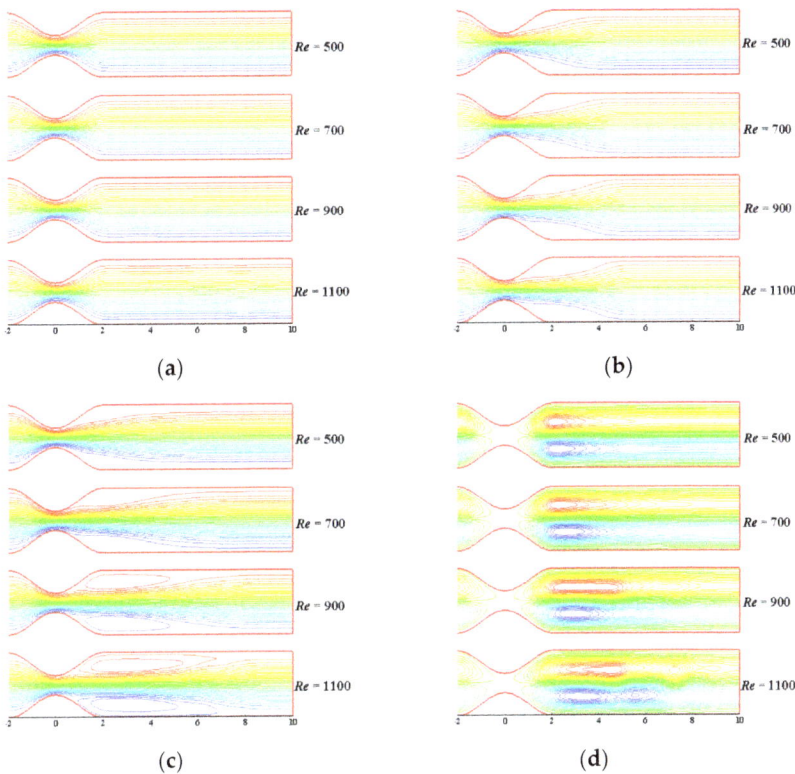

Figure 11. Streamlines of the pulsatile flow for different values of Re with $M = 5$, $St = 0.02$, and $K = 0.6$ at (**a**) $t = 0$, (**b**) $t = 0.25$, (**c**) $t = 0.50$, and (**d**) $t = 0.75$.

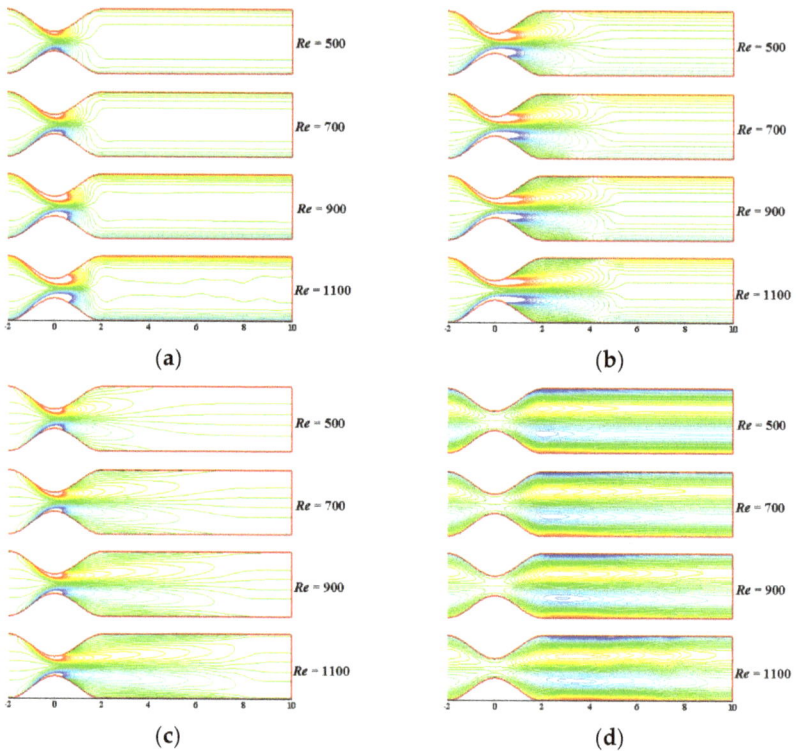

Figure 12. Vorticity contours of the pulsatile flow for different values of Re with $M = 5$, $St = 0.02$, and $K = 0.6$ at (**a**) $t = 0$, (**b**) $t = 0.25$, (**c**) $t = 0.50$, and (**d**) $t = 0.75$.

4. Conclusions

The pulsatile flow, in a constricted channel, of non-Newtonian MP fluid under the impact of the applied magnetic field was examined numerically on a Cartesian grid. The effects of M, St, Re, and K on the WSS, u, and N profiles were studied. The key outcomes of the present study are listed as follows:

- The direct relation between the WSS and M was observed, and the WSS attained its peak value at $t = 0.25$. The flow began to decelerate as the flow rate tended to be zero. The sign of the WSS changed when the net flow rate was zero;
- The WSS rose with an increasing value of St in the accelerating phase of the pulsation cycle. The WSS attained the same peak value for all values of St when $t = 0.25$;
- The WSS was reduced with the increasing values of K during a complete cycle and attained its peak value at $t = 0.25$;
- The dampening of the u profile was impacted by the magnetic field strength M. The u profile was asymmetric in the absence of an external magnetic field. However, the flow was symmetric at both ends of constrictions for $M = 0$. The flow parameters, M, Re, and K, had significant effects on N. The amplitude of the N profile had a direct relation with M and Re and an inverse relation with K;
- The streamlines became smoother as the value of M increases;
- The flow separation region was reduced by increasing M. The flow separation region had a direct relation with Re. The flow downstream of the constriction showed abrupt behavior as the value of Re increased.

The study found applications in understating the blood flow, modeled as non-Newtonian micropolar fluid, in stenotic arteries especially. The outcomes could be used in designing the biomedical devices and techniques for cardiovascular treatments.

Author Contributions: Conceptualization, A.A., M.U., and Z.A.; methodology, A.A., M.U., Z.A., Z.B., and A.S.; software, M.U. and A.S.; validation, Z.A. and G.S.; formal analysis, A.A., M.U., and Z.B.; investigation, A.A., M.U., and Z.B.; writing—original draft preparation, A.A. and M.U.; writing—review and editing, A.A. and G.S.; visualization, M.U., Z.B., and A.S.; supervision, A.A. and G.S.; project administration, A.A. and G.S. All authors have read and agreed to the published version of the manuscript.

Funding: There was no funding for the present research.

Institutional Review Board Statement: Not applicable.

Informed Consent Statement: Not applicable.

Data Availability Statement: All the required data is available in this manuscript.

Conflicts of Interest: The authors declare no conflict of interest.

Nomenclature

y_1	lower wall of the constricted channel
y_2	upper wall of the constricted channel
h_1	height of lower wall constriction
h_2	height of upper wall constriction
$\hat{u}$	velocity in $\check{x}$-direction
$\hat{v}$	velocity in $\check{y}$-direction
$\hat{p}$	pressure
ρ	density
μ	dynamic viscosity
k	vortex viscosity
E	electric field
Re	Reynolds number
M	Hartmann number
N	micro-rotation velocity
K	MP parameter
J	current density
B	magnetic field
B_0	strength of uniform magnetic field
σ	electrical conductivity
ν	kinematic viscosity
j	micro-interia density
γ	spin gradient velocity
s	switch for weak/strong concentrations/turbulence
L	length between channel walls
T	period of flow pulsation
St	Strouhal number
ψ	stream function
ω	vorticity function
ϵ	pulsating amplitude

References

1. Eringen, A.C. Theory of micropolar fluids. *J. Math. Mech.* **1966**, *16*, 1–18. [CrossRef]
2. Agarwal, R.S.; Bhargava, R.A.M.A.; Balaji, A.V. Numerical solution of flow and heat transfer of a micropolar fluid at a stagnation point on a porous stationary wall. *Indian J. Pure Appl. Math.* **1990**, *21*, 567–573.
3. Nazar, R.; Amin, N.; Filip, D.; Pop, I. Stagnation point flow of a micropolar fluid towards a stretching sheet. *Int. J. Nonlinear Mech.* **2004**, *39*, 1227–1235. [CrossRef]

4. Lok, Y.Y.; Amin, N.; Campean, D.; Pop, I. Steady mixed convection flow of a micropolar fluid near the stagnation point on a ver-tical surface. *Int. J. Numer. Methods Heat Fluid Flow* **2005**, *15*, 654–670. [CrossRef]
5. Chang, C.-L. Numerical simulation of micropolar fluid flow along a flat plate with wall conduction and buoyancy effects. *J. Phys. D: Appl. Phys.* **2006**, *39*, 1132–1140. [CrossRef]
6. Magyari, E.; Pop, I.; Valkó, P.P. Stokes' first problem for micropolar fluids. *Fluid Dyn. Res.* **2009**, *42*, 42. [CrossRef]
7. Babu, M.J.; Gupta, R.; Sandeep, N. Effect of radiation and viscous dissipation on stagnation-point flow of a micropolar fluid over a nonlinearly stretching surface with suction/injection. *Int. J. Sci. Basic Appl. Res.* **2015**, *7*, 73–82.
8. Turkyilmazoglu, M. Flow of a micropolar fluid due to a porous stretching sheet and heat transfer. *Int. J. Non-linear Mech.* **2016**, *83*, 59–64. [CrossRef]
9. Waqas, M.; Farooq, M.; Khan, M.I.; Alsaedi, A.; Hayat, T.; Yasmeen, T. Magnetohydrodynamic (MHD) mixed convection flow of micropolar liquid due to nonlinear stretched sheet with convective condition. *Int. J. Heat Mass Transf.* **2016**, *102*, 766–772. [CrossRef]
10. Ramadevi, B.; Kumar, K.A.; Sugunamma, V.; Reddy, J.V.R.; Sandeep, N. Magnetohydrodynamic mixed convective flow of mi-cropolar fluid past a stretching surface using modified Fourier's heat flux model. *J. Therm. Anal. Calorim.* **2020**, *139*, 1379–1393. [CrossRef]
11. Sheikh, N.A.; Ali, F.; Khan, I.; Saqib, M.; Khan, A. MHD Flow of Micropolar Fluid over an Oscillating Vertical Plate Embedded in Porous Media with Constant Temperature and Concentration. *Math. Probl. Eng.* **2017**, *2017*, 1–20. [CrossRef]
12. Hussanan, A.; Salleh, M.Z.; Khan, I.; Tahar, R.M. Heat and mass transfer in a micropolar fluid with Newtonian heating: An exact analysis. *Neural Comput. Appl.* **2018**, *29*, 59–67. [CrossRef]
13. Kumar, K.A.; Sugunamma, V.; Sandeep, N. Impact of nonlinear radiation on MHD non-aligned stagnation point flow of mi-cropolar fluid over a convective surface. *J. Nonequil. Thermody* **2018**, *43*, 327–345.
14. Shamsuddin, M.D.; Thumma, T. Numerical study of a dissipative micropolar fluid flow past an inclined porous plate with heat source/sink. *Propuls. Power Res.* **2019**, *8*, 56–68. [CrossRef]
15. Nadeem, S.; Malik, M.; Abbas, N. Heat transfer of three-dimensional micropolar fluid on a Riga plate. *Can. J. Phys.* **2020**, *98*, 32–38. [CrossRef]
16. Si, X.; Zheng, L.; Zhang, X.; Chao, Y. The flow of a micropolar fluid through a porous channel with expanding or contracting walls. *Open Phys.* **2011**, *9*, 825–834. [CrossRef]
17. Lu, D.; Kahshan, M.; Siddiqui, A.M. Hydrodynamical Study of Micropolar Fluid in a Porous-Walled Channel: Application to Flat Plate Dialyzer. *Symmetry* **2019**, *11*, 541. [CrossRef]
18. Fakour, M.; Vahabzadeh, A.; Ganji, D.; Hatami, M. Analytical study of micropolar fluid flow and heat transfer in a channel with permeable walls. *J. Mol. Liq.* **2015**, *204*, 198–204. [CrossRef]
19. Tutty, O.R. Pulsatile Flow in a Constricted Channel. *J. Biomech. Eng.* **1992**, *114*, 50–54. [CrossRef]
20. Mekheimer, K.S. Peristaltic Flow of a Magneto-Micropolar Fluid: Effect of Induced Magnetic Field. *J. Appl. Math.* **2008**, *2008*, 1–23. [CrossRef]
21. Hayat, T.; Ali, N. Effects of an endoscope on peristaltic flow of a micropolar fluid. *Math. Comput. Model.* **2008**, *48*, 721–733. [CrossRef]
22. Bandyopadhyay, S.; Layek, G. Study of magnetohydrodynamic pulsatile flow in a constricted channel. *Commun. Nonlinear Sci. Numer. Simul.* **2012**, *17*, 2434–2446. [CrossRef]
23. Khair, A.; Wanag, B.C.; Kuhn, D.C.S. Study of laminar-turbulent flow transition under pulsatile conditions in a constricted channel. *Int. J. Comput. Fluid D* **2015**, *29*, 447–463. [CrossRef]
24. Ali, A.; Farooq, H.; Abbas, Z.; Bukhari, Z.; Fatima, A. Impact of Lorentz force on the pulsatile flow of a non-Newtonian Casson fluid in a constricted channel using Darcy's law: A numerical study. *Sci. Rep.* **2020**, *10*, 1–15. [CrossRef]
25. Arjunon, S.; Ardana, P.H.; Saikrishnan, N.; Madhani, S.; Foster, B.; Glezer, A.; Yoganathan, A.P. Design of a Pulsatile Flow Facility to Evaluate Thrombogenic Potential of Implantable Cardiac Devices. *J. Biomech. Eng.* **2015**, *137*, 045001. [CrossRef] [PubMed]
26. Ali, A.; Syed, K.S. An Outlook of High Performance Computing Infrastructures for Scientific Computing. *Adv. Comput.* **2013**, *91*, 87–118. [CrossRef]

Article

Issues on Applying One- and Multi-Step Numerical Methods to Chaotic Oscillators for FPGA Implementation

Omar Guillén-Fernández [†], María Fernanda Moreno-López [†] and Esteban Tlelo-Cuautle [*,†]

Department of Electronics, INAOE, Tonantintla, Puebla 72840, Mexico; ing.omargufe@gmail.com (O.G.-F.); xerk.kun@gmail.com (M.F.M.-L.)
* Correspondence: etlelo@inaoep.mx; Tel.: +52-222-2470-517
† These authors contributed equally to this work.

Abstract: Chaotic oscillators have been designed with embedded systems like field-programmable gate arrays (FPGAs), and applied in different engineering areas. However, the majority of works do not detail the issues when choosing a numerical method and the associated electronic implementation. In this manner, we show the FPGA implementation of chaotic and hyper-chaotic oscillators from the selection of a one-step or multi-step numerical method. We highlight that one challenge is the selection of the time-step h to increase the frequency of operation. The case studies include the application of three one-step and three multi-step numerical methods to simulate three chaotic and two hyper-chaotic oscillators. The numerical methods provide similar chaotic time-series, which are used within a time-series analyzer (TISEAN) to evaluate the Lyapunov exponents and Kaplan–Yorke dimension (D_{KY}) of the (hyper-)chaotic oscillators. The oscillators providing higher exponents and D_{KY} are chosen because higher values mean that the chaotic time series may be more random to find applications in chaotic secure communications. In addition, we choose representative numerical methods to perform their FPGA implementation, which hardware resources are described and counted. It is highlighted that the Forward Euler method requires the lowest hardware resources, but it has lower stability and exactness compared to other one-step and multi-step methods.

Keywords: chaotic oscillator; one-step method; multi-step method; computer arithmetic; FPGA

Citation: Guillén-Fernández, O.; Moreno-López, M.F.; Tlelo-Cuautle, E. Issues on Applying One- and Multi-Step Numerical Methods to Chaotic Oscillators for FPGA Implementation. *Mathematics* 2021, 9, 151. https://doi.org/10.3390/math 9020151

Received: 8 December 2020
Accepted: 6 January 2021
Published: 12 January 2021

Publisher's Note: MDPI stays neutral with regard to jurisdictional claims in published maps and institutional affiliations.

Copyright: © 2021 by the authors. Licensee MDPI, Basel, Switzerland. This article is an open access article distributed under the terms and conditions of the Creative Commons Attribution (CC BY) license (https:// creativecommons.org/licenses/by/ 4.0/).

1. Introduction

Chaos is a nonlinear and unpredictable behavior that can be modeled by ordinary differential equations (ODEs). In continuous-time, the minimum number of ODEs for autonomous chaotic oscillators is three, as for example in [1,2]. A dynamical system modeled by four or more ODEs can generate hyper-chaotic behavior, as for example in [3]. Although sensitivity to initial conditions does not necessarily yield chaos [4], the majority of authors agree that the main characteristic of a dynamical system that generates chaos is the high sensibility to initial conditions, which is associated with a high unpredictability in the evolution of the time series of the state variables. The chaotic time series can be used to estimate Lyapunov exponents, as already shown in the seminal work [5], and by using the software for TIme SEries ANalysis (TISEAN) introduced in [6]. Lyapunov exponents are quite useful to characterize the behavior of a dynamical system, and they quantify the exponentially fast divergence or convergence of nearby orbits that can be seen in phase space.

Nowadays, it is said that a system with one positive Lyapunov exponent (LE+) is defined to be chaotic, and a system with more than one LE+ is hyper-chaotic. Some engineering applications of chaotic oscillators can be found in [7], which provides guidelines on the implementation by using field-programmable gate arrays (FPGAs), and shows the design of random number generators (RNGs) and chaotic secure communication systems [8]. The applications based on (hyper-)chaotic oscillators can be enhanced by guaranteeing higher unpredictability of the chaotic time series. One way is finding the chaotic oscillator having the highest LE+ [2], and also one must take into account other dynamical characteristics such as entropy and Kaplan–Yorke dimension (D_{KY}). For a chaotic oscillator having three ODEs, one computes three Lyapunov exponents, where one must be positive, one zero and one negative. There are methods to evaluate the Lyapunov spectrum [9,10], the seminal one was introduced in [5], and herein we apply TISEAN [6].

Recent works show the usefulness of chaotic oscillators in different engineering problems [11–13], however, there is no information on the issues related to the implementation of the numerical methods in electronic systems. In this manner, this paper uses three representative chaotic and two hyper-chaotic oscillators as case studies, which are listed in Table 1, along with their associated name, ODEs and parameter values that are used herein to generate chaotic behavior. The five chaotic oscillators are case studies to evaluate LE+ and D_{KY} from their chaotic time series that are generated by applying three one-step and three multi-step methods. Representative numerical methods are chosen to be implemented on a FPGA and their hardware resources are counted to show the challenges of minimizing hardware resources while guaranteeing the highest exactness and stability of the numerical simulations.

The three chaotic and two hyper-chaotic oscillators that are case studies in this paper are detailed in Section 2. Three one-step and three multi-step numerical methods are given in Section 3. The chaotic time series of each state variable of each (hyper-)chaotic oscillator are generated by applying all the numerical methods, and the LE+ and D_{KY} of each state variable are evaluated in Section 4. The FPGA implementation of representative numerical methods is detailed in Section 5. Finally, the conclusions are summarized in Section 6.

2. Chaotic and Hyper-Chaotic Oscillators

The three chaotic (modeled by three ODEs) and two hyper-chaotic (modeled by four ODEs) oscillators that are case studies herein are given in Table 1. In this Table CO1 is the well-known Lorenz system, introduced in 1963 as a simplified mathematical model for atmospheric convection [14], and from which was accidentally discovered the property associated to the high sensitivity to initial conditions. This originated one of the main characteristics in chaos theory and this CO1 is widely used as a work-horse to verify simulation and hardware implementation issues. In phase space, the Lorenz attractor resembles a butterfly effect, which stems from the real-world implications, i.e., in any physical system, the prediction of the evolution of the chaotic trajectories of the state variables will always fail in the absence of perfect knowledge of the initial conditions. In this manner, although physical systems can be completely deterministic, their chaotic behavior makes them inherently unpredictable (https://en.wikipedia.org/wiki/Lorenz_system#cite_note-lorenz-1).

The chaotic oscillator labeled as CO2 is another well-known system introduced by Otto Rössler in 1976, originally intended to behave similarly to the Lorenz attractor, but its dynamical behavior is simpler and has only one manifold. In the Rössler system, an orbit within the attractor follows an outward spiral around an unstable fixed point. From the mathematical model of CO2 given in Table 1, this spiral effect is seen in the x,y plane, and once the graph spirals out enough, the z-dimension shows the influence of a second fixed point causing rise and twist. After the introduction of the Rössler system, important news was that the original model was useful in modeling equilibrium in chemical reactions [15].

CO3 is based on a saturated nonlinear function series that can be approximated by a piecewise-linear (PWL) function. Considering that the PWL function has saturation levels k_i, break-points B_i and slope m, then Equation (1) can be used to generate two scrolls, and Equation (2) to generate three scrolls. In a general sense, the PWL function given in Equation (1) can be increased to generate an even number of scrolls, and Equation (2) to generate an odd number of scrolls, as shown in [16].

The hyper-chaotic oscillators labeled as HO4 and HO5, both have more than three ODEs in order to have more than one positive Lyapunov exponent, so that they present a more complex behavior than chaotic oscillators modeled by three ODEs.

Table 1. Chaotic and hyper-chaotic oscillators.

Name	ODEs	Parameters
CO1 [1]	$\dot{x} = \sigma(y - x)$ $\dot{y} = x(\rho - z) - y$ $\dot{z} = xy - \beta z$	$\sigma = 10, \beta = 8/3,$ $\rho = 28$
CO2 [1]	$\dot{x} = -y - z$ $\dot{y} = x + ay$ $\dot{z} = b + z(x - c)$	$a = b = 0.2,$ $c = 5.7$
CO3 [2]	$\dot{x} = y$ $\dot{y} = z$ $\dot{z} = -ax - by - cz$ $+ d_1 f(x, m)$	$a = 0.7, b = 0.7,$ $c = 0.7, d_1 = 0.7$
HO4 [3]	$\dot{x} = a(y - x) + yz + w$ $\dot{y} = by + cxz - px^2 + w$ $\dot{z} = xy - d$ $\dot{w} = -x - y$	$a = 16, b = 3,$ $c = 8, d = 20,$ $p = 0.1$
HO5 [17]	$\dot{x} = a(y - x) - w$ $\dot{y} = bx + 2y + xz$ $\dot{z} = c - xy$ $\dot{w} = x$	$a = 6, b = 5,$ $c = 50$

$$f(x) = \begin{cases} k_1 & \text{if } B_1 < x < B_2 \\ mx & \text{if } B_2 \leq x \leq B_3 \\ k_2 & \text{if } B_3 < x < B_4 \end{cases} \qquad (1)$$

$$f(x) = \begin{cases} k_1 & \text{if } B_1 < x < B_2 \\ m\left(x - \frac{B_2 + B_3}{2}\right) - \frac{k_1 + k_2}{2} & \text{if } B_2 \leq x \leq B_3 \\ k_2 & \text{if } B_3 < x < B_4 \\ m\left(x - \frac{B_4 + B_5}{2}\right) - \frac{k_2 + k_3}{2} & \text{if } B_4 \leq x \leq B_5 \\ k_3 & \text{if } B_5 < x < B_6 \end{cases} \qquad (2)$$

3. One-Step and Multi-Step Methods

The mathematical models of the chaotic and hyper-chaotic oscillators given in Table 1 can be formulated as initial value problems of the type $\dot{x} = f(x)$. The solution of the ODEs can be performed by applying one-step and multi-step methods. The former requires values evaluated in one step x_i to evaluate the next step denoted by x_{i+1}, while the multi-step methods require two or more previous step values denoted as $x_i, x_{i-1}, x_{i-2,...}$ to evaluate x_{i+1}. Other classifications are predictor or explicit and corrector or implicit methods. The explicit methods require past steps to evaluate the current step at iteration $i + 1$, but the implicit methods require estimation of the value at the current step $i + 1$ and past values at steps $x_i, x_{i-1}, x_{i-2,...}$. In this manner, it is common to name predictor–corrector [18] to the implicit methods, and they require an explicit method to evaluate the functions at the current iteration.

The explicit methods are faster than the implicit ones, but they may present numerical instability and lower exactness than the implicit methods. There are some rules for choosing the explicit method that is used within an implicit one to evaluate the current step x_{i+1} [18]. The step-size can also be varied during the computation or it can be constant and can be estimated from the stability analysis of the method, but one must take care of choosing the correct step-size to avoid non-convergence [19]. The explicit or predictor is the weak part in an implicit method due to the inherent truncation error, so that it puts a condition on the exactness of the initial prediction and the step-size of the corrector [18]. To enhance FPGA implementations of the numerical methods, the challenge is the selection of a method that allows a large step-size. That way, the larger the step-size of a numerical method, the higher the operating frequency of the FPGA implementation, as shown in Section 5.

The solution of the five chaotic oscillators given in the previous section are solved herein by applying the three one-step methods given in Table 2, and the three multi-step methods given in Table 3. The one-step methods are labeled as Forward Euler (FE), Backward Euler (BE) and fourth-order Runge–Kutta (RK4). The multi-step methods are labeled as sixth-order Adams–Bashforth (AB6), fourth-order Adams–Moulton (AM4) and fourth-order Gear (G4).

Table 2. One-step methods.

Method	Iterative Equation
Forward Euler (FE)	$y_{i+1} = y_i + hf(y_i, t_i)$
Backward Euler (BE)	$y_{i+1} = y_i + hf(y_{i+1}, t_{i+1})$
Runge–Kutta 4 (RK4)	$k_1 = hf(x_i, y_i)$ $k_2 = hf(x_i + \frac{1}{2}h, y_i + \frac{1}{2}k_1)$, $k_3 = hf(x_i + \frac{1}{2}h, y_i + \frac{1}{2}k_2)$, $k_4 = hf(x_i + h, y_i + k_3)$, $y_{i+1} = y_i + \frac{1}{6}(k_1 + 2k_2 + 2k_3 + k_4)$

Table 3. Multi-step methods.

Method	Iterative Equation
Adams–Bashforth 6 (AB6)	$y_{i+1} = y_i + \frac{h}{1440}(4277f(t_i, y_i)$ $-7923f(t_{i-1}, y_{i-1}) + 9982f(t_{i-2}, y_{i-2})$ $-7298f(t_{i-3}, y_{i-3}) + 2877f(t_{i-4}, y_{i-4})$ $-475f(t_{i-5}, y_{i-5}))$
Adams–Moulton 4 (AM4)	$y_{i+1} = y_i + \frac{h}{24}(9f(t_{i+1}, y_{i+1})$ $+19f(t_i, y_i) - 5f(t_{i-1}, y_{i-1})$ $+f(t_{i-2}, y_{i-2}))$
Gear 4 (G4)	$y_{i+1} = \frac{48}{25}y_i + -\frac{36}{25}y_{i-1}$ $+\frac{16}{25}y_{i-2} - \frac{3}{25}y_{i-3}$ $+\frac{12}{25}hf(y_{i+1})$

4. Chaotic Time Series, LE+ and D_{KY}

In Table 1, CO1 is the well-known Lorenz system, therefore, we show the simulation results for CO2, CO3, HO4 and HO5. The step-size h for each numerical method is given in the upper corner of each figure. One can appreciate that in some cases h is decreased to generate the same behavior provided by the majority of methods. Although the time evolution of the chaotic series is different for each method, the LE+ and D_{KY} are similar, and it can be improved by varying h, which is not a trivial task and requires the analysis of the eigenvalues associated to each Jacobian matrix of each equilibrium point of each chaotic oscillator.

Figures 1–4 show some chaotic time series of the (hyper-)chaotic oscillators simulated by applying the six numerical methods and listing the step-size h. The six methods were programmed into MatLab, and afterwards described in hardware language for FPGA implementation. In this case, a large h is desired to increase the operation frequency of an FPGA implementation, as shown in the following section.

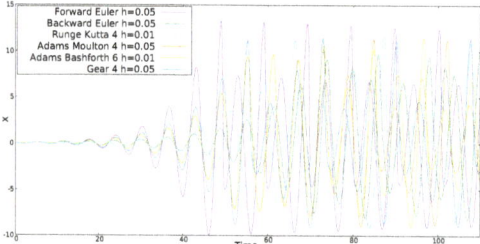

Figure 1. Time series of x of CO2 given in Table 1 with initial conditions $x_0 = y_0 = z_0 = 0.01$.

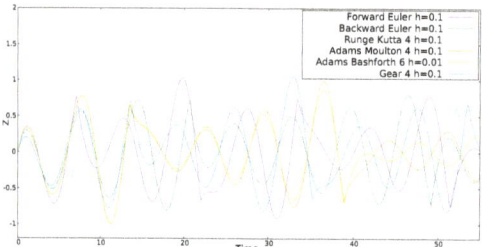

Figure 2. Time series of z of CO3 given in Table 1 with initial conditions $x_0 = y_0 = 0.1$, $z_0 = 0.0$.

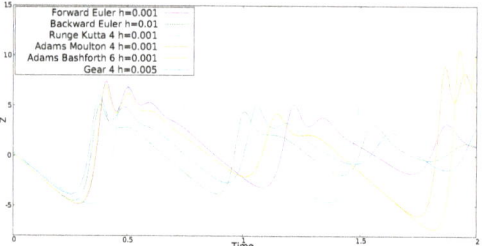

Figure 3. Time series of z of HO4 given in Table 1 with initial conditions $x_0 = y_0 = z_0 = w_0 = 0.2$.

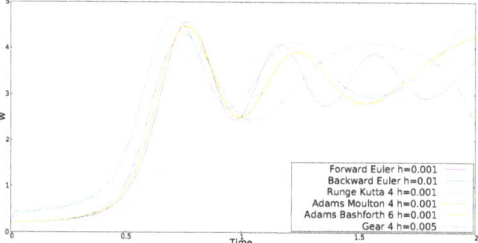

Figure 4. Time series of w of HO5 given in Table 1 with initial conditions $x_0 = y_0 = z_0 = w_0 = 0.2$.

The D_{KY} is evaluated from the Lyapunov exponents [9], and for an n-dimensional system it is evaluated by Equation (3), where $LE_1, \ldots, LE_n$ are Lyapunov exponent values ordered from the highest to the lowest value.

$$D_{KY} = (n-1) + \frac{LE_1 + LE_2 + LE_{n-1}}{|LE_n|} \quad (3)$$

The LE+ and D_{KY} were evaluated by TISEAN [6], which is based on the method introduced in [20]. The parameters for TISEAN are different for each state variable and analysis is performed using 50,000 samples for each chaotic time series. The LE+ for each state variable of each oscillator is shown in Table 4, and ordered from the highest to the lowest value. The highest LE+ is from the state variable x of HO4 and simulated with the fourth-order Runge–Kutta method, so that it is labeled as x_HO4_RK4. The same labels were adopted for the evaluation of D_{KY}, whose results are shown in Table 5.

Table 4. LE+ (ordered from the highest to the lowest) evaluated by time series analysis (TISEAN) for each state variable of the five oscillators given in Table 1, and for each numerical method.

Variable	LE+	Variable	LE+
x_HO4_RK4	0.48209	x_HO4_G4	0.05828
y_HO4_RK4	0.46388	y_HO4_BE	0.05650
w_HO4_FE	0.44930	x_HO5_BE	0.05628
z_HO4_FE	0.44903	z_HO4_BE	0.05575
x_HO4_FE	0.43356	y_HO5_FE	0.05489
w_HO4_RK4	0.43249	x_HO5_AM4	0.05380
z_HO4_AB6	0.42495	z_HO4_G4	0.05366
z_HO4_AM4	0.41541	w_HO5_BE	0.05221
w_HO4_AB6	0.41065	x_CO1_G4	0.05066
w_HO4_AM4	0.40906	x_CO1_AM4	0.04997
x_HO4_AM4	0.40645	w_HO5_G4	0.04912
x_HO4_AB6	0.40141	y_CO2_AB6	0.04814
z_HO4_RK4	0.38334	x_HO5_G4	0.04812
y_HO4_FE	0.37315	w_HO5_AM4	0.04610
y_HO4_AB6	0.36170	w_HO4_G4	0.04488
y_HO4_AM4	0.35865	y_HO5_AB6	0.04372
x_CO2_FE	0.29526	z_CO3_AM4	0.04275
z_HO5_FE	0.22398	y_HO5_AM4	0.04017
y_CO2_FE	0.20778	y_HO5_G4	0.03834
z_CO2_AM4	0.20292	z_CO3_G4	0.03809
z_CO2_G4	0.20292	y_HO4_G4	0.03730
z_HO5_RK4	0.15873	x_CO1_BE	0.03651
z_CO2_BE	0.15520	y_HO5_BE	0.03585
w_HO5_RK4	0.13860	y_CO1_RK4	0.03529
z_HO5_AB6	0.12956	z_CO3_BE	0.03187
z_CO3_RK4	0.11378	y_CO1_BE	0.02697
w_HO5_AB6	0.10937	y_CO1_FE	0.02547
y_CO2_BE	0.10362	y_CO3_FE	0.02457
w_HO4_BE	0.09907	x_CO2_G4	0.02429
x_HO5_RK4	0.09784	x_CO3_AB6	0.02294
z_HO5_G4	0.09718	y_CO3_RK4	0.02166
w_HO5_FE	0.09207	z_CO1_BE	0.02024

Table 4. Cont.

Variable	LE+	Variable	LE+
y_CO2_G4	0.09088	x_CO2_RK4	0.01952
y_CO1_AM4	0.08520	y_CO1_AB6	0.01944
z_HO5_AM4	0.08389	y_CO3_AM4	0.01918
z_CO2_FE	0.08388	x_CO2_AM4	0.01659
y_CO2_RK4	0.08253	x_CO1_AB6	0.01596
x_HO4_BE	0.08253	z_CO1_RK4	0.01593
x_HO5_AB6	0.08008	y_CO3_G4	0.01488
z_CO3_FE	0.08000	x_CO3_RK4	0.01414
z_CO1_AM4	0.07899	x_CO3_FE	0.01380
x_CO2_AB6	0.07732	x_CO2_BE	0.01352
y_CO1_G4	0.07696	z_CO1_AB6	0.01333
x_CO1_FE	0.07658	z_CO1_FE	0.01107
z_CO2_AB6	0.07475	z_CO3_AB6	0.01041
z_HO5_BE	0.07097	x_CO3_G4	0.01024
x_HO5_FE	0.06928	y_CO3_AB6	0.00988
z_CO2_RK4	0.06513	x_CO3_AM4	0.00959
x_CO1_RK4	0.06501	z_CO1_G4	0.00826
y_CO2_AM4	0.06107	x_CO3_BE	0.00788
y_HO5_RK4	0.06064	y_CO3_BE	0.00642

Table 5. D_{KY} (ordered from the highest to the lowest) evaluated by TISEAN for each state variable of the five oscillators given in Table 1, and for each numerical method.

Variable	D-KY	Variable	D-KY
x_HO5_BE	4.00000	z_CO2_RK4	3.00000
z_HO5_AM4	4.00000	x_CO1_AB6	2.98513
z_HO5_AB6	4.00000	x_CO1_BE	2.90439
z_HO5_G4	4.00000	y_HO5_BE	2.90363
w_HO5_BE	4.00000	y_CO2_BE	2.89287
z_HO4_G4	3.94577	y_HO5_AM4	2.87440
z_HO5_BE	3.92788	y_CO2_AB6	2.87170
w_HO5_G4	3.92303	y_CO1_G4	2.84422
w_HO5_AB6	3.92021	y_HO5_G4	2.81868
z_HO5_RK4	3.91847	z_CO2_FE	2.81799
w_HO5_AM4	3.90834	x_CO2_FE	2.73846
z_HO5_FE	3.90638	y_HO4_G4	2.73341
w_HO5_FE	3.88397	z_CO3_RK4	2.71019
x_HO5_AB6	3.86937	z_CO2_AM4	2.68730
x_HO5_RK4	3.86684	z_CO2_G4	2.68730
w_HO5_RK4	3.84980	y_CO2_G4	2.65563

Table 5. *Cont.*

Variable	D-KY	Variable	D-KY
x_HO5_AM4	3.80773	z_CO2_AB6	2.50008
w_HO4_RK4	3.71818	z_CO3_FE	2.37750
y_HO4_AM4	3.70668	x_CO3_AB6	2.37036
x_HO4_FE	3.68494	x_CO2_AB6	2.35488
x_HO4_AM4	3.68262	y_CO2_FE	2.28816
y_HO4_AB6	3.68177	y_CO2_AM4	2.26332
w_HO4_AB6	3.68161	z_CO3_AM4	2.24167
x_HO4_AB6	3.68044	x_CO3_RK4	2.20126
w_HO4_FE	3.67562	z_CO1_BE	2.18781
z_HO4_RK4	3.66694	z_CO1_RK4	2.17805
w_HO4_AM4	3.66422	x_CO3_FE	2.17802
z_HO4_AM4	3.65360	y_CO3_AM4	2.16888
x_HO4_G4	3.64872	z_CO3_BE	2.16674
y_HO4_FE	3.63942	z_CO3_G4	2.15179
z_HO4_AB6	3.62949	x_CO3_G4	2.14228
z_HO4_FE	3.62776	y_CO3_FE	2.13910
y_HO4_RK4	3.59111	x_CO3_AM4	2.13357
x_HO4_RK4	3.58378	y_CO1_RK4	2.12505
x_HO5_G4	3.57333	y_CO3_G4	2.11978
x_HO4_BE	3.54548	z_CO1_AB6	2.11110
y_HO5_RK4	3.53267	x_CO3_BE	2.09697
w_HO4_BE	3.44177	y_CO1_AM4	2.09374
y_HO5_FE	3.37649	z_CO3_AB6	2.08648
z_HO4_BE	3.30898	x_CO2_RK4	2.08539
y_HO5_AB6	3.30389	y_CO3_RK4	2.08183
x_HO5_FE	3.17620	z_CO1_AM4	2.06457
y_HO4_BE	3.16271	x_CO1_AM4	2.06136
w_HO4_G4	3.15787	z_CO1_FE	2.05527
x_CO1_FE	3.00000	y_CO1_FE	2.04314
x_CO1_RK4	3.00000	x_CO2_G4	2.03557
x_CO1_G4	3.00000	y_CO3_AB6	2.03262
y_CO1_BE	3.00000	z_CO1_G4	2.02645
y_CO1_AB6	3.00000	x_CO2_AM4	2.00114
y_CO2_RK4	3.00000	y_CO3_BE	2.00079
z_CO2_BE	3.00000	x_CO2_BE	1.98573

5. FPGA Implementation Issues

The development of engineering applications like chaotic secure communication systems and lightweight cryptography have positioned chaotic oscillators as a hot topic for research in this century. Nowadays, one can find implementations of chaotic systems using either analog or digital electronics, as already shown in [21]. This paper shows the implementation of (hyper-)chaotic oscillators from the selection of a numerical method, and by using FPGAs, which can be programmed/configured in the field after manufacture, and allow fast prototyping at relatively low development cost while providing good performance, computational power and programming flexibility.

Lets us consider the Lorenz oscillator (CO1) given in Table 1. The ODEs can be discretized by applying the most simple method known as Forward Euler (FE), to give the equations given in Equation (4). It is easy to see that these equations can be implemented by using multipliers, adders and subtractors. In addition, each block can be implemented including a clock (clk) and a reset (rst) pin to control the iterative process. As the multiplier consumes more power, if the multiplication includes a constant, as h, σ, ρ, β, one can design single-constant-multipliers (SCMs), as shown in [21], which use shift registers and adders to reduce power consumption and hardware resources. In this manner, the block description of Equation (4) is shown in Figure 5. The registers have an enable (en) pin and the description is divided into the macro-blocks labeled as Function Evaluation and Integrator FE. A counter is added to control the number of clks required in the FPGA implementation to evaluate the current iteration at $n + 1$, which is saved in the registers to process the next iteration.

$$\begin{aligned} x_{fe_{n+1}} &= x_n + h[\sigma(y_n - x_n)] \\ y_{fe_{n+1}} &= y_n + h[-x_n z_n + \rho x_n - y_n] \\ z_{fe_{n+1}} &= z_n + h[x_n y_n - \beta z_n] \end{aligned} \quad (4)$$

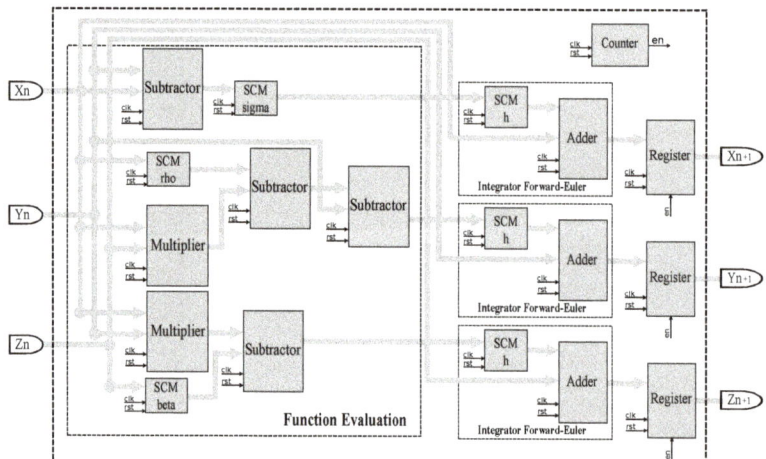

Figure 5. Block diagram of Equation (4) for field-programmable gate array (FPGA) implementation applying Forward Euler (FE).

The discretization of CO1 by applying an implicit method like Backward Euler (BE) is given in Equation (5), where it can be appreciated that the predictor is the FE given in Equation (4) to evaluate $x_{fe_{n+1}}, x_{fe_{n+1}}, x_{fe_{n+1}}$. The block description for FPGA implementation is more complex and it embeds the FE method as shown in Figure 6. One can infer that the hardware resources for the BE method almost double compared to FE.

$$\begin{aligned} x_{n+1} &= x_n + h[\sigma(y_{fe_{n+1}} - x_{fe_{n+1}})] \\ y_{n+1} &= y_n + h[-x_{fe_{n+1}} z_{fe_{n+1}} + \rho x_{fe_{n+1}} - y_{fe_{n+1}}] \\ z_{n+1} &= z_n + h[x_{fe_{n+1}} y_{fe_{n+1}} - \beta z_{fe_{n+1}}] \end{aligned} \quad (5)$$

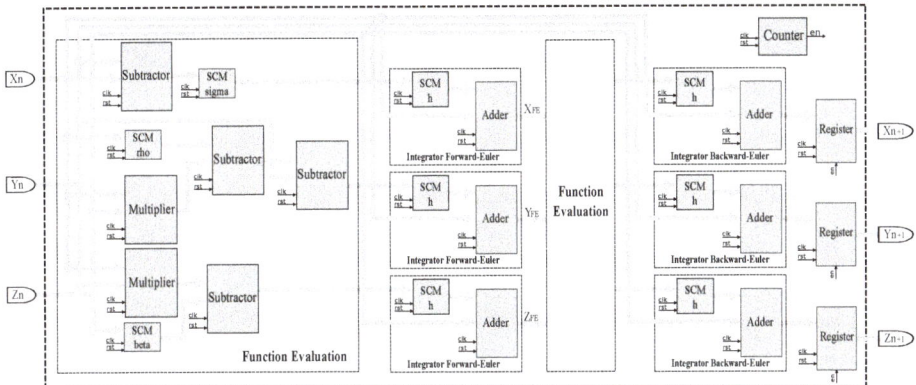

Figure 6. Block diagram description of Equation (5) applying Backward Euler (BE) highlighting function evaluation, integrator Forward Euler and integrator Backward Euler blocks.

The application of other one-step and multi-step methods to discretize a (hyper-)chaotic oscillator is performed in a similar manner as for the FE and BE methods. For example, the application of the multi-step sixth-order Adams–Bashforth (AB6) method is more complex than FE or BE. It requires five past steps associated to $f(n), f(n-1)$, $f(n-2), f(n-3), f(n-4), f(n-5)$ that can be evaluated by the 4th-order Runge–Kutta (RK4) method. In this manner, using the iterative equation associated to AB6 that is given in Table 3, the discrete equations of CO1 are given in Equation (6). Figure 7 shows the block description for the FPGA implementation of CO1. One can see the predictor RK4, function evaluation and integrator Adams–Bashforth blocks, which are designed as for the FE and BE methods. The evaluation of Equation (6) also requires a finite-state machine (FSM) to control the iterative process, a cumulative sum block to process the RK4 method and random access memories (RAMs) to save the past steps $f(n), f(n-1), f(n-2), f(n-3), f(n-4), f(n-5)$ that are required for the next iteration, and they are controlled by STP (StarT Prediction) and EOP (End Of Prediction). The predictor RK4 is disconnected after the first iteration, which is controlled by the FSM.

$$\begin{aligned} x_{ab6_{n+1}} &= x_n + h/1440[4277f(n) - 7923f(n-1) + 9982f(n-2) - 7298f(n-3) + 2877f(n-4) - 475f(n-5)] \\ y_{ab6_{n+1}} &= y_n + h/1440[4277f(n) - 7923f(n-1) + 9982f(n-2) - 7298f(n-3) + 2877f(n-4) - 475f(n-5)] \\ z_{ab6_{n+1}} &= z_n + h/1440[4277f(n) - 7923f(n-1) + 9982f(n-2) - 7298f(n-3) + 2877f(n-4) - 475f(n-5)] \end{aligned} \quad (6)$$

In all the previous cases, the FPGA synthesis can be performed by adopting computer arithmetic of fixed-point notation, where the number of bits depends on the amplitudes of the state variables, as detailed in [7], where one can also find guidelines on Very High Speed Integrated Circuit Hardware Description Language (VHDL) programming. In this paper, the fixed-point notation has the format 12.20. Table 6 summarizes the hardware resources for the implementation of CO1, CO2, CO3, HO4 and HO5 applying FE and using FPGA Cyclone IV EP4CGX150DF31C7 under the synthesizer of software "Quartus II 13.0". In the same table, the last two rows provide the number of clk cycles that are required to evaluate one iteration n, and the latency is given in nanoseconds when using a 50 MHz clk signal.

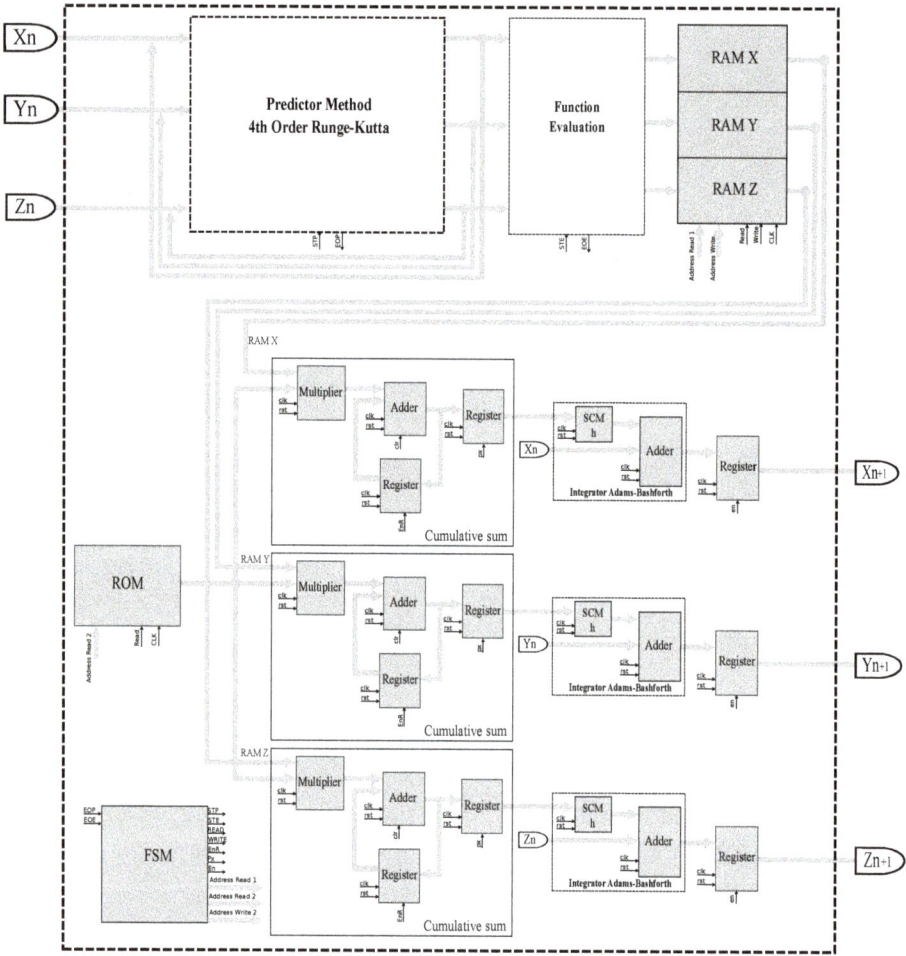

Figure 7. Block diagram description of Equation (6) for FPGA implementation applying AB6 highlighting function evaluation, integrator Adams–Bashforth, cumulative sum, finite-state machine (FSM) and RK4.

Table 6. Hardware resources using FPGA Cyclone IV EP4CGX150DF31C7 and applying FE to CO1, CO2, CO3, HO4 and HO5.

Resources	CO1	CO2	CO3	HO4	HO5	Available
Logic elements	1295	1083	2567	2554	1707	149,760
Registers	654	565	588	1591	1045	149,760
9*9 bit multipliers	16	8	8	135	92	720
Max freq (MHz)	90.88	102.75	58.55	79.77	82.7	50
Clock cycles by iteration	5	7	9	12	9	-
Latency (ns)	100	140	180	240	180	-

Table 7 shows the hardware resources for CO1 using FPGA Cyclone IV EP4CGX150DF31C7 under "Quartus II 13.0" and by applying the three one-step (FE, BE, RK4) and three multi-step (AB6, AM4, G4) methods. As supposed, FE requires the lowest hardware resources and clks to process one iteration. The use of SCMs makes a considerable reduction on the number of multipliers. Although RK4 requires almost four times the hardware resources than FE, it is more exact and allows a higher h [7]. AB6 requires the higher number of hardware resources compared to the other five methods. If one does not design an SCM, the VHDL description of AB6 will require more than the 720 available multipliers in the FPGA Cyclone IV, and therefore it may not be implemented on this FPGA, so that one must use an FPGA with more density resources.

Table 7. Hardware resources for CO1 using FPGA Cyclone IV and applying different methods.

Resources	FE	BE	RK4	AB6	AM4	G4	Available
Logic elements	1295	1988	4708	8512	7684	7220	149,760
Registers	654	1160	2662	4232	3856	3484	149,760
Multipliers	16	32	208	325	290	274	720
Freq (MHz)	90.88	92.59	84.77	83.53	84.18	82.73	50
Clks/iteration	5	11	32	190	130	100	-
Latency (ns)	100	220	640	3800	2600	2000	-

The hardware resources for the FPGA implementation of the remaining chaotic systems labeled as CO2, CO3, HO4 and HO5, have similar increases for each numerical method, the main difference being due to the number of ODEs and nonlinear functions.

Figure 8 shows a representative case of the FPGA implementation of CO1 applying the one-step method BE, and Figure 9 shows the application of the multi-step method G4, considering $h = 0.001$ in both cases.

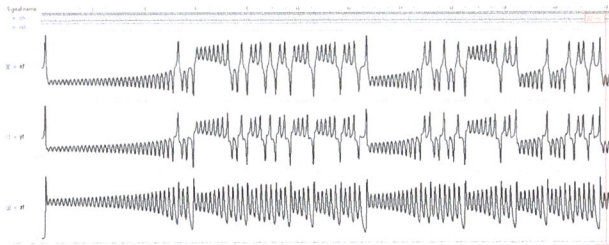

Figure 8. FPGA simulation of CO1 applying BE.

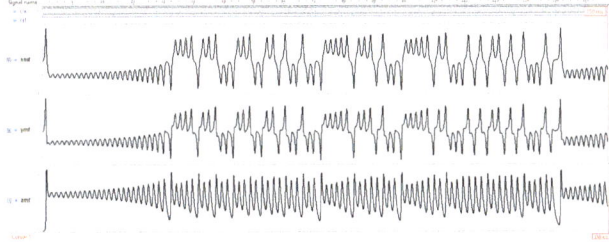

Figure 9. FPGA simulation of CO1 applying G4.

6. Conclusions

We have shown the issues with the FPGA implementation of chaotic and hyperchaotic oscillators from the selection of a one-step and multi-step numerical method. The challenge is the selection of the time-step h to increase the frequency of operation of the FPGA design. It was appreciated that each one-step or multi-step method requires different hardware resources, so that trade-offs arise among reducing hardware resources, improving exactness and maximum operation frequency. Another open problem is the selection of the best chaotic oscillator, which can be done by evaluating the LE+ and D_{KY}. This last characteristic increases as the number of ODEs increases, so that according to the results provided by TISEAN, the hyper-chaotic oscillators have the higher LE+ and D_{KY} values. The FPGA implementation of the Lorenz system CO1 showed good agreement on the time series generated by applying BE and G4 methods, and using 32 bits in fixed-point notation of 12.30. The exactness can also be accomplished through using more bits, so that one can enhance applications in chaotic secure communications and the Internet of Things (IoT) to guarantee security and privacy. In particular, the IoT application requires a connectivity protocol in which chaotic oscillators can be synchronized to mask the data being transmitted, like in the extremely lightweight publish/subscribe messaging transport known as MQTT (mqtt.org), which is ideal for connecting remote devices with a small code footprint and minimal network bandwidth.

Author Contributions: Conceptualization O.G.-F., M.F.M.-L. and E.T.-C.; software O.G.-F. and M.F.M.-L.; validation O.G.-F., M.F.M.-L. and E.T.-C.; investigation O.G.-F. and M.F.M.-L.; resources O.G.-F. and M.F.M.-L.; writing—original draft preparation O.G.-F. and M.F.M.-L.; writing—review and editing O.G.-F., M.F.M.-L. and E.T.-C.; supervision E.T.-C. All authors have read and agreed to the published version of the manuscript.

Funding: This research received no external funding.

Conflicts of Interest: The authors declare no conflicts of interest.

References

1. Fuchs, A. *Nonlinear Dynamics in Complex Systems*, 1st ed.; Springer: Berlin/Heidelberg, Germany, 2013. [CrossRef]
2. Muñoz-Pacheco, J.M.; Guevara-Flores, D.K.; Félix-Beltrán, O.G.; Tlelo-Cuautle, E.; Barradas-Guevara, J.E.; Volos, C.K. Experimental Verification of Optimized Multiscroll Chaotic Oscillators Based on Irregular Saturated Functions. *Complexity* **2018**, *2018*, 3151840. [CrossRef]
3. Vaidyanathan, S.; Lien, C.H.; Fuadi, W.; Mujiarto; Mamat, M.; Subiyanto. A New 4-D Multi-Stable Hyperchaotic Two-Scroll System with No-Equilibrium and its Hyperchaos Synchronization. *J. Phys. Conf. Ser.* **2020**, *1477*, 022018. [CrossRef]
4. Shang, Y. Deffuant model with general opinion distributions: First impression and critical confidence bound. *Complexity* **2013**, *19*, 38–49. [CrossRef]
5. Wolf, A.; Swift, J.B.; Swinney, H.L.; Vastano, J.A. Determining Lyapunov exponents from a time series. *Phys. D Nonlinear Phenom.* **1985**, *16*, 285–317. [CrossRef]
6. Hegger, R.; Kantz, H.; Schreiber, T. Practical implementation of nonlinear time series methods: The TISEAN package. *Chaos Interdiscip. J. Nonlinear Sci.* **1999**, *9*, 413–435. [CrossRef] [PubMed]
7. Tlelo-Cuautle, E.; Rangel-Magdaleno, J.; de la Fraga, L.G. *Engineering Applications of FPGAs*; Springer International Publishing: Cham, Switzerland, 2016; pp. 1–222. [CrossRef]
8. Fountain, D.M.; Kolias, A.G.; Laing, R.J.; Hutchinson, P.J. The financial outcome of traumatic brain injury: A single centre study. *Br. J. Neurosurg.* **2017**, *31*, 350–355. [CrossRef] [PubMed]
9. Chen, H.; Bayani, A.; Akgul, A.; Jafari, M.A.; Pham, V.T.; Wang, X.; Jafari, S. A flexible chaotic system with adjustable amplitude, largest Lyapunov exponent, and local Kaplan—Yorke dimension and its usage in engineering applications. *Nonlinear Dyn.* **2018**, *92*, 1791–1800. [CrossRef]
10. Yakovleva, T.V.; Kutepov, I.E.; Karas, A.Y.; Yakovlev, N.M.; Dobriyan, V.V.; Papkova, I.V.; Zhigalov, M.V.; Saltykova, O.A.; Krysko, A.V.; Yaroshenko, T.Y.; et al. EEG Analysis in Structural Focal Epilepsy Using the Methods of Nonlinear Dynamics (Lyapunov Exponents, Lempel-Ziv Complexity, and Multiscale Entropy). *Sci. World J.* **2020**, *2020*, 8407872. [CrossRef] [PubMed]
11. Akhmet, M.; Tleubergenova, M.; Zhamanshin, A. Inertial Neural Networks with Unpredictable Oscillations. *Mathematics* **2020**, *8*, 1797. [CrossRef]
12. Jia, H.; Guo, C. The Application of Accurate Exponential Solution of a Differential Equation in Optimizing Stability Control of One Class of Chaotic System. *Mathematics* **2020**, *8*, 1740. [CrossRef]
13. Lin, C.H.; Hu, G.H.; Yan, J.J. Chaos Suppression in Uncertain Generalized Lorenz–Stenflo Systems via a Single Rippling Controller with Input Nonlinearity. *Mathematics* **2020**, *8*, 327. [CrossRef]
14. Lorenz, E.N. Deterministic nonperiodic flow. *J. Atmos. Sci.* **1963**, *20*, 130–141. [CrossRef]
15. Gosar, Z. Chaotic Dynamics—Rössler System. Unpublished, 2011. [CrossRef]
16. Tlelo-Cuautle, E.; Quintas-Valles, A.D.J.; de la Fraga, L.G.; Rangel-Magdaleno, J.D.J. VHDL Descriptions for the FPGA Implementation of PWL-Function-Based Multi-Scroll Chaotic Oscillators. *PLoS ONE* **2016**, *11*, e0168300. [CrossRef] [PubMed]
17. Vaidyanathan, S.; Tlelo-Cuautle, E.; Munoz-Pacheco, J.M.; Sambas, A. A new four-dimensional chaotic system with hidden attractor and its circuit design. In Proceedings of the 2018 IEEE 9th Latin American Symposium on Circuits & Systems (LASCAS), Puerto Vallarta, Mexico, 25–28 February 2018; pp. 1–4. [CrossRef]
18. Chapra, S.C.; Canale, R.P. *Numerical Methods for Engineers*, 5th ed.; McGraw-Hill/Interamericana: New York, NY, USA, 2006.

19. Tannehill, J.C.; Anderson, D.A.; Pletcher, R.H. *Computational Fluid Mechanics and Heat Transfer*, 2nd ed.; Taylor & Francis: Boca Raton, FL, USA, 1997; pp. 1–740.
20. Sano, M.; Sawada, Y. Measurement of the Lyapunov Spectrum from a Chaotic Time Series. *Phys. Rev. Lett.* **1985**, *55*, 1082–1085. [CrossRef] [PubMed]
21. Tlelo-Cuautle, E.; Pano-Azucena, A.D.; Guillén-Fernández, O.; Silva-Juárez, A. *Analog/Digital Implementation of Fractional Order Chaotic Circuits and Applications*; Springer: Berlin/Heidelberg, Germany, 2020

Article

Hybrid Nanofluid Flow over a Permeable Non-Isothermal Shrinking Surface

Iskandar Waini [1,2], Anuar Ishak [2,*] and Ioan Pop [3]

[1] Fakulti Teknologi Kejuruteraan Mekanikal dan Pembuatan, Universiti Teknikal Malaysia Melaka, Hang Tuah Jaya, Durian Tunggal 76100, Melaka, Malaysia; iskandarwaini@utem.edu.my

[2] Department of Mathematical Sciences, Faculty of Science and Technology, Universiti Kebangsaan Malaysia, UKM Bangi 43600, Selangor, Malaysia

[3] Department of Mathematics, Babeş-Bolyai University, 400084 Cluj-Napoca, Romania; ipop@math.ubbcluj.ro

* Correspondence: anuar_mi@ukm.edu.my

Abstract: In this paper, we examine the influence of hybrid nanoparticles on flow and heat transfer over a permeable non-isothermal shrinking surface and we also consider the radiation and the magnetohydrodynamic (MHD) effects. A hybrid nanofluid consists of copper (Cu) and alumina (Al_2O_3) nanoparticles which are added into water to form Cu-Al_2O_3/water. The similarity equations are obtained using a similarity transformation and numerical results are obtained via bvp4c in MATLAB. The results show that dual solutions are dependent on the suction strength of the shrinking surface; in addition, the heat transfer rate is intensified with an increase in the magnetic parameter and the hybrid nanoparticles volume fractions for higher values of the radiation parameter. Furthermore, the heat transfer rate is higher for isothermal surfaces as compared with non-isothermal surfaces. Further analysis proves that the first solution is physically reliable and stable.

Keywords: hybrid nanofluid; heat transfer; non-isothermal; shrinking surface; MHD; radiation

Citation: Waini, I.; Ishak, A.; Pop, I. Hybrid Nanofluid Flow over a Permeable Non-Isothermal Shrinking Surface. *Mathematics* **2021**, *9*, 538. https://doi.org/10.3390/math9050538

Academic Editor: Aihua Wood

Received: 2 February 2021
Accepted: 28 February 2021
Published: 4 March 2021

Publisher's Note: MDPI stays neutral with regard to jurisdictional claims in published maps and institutional affiliations.

Copyright: © 2021 by the authors. Licensee MDPI, Basel, Switzerland. This article is an open access article distributed under the terms and conditions of the Creative Commons Attribution (CC BY) license (https://creativecommons.org/licenses/by/4.0/).

1. Introduction

In the history of fluid mechanics, flow development over stretching and shrinking surfaces was first described by Crane [1] and Wang [2], respectively. Meanwhile, Miklavčič and Wang [3] reported the existence of non-unique solutions for flow over a shrinking sheet. Since then, many studies have considered the effect of several physical parameters such as magnetohydrodynamic (MHD) and radiation on stretching and shrinking surfaces [4–12]. The effect of the MHD parameter is an important factor in many industrial and engineering applications, for example, MHD power generators, metallurgical process, crystal growth, metal casting, and cooling of nuclear reactors [13]. Thermal radiation is also important in designing innovative energy conversion systems operational at high temperatures [14].

In general, most previous studies have considered isothermal surface conditions; however, heating or cooling can occur under non-isothermal conditions for many practical applications such as in microelectromechanical (MEM) condensation applications, a thin-film solar energy collector device, the cooling of metallic plate in a cooling bath, metal spinning, paper production, and aerodynamic extrusion of plastic sheets [15,16]. In this respect, Soundalgekar and Ramana Murty [17], and Grubka and Bobba [18] considered flow over moving and stretching surfaces under non-isothermal conditions, respectively. This type of heating condition also has been reported by several researchers [19–22].

In 1995, Choi and Eastman [23] introduced nanofluids, which are a mixture of a base fluid and a single type of nanoparticle, to enhance thermal conductivity. Various studies on such fluids have been conducted [24–29]. Recently, some studies have found that advanced nanofluid consists of another type of nanoparticle that is mixed in with the regular nanofluid and improves its thermal properties, namely a "hybrid nanofluid". Prior experimental studies using hybrid nanoparticles have been conducted by several researchers [30–32] and numerical studies on the flow of hybrid nanofluids were studied by Takabi and Salehi [33]. Moreover, dual solutions of hybrid nanofluid flow were examined by Waini et al. [34–39]. Other physical aspects have been considered by several authors [40–49] and review papers are also available [50–55].

In this study, we aim at investigating the effects of Cu-Al$_2$O$_3$ hybrid nanoparticles on the radiative MHD flow over a permeable non-isothermal shrinking surface. The simultaneous effects of radiation and the hybrid nanoparticles are examined and the influence of magnetic field and variation of the temperature index is also considered. To the best of our knowledge, based on the above studies, the flow of hybrid nanofluids over non-isothermal shrinkage surfaces is not yet available in the literature, and therefore the results of this study are new. Most importantly, in this study, two solutions are discovered and the long-term stability of these solutions is investigated.

2. Mathematical Formulation

Let us consider the two-dimensional, laminar, and incompressible flow of a hybrid nanofluid over a permeable non-isothermal shrinking surface, as shown in Figure 1. The surface velocity is represented by $u_w(x) = ax$ where $a > 0$ is constant and v_0 is the constant mass flux velocity. The flow is subjected to the combined effect of a transverse magnetic field of strength B_0 and the radiative heat flux q_r, which is assumed to be applied normal to the surface in the positive y-direction. Accordingly, the hybrid nanofluid Equations (see Grubka and Bobba [18], Rashid et al. [20], Waini et al. [34]) are:

$$\frac{\partial u}{\partial x} + \frac{\partial v}{\partial y} = 0 \tag{1}$$

$$u\frac{\partial u}{\partial x} + v\frac{\partial u}{\partial y} = \frac{\mu_{hnf}}{\rho_{hnf}}\frac{\partial^2 u}{\partial y^2} - \frac{\sigma_{hnf}}{\rho_{hnf}}B_0^2 u \tag{2}$$

$$u\frac{\partial T}{\partial x} + v\frac{\partial T}{\partial y} = \frac{k_{hnf}}{(\rho C_p)_{hnf}}\frac{\partial^2 T}{\partial y^2} - \frac{1}{(\rho C_p)_{hnf}}\frac{\partial q_r}{\partial y} \tag{3}$$

subject to:

$$\begin{array}{c} v = v_0,\ u = \lambda u_w(x),\ T = T_w(x)\ \text{at}\ y = 0 \\ u \to 0,\ T \to T_\infty\ \text{as}\ y \to \infty \end{array} \tag{4}$$

where u and v represent the velocity components along the x- and y-axes and the temperature of the hybrid nanofluid is given by T.

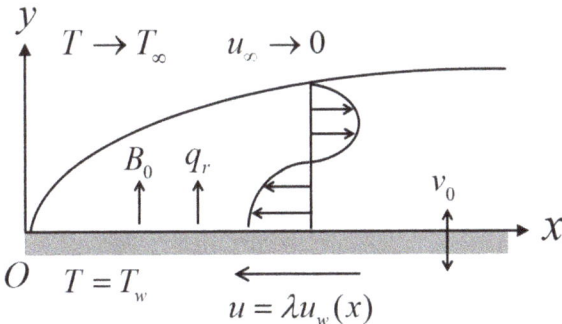

Figure 1. The flow configuration.

The expression of the radiative heat flux is as follows [9]:

$$q_r = -\frac{4\sigma^*}{3k^*}\frac{\partial T^4}{\partial y} \tag{5}$$

where σ^* and k^* denote the Stefan–Boltzmann constant and the mean absorption coefficient, respectively. Following Rosseland [56], after employing a Taylor series, one gets $T^4 \cong 4 T_\infty^3 T - 3T_\infty^4$. Then, Equation (3) becomes the following:

$$u\frac{\partial T}{\partial x} + v\frac{\partial T}{\partial y} = \frac{1}{(\rho C_p)_{hnf}}\left[k_{hnf} + \frac{16\sigma^* T_\infty^3}{3k^*}\right]\frac{\partial^2 T}{\partial y^2} \quad (6)$$

Furthermore, the thermophysical properties can be referred to in Tables 1 and 2. Data from these tables are adapted from previous studies [26,33,34,57]. Note that φ_1 (Al$_2$O$_3$) and φ_2 (Cu) are nanoparticles volume fractions, and the subscripts $n1$ and $n2$ correspond to their solid components, while the subscripts hnf and f represent the hybrid nanofluid and the base fluid, respectively.

Table 1. Thermophysical properties of nanoparticles and water.

Properties	Nanoparticles		Base Fluid
	Cu	Al$_2$O$_3$	Water
ρ (kg/m^3)	8933	3970	997.1
C_p (J/kgK)	385	765	4179
k (W/mK)	400	40	0.613
σ (S/m)	5.96×10^7	3.69×10^7	0.05
Prandtl number, Pr			6.2

Table 2. Thermophysical properties of nanofluid and hybrid nanofluid.

Thermophysical Properties	Correlations
Dynamic viscosity	$\mu_{hnf} = \frac{\mu_f}{(1-\varphi_{hnf})^{2.5}}$
Density	$\rho_{hnf} = (1-\varphi_{hnf})\rho_f + \varphi_1\rho_{n1} + \varphi_2\rho_{n2}$
Heat capacity	$(\rho C_p)_{hnf} = (1-\varphi_{hnf})(\rho C_p)_f + \varphi_1(\rho C_p)_{n1} + \varphi_2(\rho C_p)_{n2}$
Thermal conductivity	$\frac{k_{hnf}}{k_f} = \frac{\frac{\varphi_1 k_{n1} + \varphi_2 k_{n2}}{\varphi_{hnf}} + 2k_f + 2(\varphi_1 k_{n1} + \varphi_2 k_{n2}) - 2\varphi_{hnf}k_f}{\frac{\varphi_1 k_{n1} + \varphi_2 k_{n2}}{\varphi_{hnf}} + 2k_f - (\varphi_1 k_{n1} + \varphi_2 k_{n2}) + \varphi_{hnf}k_f}$
Electrical conductivity	$\frac{\sigma_{hnf}}{\sigma_f} = \frac{\frac{\varphi_1 \sigma_{n1} + \varphi_2 \sigma_{n2}}{\varphi_{hnf}} + 2\sigma_f + 2(\varphi_1 \sigma_{n1} + \varphi_2 \sigma_{n2}) - 2\varphi_{hnf}\sigma_f}{\frac{\varphi_1 \sigma_{n1} + \varphi_2 \sigma_{n2}}{\varphi_{hnf}} + 2\sigma_f - (\varphi_1 \sigma_{n1} + \varphi_2 \sigma_{n2}) + \varphi_{hnf}\sigma_f}$

For the similarity solution of Equations (1), (2), and (6), the surface temperature is taken as follows (see Grubka and Bobba [18], Rashid et al. [20]):

$$T_w(x) = T_\infty + T_0(x/L)^m \quad (7)$$

where L is a characteristic length of the sheet and T_0 is a temperature characteristic. The ambient temperature T_∞ is assumed to be constant and m represents the temperature power-law index, with $m = 0$ indicating an isothermal surface and $m > 0$ indicating a non-isothermal surface.

Now, using the following similarity transformation:

$$\psi = \sqrt{av_f}xf(\eta), \ \theta(\eta) = \frac{T - T_\infty}{T_w - T_\infty}, \ \eta = y\sqrt{\frac{a}{v_f}} \quad (8)$$

with the stream function ψ. Here, $u = \partial\psi/\partial y$ and $v = -\partial\psi/\partial x$, then:

$$u = axf'(\eta), \ v = -\sqrt{av_f}f(\eta) \quad (9)$$

From Equation (9), by setting $\eta = 0$, one obtains:

$$v_0 = -\sqrt{av_f}\, S \tag{10}$$

where $f(0) = S$ is the constant mass flux parameter which determines the permeability of the surface. Here, $S < 0$ and $S > 0$ are for injection and suction cases, respectively, while $S = 0$ represents an impermeable case.

On using Equations (8) and (9), Equation (1) is identically fulfilled. Now, Equations (2) and (6) are reduced to:

$$\frac{\mu_{hnf}/\mu_f}{\rho_{hnf}/\rho_f} f''' + ff'' - f'^2 - \frac{\sigma_{hnf}/\sigma_f}{\rho_{hnf}/\rho_f} Mf' = 0 \tag{11}$$

$$\frac{1}{\Pr} \frac{1}{(\rho C_p)_{hnf}/(\rho C_p)_f} \left(\frac{k_{hnf}}{k_f} + \frac{4}{3}R\right)\theta'' + f\theta' - mf'\theta = 0 \tag{12}$$

subject to the following:

$$\begin{array}{c} f(0) = S,\; f'(0) = \lambda,\; \theta(0) = 1, \\ f'(\eta) \to 0,\; \theta(\eta) \to 0 \text{ as } \eta \to \infty \end{array} \tag{13}$$

where primes denote differentiation with respect to η. Note that $\lambda < 0$ and $\lambda > 0$ represent the shrinking and stretching surfaces, while $\lambda = 0$ is a rigid surface. In addition, Pr is the Prandtl number, while R and M are the radiation and the magnetic parameters, respectively, which are defined as follows:

$$\Pr = \frac{(\mu C_p)_f}{k_f},\; R = \frac{4\sigma^* T_\infty^3}{k^* k_f},\; M = \frac{\sigma_f}{\rho_f a} B_0^2 \tag{14}$$

The coefficient of the skin friction C_f and the local Nusselt number Nu_x are given as follows [9]:

$$C_f = \frac{\mu_{hnf}}{\rho_f u_w^2}\left(\frac{\partial u}{\partial y}\right)_{y=0},\; Nu_x = \frac{x}{k_f(T_w - T_\infty)}\left(-k_{hnf}\left(\frac{\partial T}{\partial y}\right)_{y=0} + (q_r)_{y=0}\right) \tag{15}$$

Using Equations (8) and (15), one obtains:

$$\mathrm{Re}_x^{1/2} C_f = \frac{\mu_{hnf}}{\mu_f} f''(0),\; \mathrm{Re}_x^{-1/2} Nu_x = -\left(\frac{k_{hnf}}{k_f} + \frac{4}{3}R\right)\theta'(0) \tag{16}$$

where $\mathrm{Re}_x = u_w(x)x/v_f$ defines the local Reynolds number.

It should be noted that for $\varphi_{hnf} = S = M = R = 0$, Equations (11) and (12) reduce to Equations (5) and (6) from Grubka and Bobba [18] when $\lambda = 1$.

3. Stability Analysis

The temporal stability of the dual solutions as time evolves is studied. This analysis was first introduced by Merkin [58], and then followed by Weidman et al. [59]. Firstly, consider the new variables as follows:

$$\psi = \sqrt{av_f}\, xf(\eta),\; \theta(\eta) = \frac{T - T_\infty}{T_w - T_\infty},\; \eta = y\sqrt{\frac{a}{v_f}},\; \tau = at \tag{17}$$

Now, the unsteady form of Equations (2) and (3) are employed, while Equation (1) remains unchanged. On using (17), one obtains:

$$\frac{\mu_{hnf}/\mu_f}{\rho_{hnf}/\rho_f}\frac{\partial^3 f}{\partial \eta^3} + f\frac{\partial^2 f}{\partial \eta^2} - \left(\frac{\partial f}{\partial \eta}\right)^2 - \frac{\sigma_{hnf}/\sigma_f}{\rho_{hnf}/\rho_f} M\frac{\partial f}{\partial \eta} - \frac{\partial^2 f}{\partial \eta \partial \tau} = 0 \tag{18}$$

$$\frac{1}{\Pr}\frac{1}{(\rho C_p)_{hnf}/(\rho C_p)_f}\left(\frac{k_{hnf}}{k_f}+\frac{4}{3}R\right)\frac{\partial^2\theta}{\partial\eta^2}+f\frac{\partial\theta}{\partial\eta}-m\frac{\partial f}{\partial\eta}\theta-\frac{\partial\theta}{\partial\tau}=0 \qquad (19)$$

subject to the following:

$$f(0,\tau)=S,\ \frac{\partial f}{\partial\eta}(0,\tau)=\lambda,\ \theta(0,\tau)=1,$$
$$\frac{\partial f}{\partial\eta}(\infty,\tau)=0,\ \theta(\infty,\tau)=0 \qquad (20)$$

Then, consider the following perturbation functions [59]:

$$f(\eta,\tau)=f_0(\eta)+e^{-\gamma\tau}F(\eta),\ \theta(\eta,\tau)=\theta_0(\eta)+e^{-\gamma\tau}G(\eta) \qquad (21)$$

Here, Equation (21) is used to apply a small disturbance on the steady solutions $f=f_0(\eta)$ and $\theta=\theta_0(\eta)$ of Equations (11)–(13). The functions $F(\eta)$ and $G(\eta)$ in Equation (19) are relatively small as compared with $f_0(\eta)$ and $\theta_0(\eta)$. The sign (positive or negative) of the eigenvalue γ determines the stability of the solutions. By employing Equation (21), Equations (18) to (20) become:

$$\frac{\mu_{hnf}/\mu_f}{\rho_{hnf}/\rho_f}F'''+f_0F''+f_0''F-2f_0'F'-\frac{\sigma_{hnf}/\sigma_f}{\rho_{hnf}/\rho_f}MF'+\gamma F'=0 \qquad (22)$$

$$\frac{1}{\Pr}\frac{1}{(\rho C_p)_{hnf}/(\rho C_p)_f}\left(\frac{k_{hnf}}{k_f}+\frac{4}{3}R\right)G''+f_0G'+\theta_0'F-m(f_0'G+\theta_0F')+\gamma G=0 \qquad (23)$$

subject to the following:

$$F(0)=0,\ F'(0)=0,\ G(0)=0,$$
$$F'(\infty)=0,\ G(\infty)=0 \qquad (24)$$

Without loss of generality, we set $F''(0)=1$ [60] to get the eigenvalues γ in Equations (22) and (23).

4. Results and Discussion

By utilising the package bvp4c in MATLAB software, Equations (11)–(13) were solved numerically. This solver employs the three-stage Lobatto IIIa formula [61]. The effect of several physical parameters on the flow behaviour is examined. The total composition of Al_2O_3 and Cu volume fractions are applied in a one-to-one ratio. For instance, 1% of Al_2O_3 ($\varphi_1=1\%$) and 1% of Cu ($\varphi_2=1\%$) are mixed to produce 2% of Al_2O_3-Cu hybrid nanoparticles volume fractions, i.e., $\varphi_{hnf}=2\%$. Meanwhile, $\varphi_{hnf}=0$ indicates a regular viscous fluid.

The values of $-\theta'(0)$ for various values of m and $\Pr$ when $\varphi_{hnf}=S=M=R=0$ and $\lambda=1$ (stretching sheet) are compared with Grubka and Bobba [18], and Ishak et al. [15] and the results for each m and $\Pr$ considered are comparable, as shown in Table 3. In addition, it should be noted that the values of $-\theta'(0)$ increase for higher values of m and $\Pr$. Furthermore, Table 4 provides the values of $Re_x^{1/2}C_f$ and $Re_x^{-1/2}Nu_x$ when $\varphi_{hnf}=2\%$, $S=2$, and $\lambda=-1$ (shrinking sheet) for different physical parameters. The consequence of increasing m and R values is to reduce the local Nusselt number $Re_x^{-1/2}Nu_x$ for both branch solutions. However, the skin friction coefficient $Re_x^{1/2}C_f$ is not affected by these parameters. Moreover, the values of $Re_x^{1/2}C_f$ and $Re_x^{-1/2}Nu_x$ for the first solution increase, but they decrease for the second solution as M increases.

Table 3. Values of $-\theta'(0)$ under different values of m and Pr when $\varphi_{hnf} = S = M = R = 0$ and $\lambda = 1$ (stretching sheet).

m	Pr	Grubka and Bobba [18]	Ishak et al. [15]	Present Results
0	1	0.5820	-	0.5820
1	-	1.0000	-	1.0000
2	-	1.3333	-	1.3333
3	-	1.6154	-	1.6154
1	0.72	0.8086	0.8086	0.8086
-	1	1.0000	1.0000	1.0000
-	3	1.9237	1.9237	1.9237
-	10	3.7207	3.7207	3.7207

Table 4. Values of $Re_x^{1/2}C_f$ and $Re_x^{-1/2}Nu_x$ when $\varphi_{hnf} = 2\%$, $S = 2$ and $\lambda = -1$ (shrinking sheet) for different physical parameters.

m	R	M	First Solution		Second Solution	
			$Re_x^{1/2}C_f$	$Re_x^{-1/2}Nu_x$	$Re_x^{1/2}C_f$	$Re_x^{-1/2}Nu_x$
0	0	0	1.3622	11.8319	0.8566	11.8066
0.5	-	-	1.3622	11.5596	0.8566	11.5177
1	-	-	1.3622	11.2748	0.8566	11.2126
1	1	-	1.3622	9.9890	0.8566	9.5366
-	2	-	1.3622	8.8910	0.8566	7.6301
-	3	-	1.3622	8.0105	0.8566	5.7594
-	3	0.01	1.3834	8.0575	0.8354	5.5222
-	-	0.05	1.4554	8.2064	0.7634	4.2952
-	-	0.1	1.5284	8.3426	0.6904	0.4505

The variations of $Re_x^{-1/2}Nu_x$ against R when $\lambda = -1$, $S = 2$, $M = 0.1$, $\varphi_{hnf} = 2\%$, and Pr = 6.2 for various values of m are presented in Figure 2. Reductions in the values of $Re_x^{-1/2}Nu_x$ on both solutions are observed with an increase in R and m. Moreover, the simultaneous effect of R and φ_{hnf} on $Re_x^{-1/2}Nu_x$ when $\lambda = -1$, $S = 2$, $M = 0.1$, $m = 1$, and Pr = 6.2 can be observed in Figure 3. The values of $Re_x^{-1/2}Nu_x$ on the first solution decrease with a high percentage of φ_{hnf} for smaller values of R. This finding seems to contradict the fact that the added hybrid nanoparticles improve the heat transfer rate due to synergistic effects as discussed by Sarkar et al. [50]. However, it is interesting to note that this behaviour is opposite when higher values of R are applied to the system where the enhancement in the values of $Re_x^{-1/2}Nu_x$ are observed with a high percentage of φ_{hnf}. From these observations, we conclude that the rate of heat transfer could be controlled by manipulating the values of R and φ_{hnf}.

Next, the variations of $Re_x^{1/2}C_f$ and $Re_x^{-1/2}Nu_x$ against S for various values of φ_{hnf} and M are presented in Figures 4–7, respectively. The enhancement in the values of $Re_x^{1/2}C_f$ and $Re_x^{-1/2}Nu_x$ on the first solution are observed with an increase in S, φ_{hnf} and M values. The dual solutions are also obtained when a suitable suction strength is imposed on the shrinking surface. The flow is unlikely to exist since the vorticity could not be confined in the boundary layer. These figures reveal that a sufficient suction strength is needed to preserve the flow over a shrinking sheet. The similarity solutions are terminated at $S = S_c$ (critical value) and this point is known as the bifurcation point of the solutions. The boundary layer separation is also delayed with an increase in φ_{hnf} and M by expanding the domain of S. Here, the critical values are $S_{c1} = 1.8974$, $S_{c2} = 1.8733$, and $S_{c3} = 1.8519$ for $\varphi_{hnf} = 0\%$, 1%, and 2%, respectively. Meanwhile, for $M = 0$, 0.05, and, 0.1, the critical values are $S_{c1} = 1.9474$, $S_{c2} = 1.9003$, and $S_{c3} = 1.8519$, respectively. It can be seen that the presence of those parameters suppressed the vorticity generation due to the shrinking of the sheet and the steady boundary layer flow is maintained.

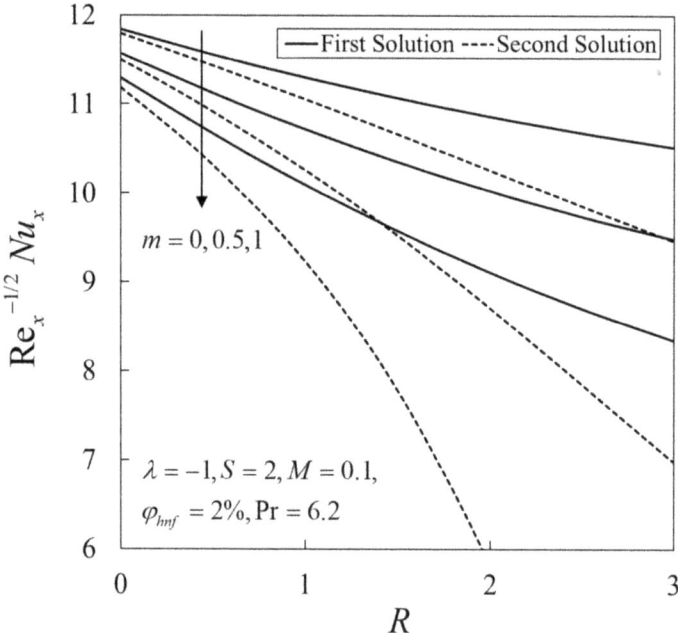

Figure 2. Variations of the local Nusselt number $Re_x^{-1/2} Nu_x$ against the radiation parameter R for different values of m.

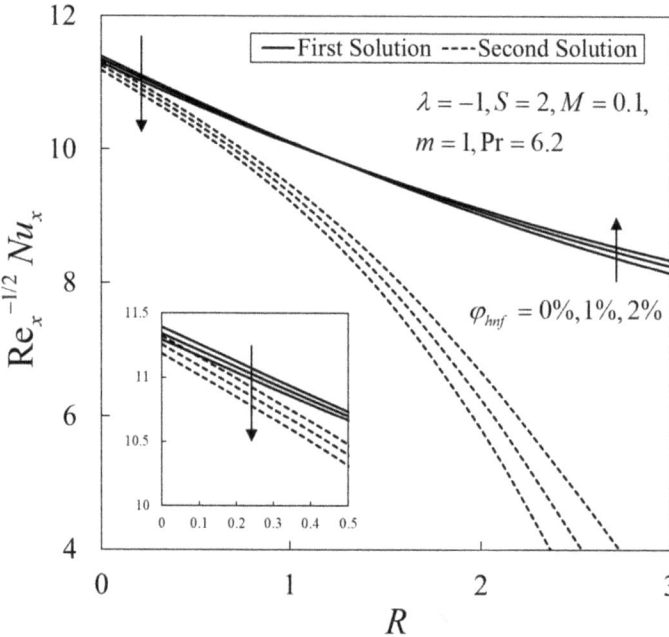

Figure 3. Variations of the local Nusselt number $Re_x^{-1/2} Nu_x$ against against the radiation parameter R and for different values of φ_{hnf}.

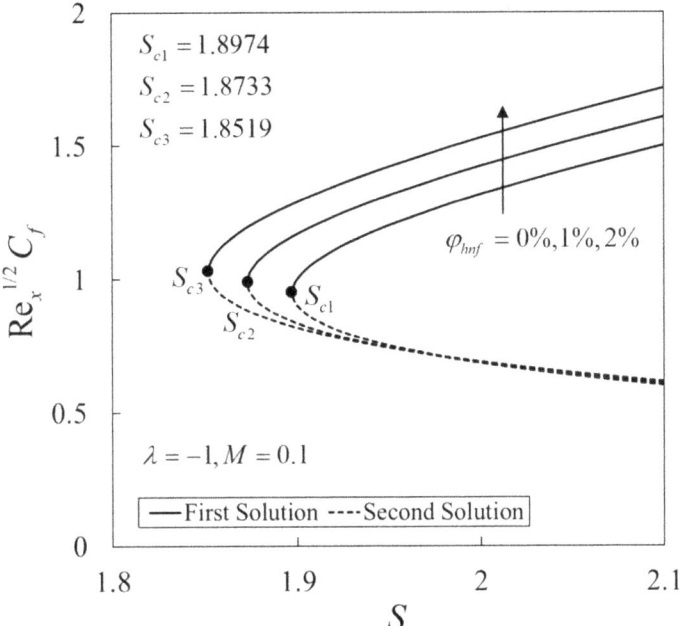

Figure 4. Variations of the skin friction coefficient $Re_x^{1/2}C_f$ against suction parameter S for different values of φ_{hnf}.

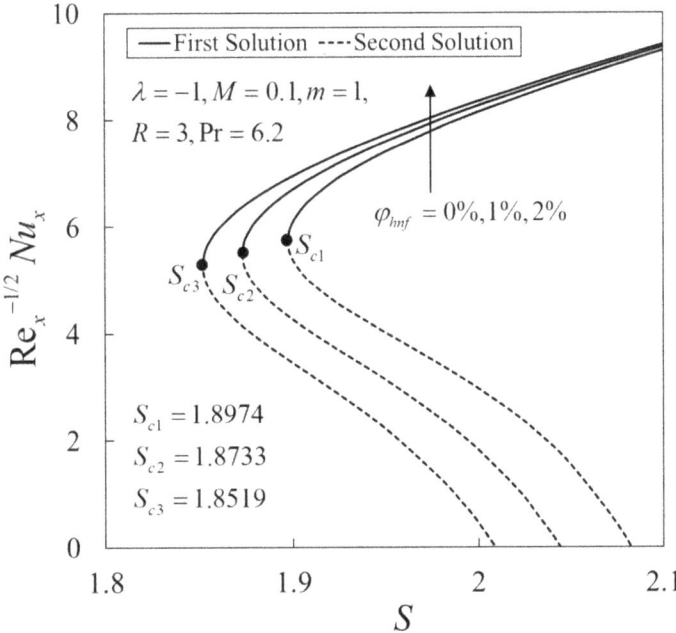

Figure 5. Variations of the local Nusselt number $Re_x^{-1/2}Nu_x$ against suction parameter S for different values of φ_{hnf}.

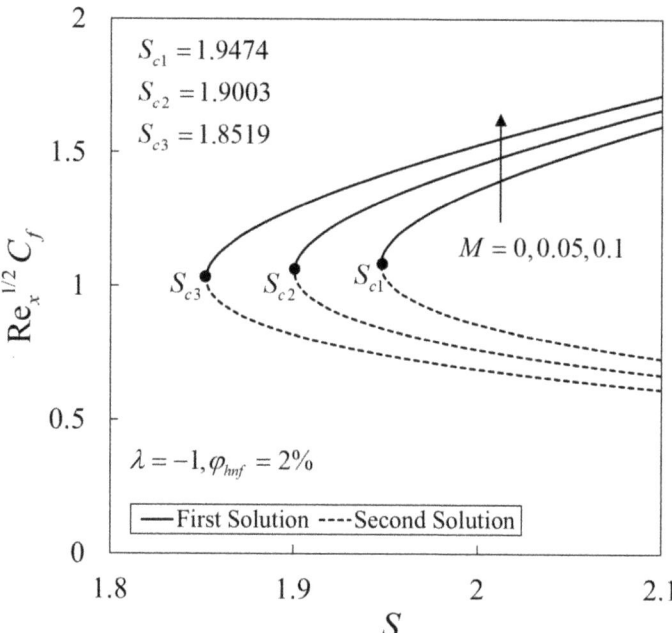

Figure 6. Variations of the local Nusselt number $\mathrm{Re}_x^{1/2} C_f$ against suction parameter S for different values of M.

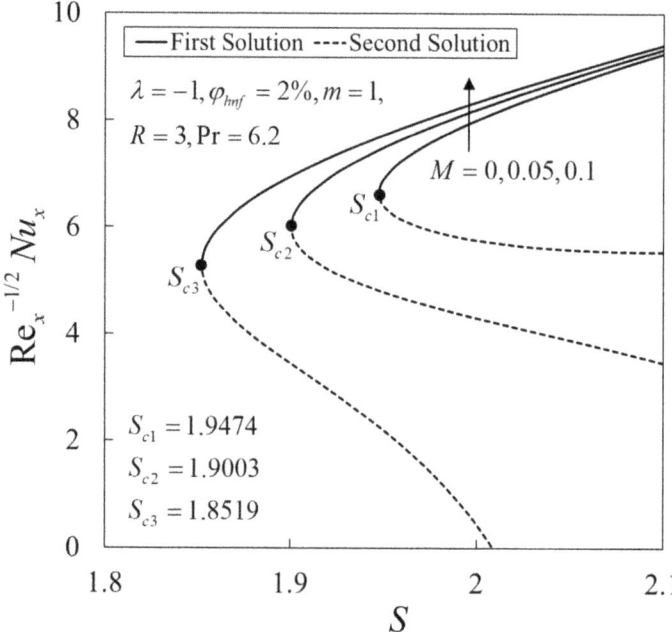

Figure 7. Variations of the local Nusselt number $\mathrm{Re}_x^{-1/2} Nu_x$ against suction parameter S for different values of M.

The influence of m and R on the variations $Re_x^{-1/2} Nu_x$ against S are given in Figures 8 and 9, respectively. The heat transfer rate is higher for the isothermal surface ($m = 0$) as compared with the non-isothermal surface ($m > 0$). An increase in R leads to a reduction in the values of $Re_x^{-1/2} Nu_x$. In addition, the boundary layer separation occurs at the same point where the critical value is $S_c = 1.8519$ for all values of m and R considered.

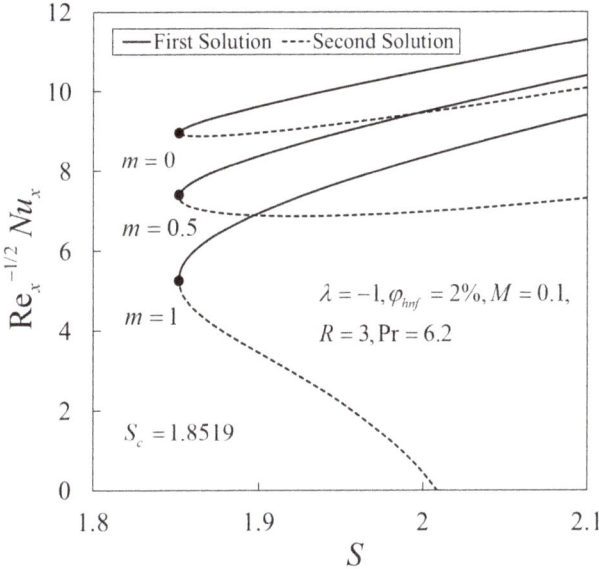

Figure 8. Variations of the local Nusselt number $Re_x^{-1/2} Nu_x$ against suction parameter S for different values of m.

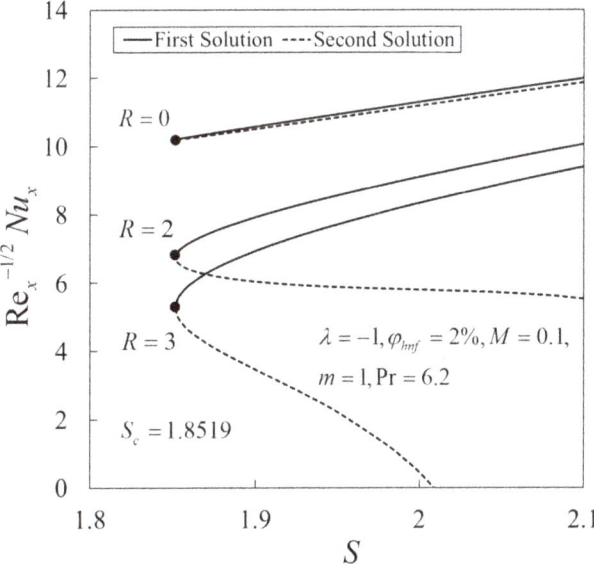

Figure 9. Variations of the local Nusselt number $Re_x^{-1/2} Nu_x$ against suction parameter S for different values of R.

The profiles of the velocity $f'(\eta)$ and the temperature $\theta(\eta)$ for several pertinent parameters are presented in Figures 10–17. There are dual solutions for $f'(\eta)$ and $\theta(\eta)$ which satisfy the infinity boundary conditions (13) asymptotically. For more detail, the profiles of $f'(\eta)$ and $\theta(\eta)$ for several values of S when $\lambda = -1$, $M = 0.1$, $\varphi_{hnf} = 2\%$, $m = 1$, $R = 3$, and $Pr = 6.2$ are given in Figures 10 and 11. Note that the profiles of the first and the second solutions are merged towards some values of S. This behaviour can also be seen in Figures 2–9 where the similarity solutions ended at $S = S_c$.

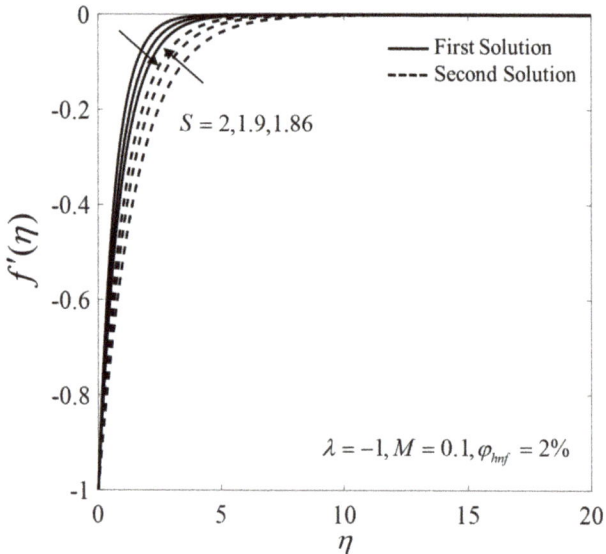

Figure 10. Velocity profiles $f'(\eta)$ or different values of suction strength S.

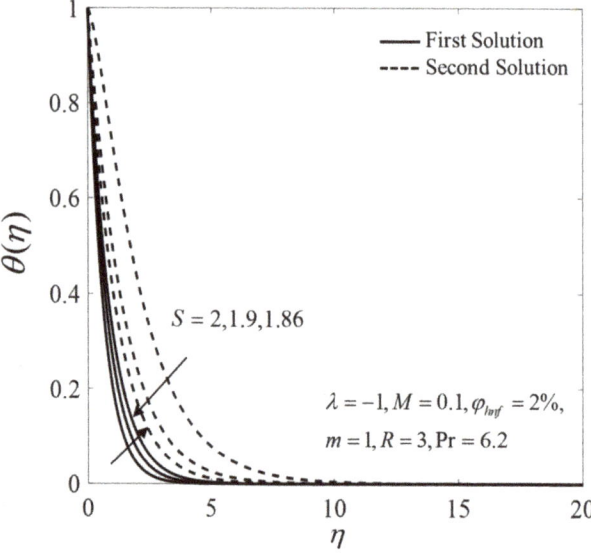

Figure 11. Temperature profiles $\theta(\eta)$ for different values of S.

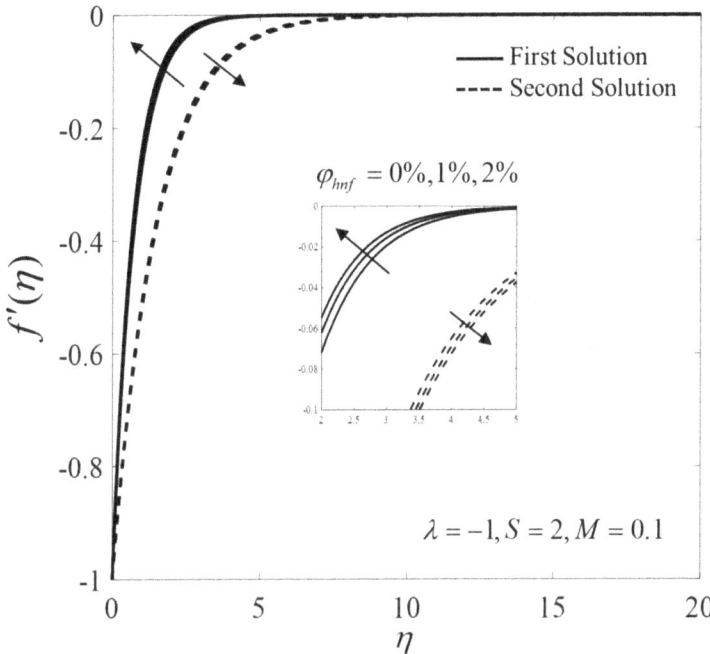

Figure 12. Velocity profiles $f'(\eta)$ for different values of φ_{hnf}.

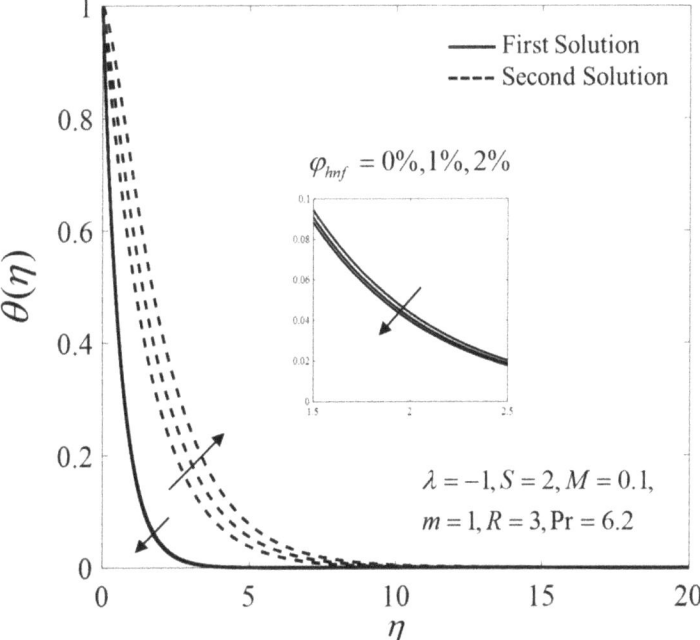

Figure 13. Temperature profiles $\theta(\eta)$ for different values of φ_{hnf}.

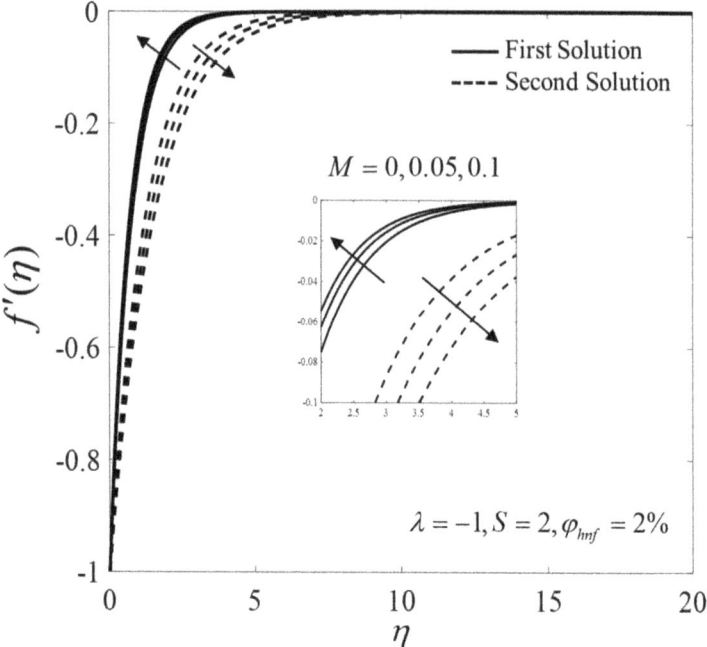

Figure 14. Velocity profiles $f'(\eta)$ for different values of M.

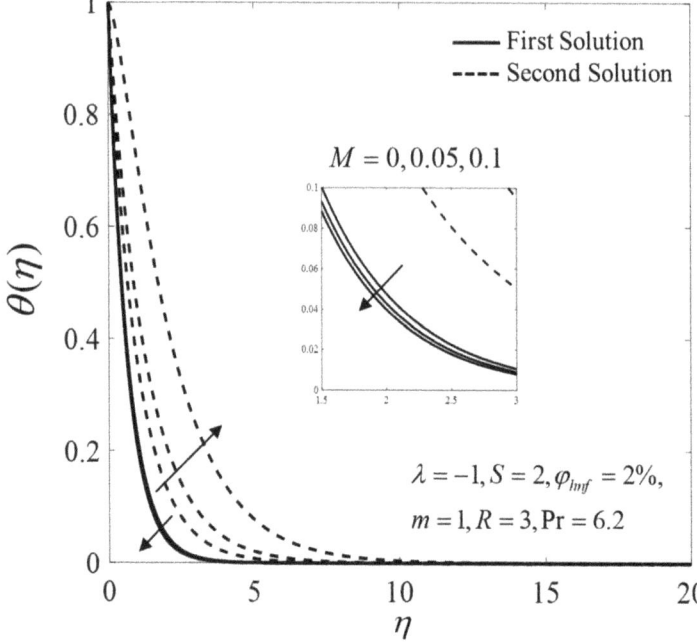

Figure 15. Temperature profiles $\theta(\eta)$ for different values of M.

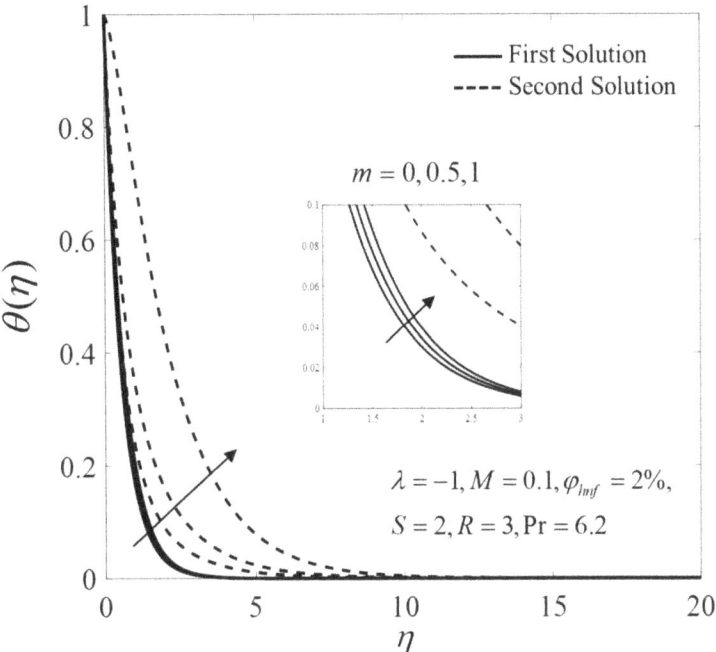

Figure 16. Temperature profiles $\theta(\eta)$ for different values of m.

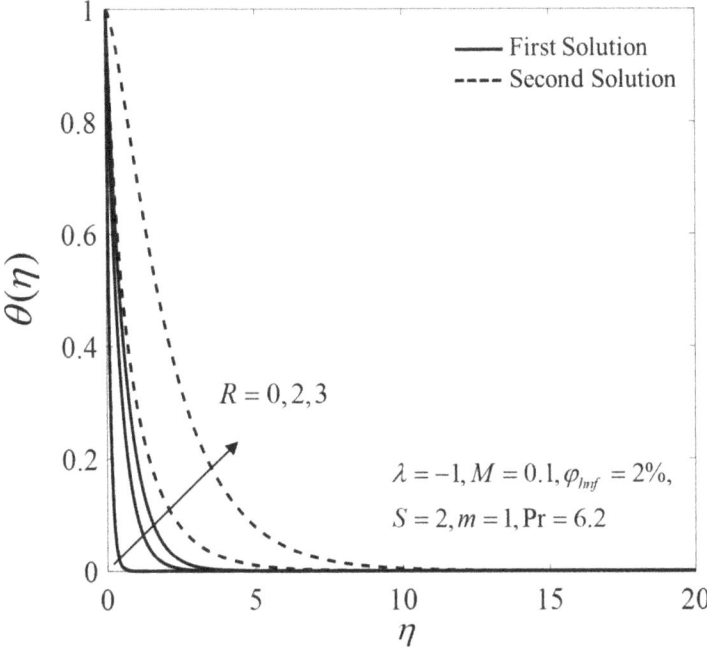

Figure 17. Temperature profiles $\theta(\eta)$ for different values of R.

Next, an increase in φ_{hnf} and M values lead to an upsurge in the velocity $f'(\eta)$ but reduces the temperature $\theta(\eta)$ on the first solution, as shown in Figures 12–15, respectively. Physically, the addition of the nanoparticles makes the fluid more viscous, and thus slows down the flow; the added nanoparticles also dissipate energy in the form of heat and consequently exert more energy which enhances the temperature. However, in this study, we discover that the velocity increases, but the temperature decreases, as φ_{hnf} increases. Furthermore, an increase in magnetic strength enhances the magnitude of Lorentz force and results in an increment in the velocity and a reduction in the temperature for the shrinking sheet case.

Moreover, Figures 16 and 17 show the consequence effects of m and R on the temperature $\theta(\eta)$. It is seen that both branch solutions of $\theta(\eta)$ show an increasing pattern for larger values of m and R; in addition, the boundary layer thickness of the first and the second solutions expand as m and R increase. For $m > 0$, the temperature in the flow field increases due to direct variation of the wall temperature along the shrinking surface. Moreover, the radiation is dominant over conduction with an increase in R. Therefore, the temperature $\theta(\eta)$ increases due to the high radiation energy presence in the flow field.

The variations of γ against S when $\lambda = -1$, $\varphi_{hnf} = 2\%$, and $M = 0.1$ are described in Figure 18. For the positive value of γ, it is noted that $e^{-\gamma\tau} \to 0$ as time evolves ($\tau \to \infty$). In the meantime, for the negative value of γ, $e^{-\gamma\tau} \to \infty$. These behaviours show that the first solution is stable and physically reliable, while the second solution becomes unstable over time.

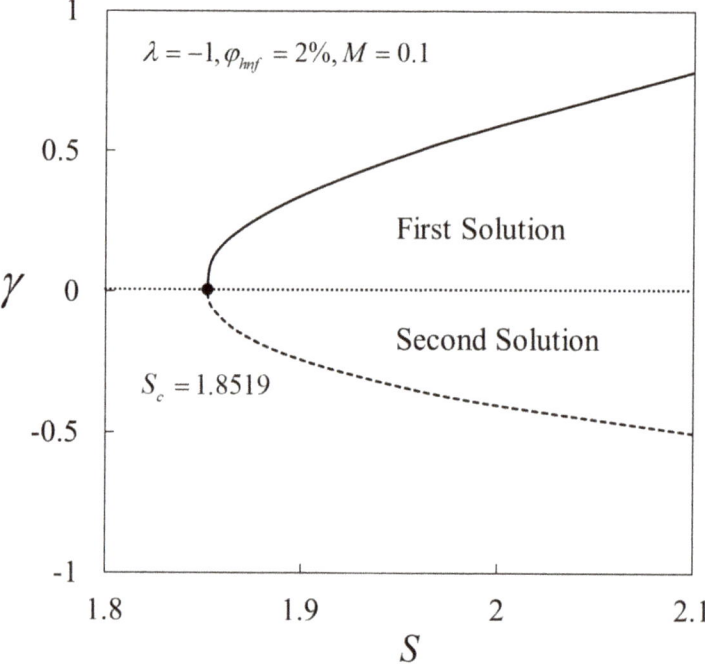

Figure 18. Variations of the minimum eigenvalues γ against suction S.

5. Conclusions

The flow and heat transfer over a permeable non-isothermal shrinking surface with radiation and magnetohydrodynamic (MHD) effects were examined in this paper. The findings revealed that dual solutions appeared when satisfactory suction strength was applied on the shrinking surface. Moreover, the heat transfer rate was enhanced with a high percentage of φ_{hnf} when higher values of the radiation parameter, R, were applied to the system; additionally, the heat transfer rate was higher for the isothermal surface ($m = 0$) as compared with the non-isothermal surface ($m > 0$). Increased φ_{hnf} and M values also enhanced the skin friction coefficient $\text{Re}_x^{1/2} C_f$ and the local Nusselt number $\text{Re}_x^{-1/2} Nu_x$. The effect of m, as well as R, was to increase the temperature $\theta(\eta)$ inside the boundary layer. Lastly, it was discovered that the first solution was stable, and thus physically reliable in the long run.

Author Contributions: Conceptualization, I.P.; funding acquisition, A.I.; methodology, I.W.; Project administration, A.I.; supervision, A.I. and I.P.; validation, I.P.; writing—original draft, I.W.; writing—review and editing, A.I., I.P. All authors have read and agreed to the published version of the manuscript.

Funding: This research was funded by the Universiti Kebangsaan Malaysia (project code DIP-2020-001).

Acknowledgments: The authors would like to thank the anonymous reviewers for their constructive comments and suggestions. The financial supports received from the Universiti Kebangsaan Malaysia (project code: DIP-2020-001) and the Universiti Teknikal Malaysia Melaka are gratefully acknowledged.

Conflicts of Interest: The authors declare no conflict of interest.

References

1. Crane, L.J. Flow past a stretching plate. *Z. Angew. Math. Phys. ZAMP* **1970**, *21*, 645–647. [CrossRef]
2. Andersson, H.I.; Aarseth, J.B.; Dandapat, B.S. Heat transfer in a liquid film on an unsteady stretching surface. *Int. J. Heat Mass Transf.* **2000**, *43*, 69–74. [CrossRef]
3. Miklavčič, M.; Wang, C.Y. Viscous flow due to a shrinking sheet. *Q. Appl. Math.* **2006**, *64*, 283–290. [CrossRef]
4. Fang, T.; Zhang, J. Closed-form exact solutions of MHD viscous flow over a shrinking sheet. *Commun. Nonlinear Sci. Numer. Simul.* **2009**, *14*, 2853–2857. [CrossRef]
5. Bhattacharyya, K.; Pop, I. MHD boundary layer flow due to an exponentially shrinking sheet. *Magnetohydrodynamics* **2011**, *47*, 337–344.
6. Bhattacharyya, K. Effects of radiation and heat source/sink on unsteady MHD boundary layer flow and heat transfer over a shrinking sheet with suction/injection. *Front. Chem. Eng. China* **2011**, *5*, 376–384. [CrossRef]
7. Ishak, A. MHD boundary layer flow due to an exponentially stretching sheet with radiation effect. *Sains Malays.* **2011**, *40*, 391–395.
8. Ishak, A.; Yacob, N.A.; Bachok, N. Radiation effects on the thermal boundary layer flow over a moving plate with convective boundary condition. *Meccanica* **2011**, *46*, 795–801. [CrossRef]
9. Cortell, R. Heat and fluid flow due to non-linearly stretching surfaces. *Appl. Math. Comput.* **2011**, *217*, 7564–7572. [CrossRef]
10. Vyas, P.; Srivastava, N. Radiative boundary layer flow in porous medium due to exponentially shrinking permeable sheet. *ISRN Thermodyn.* **2012**, *2012*, 214362. [CrossRef]
11. Yasin, M.H.M.; Ishak, A.; Pop, I. MHD heat and mass transfer flow over a permeable stretching/shrinking sheet with radiation effect. *J. Magn. Magn. Mater.* **2016**, *407*, 235–240. [CrossRef]
12. Mabood, F.; Khan, W.A.; Ismail, A.I.M. MHD flow over exponential radiating stretching sheet using homotopy analysis method. *J. King Saud Univ. Eng. Sci.* **2017**, *29*, 68–74. [CrossRef]
13. Zainal, N.A.; Nazar, R.; Naganthran, K.; Pop, I. MHD flow and heat transfer of hybrid nanofluid over a permeable moving surface in the presence of thermal radiation. *Int. J. Numer. Methods Heat Fluid Flow* **2020**, in press. [CrossRef]
14. Chamkha, A.J.; Mujtaba, M.; Quadri, A.; Issa, C. Thermal radiation effects on MHD forced convection flow adjacent to a non-isothermal wedge in the presence of a heat source or sink. *Heat Mass Transf.* **2003**, *39*, 305–312. [CrossRef]
15. Ishak, A.; Nazar, R.; Pop, I. Heat transfer over an unsteady stretching permeable surface with prescribed wall temperature. *Nonlinear Anal. Real World Appl.* **2009**, *10*, 2909–2913. [CrossRef]
16. Muthtamilselvan, M.; Prakash, D. Unsteady hydromagnetic slip flow and heat transfer of nanofluid over a moving surface with prescribed heat and mass fluxes. *Proc. Inst. Mech. Eng. Part C J. Mech. Eng. Sci.* **2015**, *229*, 703–715. [CrossRef]
17. Soundalgekar, V.M.; Ramana Murty, T.V. Heat transfer in flow past a continuous moving plate with variable temperature. *Wärme Stoffübertragung* **1980**, *14*, 91–93. [CrossRef]
18. Grubka, L.J.; Bobba, K.M. Heat Transfer Characteristics of a Continuous, Stretching Surface With Variable Temperature. *J. Heat Transf.* **1985**, *107*, 248–250. [CrossRef]
19. Bhattacharyya, K.; Uddin, M.S.; Layek, G.C. Exact solution for thermal boundary layer in Casson fluid flow over permeable shrinking sheet with variable wall temperature and thermal radiation. *Alex. Eng. J.* **2016**, *55*, 1703–1712. [CrossRef]
20. Rashid, I.; Haq, R.U.; Khan, Z.H.; Al-Mdallal, Q.M. Flow of water based alumina and copper nanoparticles along a moving surface with variable temperature. *J. Mol. Liq.* **2017**, *246*, 354–362. [CrossRef]

21. Seth, G.S.; Singha, A.K.; Mandal, M.S.; Banerjee, A.; Bhattacharyya, K. MHD stagnation-point flow and heat transfer past a non-isothermal shrinking/stretching sheet in porous medium with heat sink or source effect. *Int. J. Mech. Sci.* **2017**, *134*, 98–111. [CrossRef]
22. Uddin, M.S.; Bhattacharyya, K. Thermal boundary layer in stagnation-point flow past a permeable shrinking sheet with variable surface temperature. *Propuls. Power Res.* **2017**, *6*, 186–194. [CrossRef]
23. Choi, S.U.S.; Eastman, J.A. Enhancing thermal conductivity of fluids with nanoparticles. In Proceedings of the 1995 International Mechanical Engineering Congress and Exhibition, San Francisco, CA, USA, 12–17 November 1995; Volume 66, pp. 99–105.
24. Khanafer, K.; Vafai, K.; Lightstone, M. Buoyancy-driven heat transfer enhancement in a two-dimensional enclosure utilizing nanofluids. *Int. J. Heat Mass Transf.* **2003**, *46*, 3639–3653. [CrossRef]
25. Tiwari, R.K.; Das, M.K. Heat transfer augmentation in a two-sided lid-driven differentially heated square cavity utilizing nanofluids. *Int. J. Heat Mass Transf.* **2007**, *50*, 2002–2018. [CrossRef]
26. Oztop, H.F.; Abu-Nada, E. Numerical study of natural convection in partially heated rectangular enclosures filled with nanofluids. *Int. J. Heat Fluid Flow* **2008**, *29*, 1326–1336. [CrossRef]
27. Bachok, N.; Ishak, A.; Pop, I. Stagnation-point flow over a stretching/shrinking sheet in a nanofluid. *Nanoscale Res. Lett.* **2011**, *6*, 623. [CrossRef] [PubMed]
28. Yacob, N.A.; Ishak, A.; Pop, I.; Vajravelu, K. Boundary layer flow past a stretching/shrinking surface beneath an external uniform shear flow with a convective surface boundary condition in a nanofluid. *Nanoscale Res. Lett.* **2011**, *6*, 314. [CrossRef] [PubMed]
29. Waini, I.; Ishak, A.; Pop, I. Dufour and Soret effects on Al_2O_3-water nanofluid flow over a moving thin needle: Tiwari and Das model. *Int. J. Numer. Methods Heat Fluid Flow* **2020**, in press. [CrossRef]
30. Turcu, R.; Darabont, A.; Nan, A.; Aldea, N.; Macovei, D.; Bica, D.; Vekas, L.; Pana, O.; Soran, M.L.; Koos, A.A.; et al. New polypyrrole-multiwall carbon nanotubes hybrid materials. *J. Optoelectron. Adv. Mater.* **2006**, *8*, 643–647.
31. Jana, S.; Salehi-Khojin, A.; Zhong, W.H. Enhancement of fluid thermal conductivity by the addition of single and hybrid nano-additives. *Thermochim. Acta* **2007**, *462*, 45–55. [CrossRef]
32. Suresh, S.; Venkitaraj, K.P.; Selvakumar, P.; Chandrasekar, M. Synthesis of Al_2O_3-Cu/water hybrid nanofluids using two step method and its thermo physical properties. *Colloids Surf. A Physicochem. Eng. Asp.* **2011**, *388*, 41–48. [CrossRef]
33. Takabi, B.; Salehi, S. Augmentation of the heat transfer performance of a sinusoidal corrugated enclosure by employing hybrid nanofluid. *Adv. Mech. Eng.* **2014**, *6*, 147059. [CrossRef]
34. Waini, I.; Ishak, A.; Pop, I. Hybrid nanofluid flow induced by an exponentially shrinking sheet. *Chin. J. Phys.* **2020**, *68*, 468–482. [CrossRef]
35. Waini, I.; Ishak, A.; Pop, I. Hybrid nanofluid flow past a permeable moving thin needle. *Mathematics* **2020**, *8*, 612. [CrossRef]
36. Waini, I.; Ishak, A.; Pop, I. Squeezed hybrid nanofluid flow over a permeable sensor surface. *Mathematics* **2020**, *8*, 898. [CrossRef]
37. Waini, I.; Ishak, A.; Pop, I. Mixed convection flow over an exponentially stretching/shrinking vertical surface in a hybrid nanofluid. *Alex. Eng. J.* **2020**, *59*, 1881–1891. [CrossRef]
38. Waini, I.; Ishak, A.; Pop, I. Hiemenz flow over a shrinking sheet in a hybrid nanofluid. *Results Phys.* **2020**, *19*, 103351. [CrossRef]
39. Waini, I.; Ishak, A.; Pop, I. Hybrid nanofluid flow towards a stagnation point on an exponentially stretching/shrinking vertical sheet with buoyancy effects. *Int. J. Numer. Methods Heat Fluid Flow* **2021**, *31*, 216–235. [CrossRef]
40. Aly, E.H.; Pop, I. MHD flow and heat transfer over a permeable stretching/shrinking sheet in a hybrid nanofluid with a convective boundary condition. *Int. J. Numer. Methods Heat Fluid Flow* **2019**, *29*, 3012–3038. [CrossRef]
41. Khan, U.; Zaib, A.; Khan, I.; Baleanu, D.; Nisar, K.S. Enhanced heat transfer in moderately ionized liquid due to hybrid MoS_2/SiO_2 nanofluids exposed by nonlinear radiation: Stability analysis. *Crystals* **2020**, *10*, 142. [CrossRef]
42. Khan, U.; Zaib, A.; Khan, I.; Baleanu, D.; Sherif, E.S.M. Comparative investigation on MHD nonlinear radiative flow through a moving thin needle comprising two hybridized AA7075 and AA7072 alloys nanomaterials through binary chemical reaction with activation energy. *J. Mater. Res. Technol.* **2020**, *9*, 3817–3828. [CrossRef]
43. Khashi'ie, N.S.; Arifin, N.M.; Pop, I.; Wahid, N.S. Flow and heat transfer of hybrid nanofluid over a permeable shrinking cylinder with Joule heating: A comparative analysis. *Alex. Eng. J.* **2020**, *59*, 1787–1798. [CrossRef]
44. Khashi'ie, N.S.; Arifin, N.M.; Pop, I.; Nazar, R.; Hafidzuddin, E.H.; Wahi, N. Non-axisymmetric Homann stagnation point flow and heat transfer past a stretching/shrinking sheet using hybrid nanofluid. *Int. J. Numer. Methods Heat Fluid Flow* **2020**, in press. [CrossRef]
45. Khashi'ie, N.S.; Arifin, N.M.; Wahi, N.; Pop, I.; Nazar, R.; Hafidzuddin, E.H. Thermal marangoni flow past a permeable stretching/shrinking sheet in a hybrid $Cu-Al_2O_3$/water nanofluid. *Sains Malays.* **2020**, *49*, 211–222. [CrossRef]
46. Khashi'ie, N.S.; Arifin, N.M.; Pop, I. Mixed convective stagnation point flow towards a vertical Riga plate in hybrid $Cu-Al_2O_3$/water nanofluid. *Mathematics* **2020**, *8*, 912. [CrossRef]
47. Zainal, N.A.; Nazar, R.; Naganthran, K.; Pop, I. Unsteady three-dimensional MHD nonaxisymmetric Homann stagnation point flow of a hybrid nanofluid with stability analysis. *Mathematics* **2020**, *8*, 784. [CrossRef]
48. Zainal, N.A.; Nazar, R.; Naganthran, K.; Pop, I. Impact of anisotropic slip on the stagnation-point flow past a stretching/shrinking surface of the Al_2O_3-Cu/H_2O hybrid nanofluid. *Appl. Math. Mech.* **2020**, *41*, 1401–1416. [CrossRef]
49. Anuar, N.S.; Bachok, N.; Pop, I. $Cu-Al_2O_3$/water hybrid nanofluid stagnation point flow past MHD stretching/shrinking sheet in presence of homogeneous-heterogeneous and convective boundary conditions. *Mathematics* **2020**, *8*, 1237. [CrossRef]

50. Sarkar, J.; Ghosh, P.; Adil, A. A review on hybrid nanofluids: Recent research, development and applications. *Renew. Sustain. Energy Rev.* **2015**, *43*, 164–177. [CrossRef]
51. Sidik, N.A.C.; Adamu, I.M.; Jamil, M.M.; Kefayati, G.H.R.; Mamat, R.; Najafi, G. Recent progress on hybrid nanofluids in heat transfer applications: A comprehensive review. *Int. Commun. Heat Mass Transf.* **2016**, *78*, 68–79. [CrossRef]
52. Babu, J.A.R.; Kumar, K.K.; Rao, S.S. State-of-art review on hybrid nanofluids. *Renew. Sustain. Energy Rev.* **2017**, *77*, 551–565. [CrossRef]
53. Sajid, M.U.; Ali, H.M. Thermal conductivity of hybrid nanofluids: A critical review. *Int. J. Heat Mass Transf.* **2018**, *126*, 211–234. [CrossRef]
54. Huminic, G.; Huminic, A. Entropy generation of nanofluid and hybrid nanofluid flow in thermal systems: A review. *J. Mol. Liq.* **2020**, *302*, 112533. [CrossRef]
55. Yang, L.; Ji, W.; Mao, M.; Huang, J. An updated review on the properties, fabrication and application of hybrid-nanofluids along with their environmental effects. *J. Clean. Prod.* **2020**, *257*, 120408. [CrossRef]
56. Rosseland, S. *Astrophysik und Atom-Theoretische Grundlagen*; Springer: Berlin/Heidelberg, Germany, 1931.
57. Hussain, S.; Ahmed, S.E.; Akbar, T. Entropy generation analysis in MHD mixed convection of hybrid nanofluid in an open cavity with a horizontal channel containing an adiabatic obstacle. *Int. J. Heat Mass Transf.* **2017**, *114*, 1054–1066. [CrossRef]
58. Merkin, J.H. On dual solutions occurring in mixed convection in a porous medium. *J. Eng. Math.* **1986**, *20*, 171–179. [CrossRef]
59. Weidman, P.D.; Kubitschek, D.G.; Davis, A.M.J. The effect of transpiration on self-similar boundary layer flow over moving surfaces. *Int. J. Eng. Sci.* **2006**, *44*, 730–737. [CrossRef]
60. Harris, S.D.; Ingham, D.B.; Pop, I. Mixed convection boundary-layer flow near the stagnation point on a vertical surface in a porous medium: Brinkman model with slip. *Transp. Porous Media* **2009**, *77*, 267–285. [CrossRef]
61. Shampine, L.F.; Gladwell, I.; Thompson, S. *Solving ODEs with MATLAB*; Cambridge University Press: Cambridge, UK, 2003.

Article

Node Generation for RBF-FD Methods by QR Factorization

Tony Liu [1,*] and Rodrigo B. Platte [2]

[1] Department of Mathematics and Statistics, Air Force Institute of Technology, Dayton, OH 45433, USA
[2] School of Mathematical and Statistical Sciences, Arizona State University, Tempe, AZ 85281, USA; rbp@asu.edu
* Correspondence: Tony.Liu@afit.edu

Abstract: Polyharmonic spline (PHS) radial basis functions (RBFs) have been used in conjunction with polynomials to create RBF finite-difference (RBF-FD) methods. In 2D, these methods are usually implemented with Cartesian nodes, hexagonal nodes, or most commonly, quasi-uniformly distributed nodes generated through fast algorithms. We explore novel strategies for computing the placement of sampling points for RBF-FD methods in both 1D and 2D while investigating the benefits of using these points. The optimality of sampling points is determined by a novel piecewise-defined Lebesgue constant. Points are then sampled by modifying a simple, robust, column-pivoting QR algorithm previously implemented to find sets of near-optimal sampling points for polynomial approximation. Using the newly computed sampling points for these methods preserves accuracy while reducing computational costs by mitigating stencil size restrictions for RBF-FD methods. The novel algorithm can also be used to select boundary points to be used in conjunction with fast algorithms that provide quasi-uniformly distributed nodes.

Keywords: radial basis functions; RBF-FD; node sampling; lebesgue constant; complex regions; finite-difference methods

Citation: Liu, T.; Platte, R.B. Node Generation for RBF-FD Methods by QR Factorization. *Mathematics* **2021**, *9*, 1845. https://doi.org/10.3390/math9161845

Academic Editor: Alicia Cordero Barbero

Received: 29 June 2021
Accepted: 30 July 2021
Published: 5 August 2021

Publisher's Note: MDPI stays neutral with regard to jurisdictional claims in published maps and institutional affiliations.

Copyright: © 2021 by the authors. Licensee MDPI, Basel, Switzerland. This article is an open access article distributed under the terms and conditions of the Creative Commons Attribution (CC BY) license (https://creativecommons.org/licenses/by/4.0/).

MSC: 65D12; 65D25

1. Introduction

In [1–6], Polyharmonic Splines (PHSs) and polynomials were combined to generate radial basis function finite-difference (RBF-FD) methods. One of the key benefits of combining PHSs with polynomials was the fact that high-order accuracy could be obtained from resulting RBF-FD differentiation matrices. Another improvement was the elimination of the requirement to select optimal shape parameters. When implementing RBF-FD methods, the choice of shape parameter plays a crucial role in the conditioning of interpolation matrices as well as accuracy [7,8]. As a result, the need to balance accuracy and conditioning through the tuning of the shape parameter becomes a problem itself. The use of PHSs with polynomials eliminates this requirement. Instead of having to select shape parameters to handle different resolutions, the only parameter selection required is the degree of the PHS and polynomials used, which is pre-selected and remains constant.

The need to tune the shape parameter can be observed in the stagnation error of RBF-FD methods strictly using RBFs. These methods encountered convergence, which plateaued or worsened as the number of sampling points increased. This was directly due to the fact that as the resolution increases, the shape parameter needed to be tuned. As a result, accuracy was traded off in order to maintain the conditioning of interpolation matrices. The implementation of RBF-FD matrices using PHSs and polynomials eliminated such stagnation error. These methods maintain accurate approximations while also eliminating the complexities of shape parameter selection.

Along with these advantages for using PHSs with polynomials came one key constraint: the number of nodes used in each stencil was required to be approximately twice the size of the number of polynomial basis functions appended. For example, in [1,2], stencils comprising of 37 nodes were used. In this case, only polynomials up to degree 4 could be appended to the RBFs. The accuracy of the resulting approximation depends on the degree of the polynomials appended. Thus, for higher-order methods, larger stencils are required. This results in increased computational costs as the differentiation matrices used became less sparse since derivative calculations at each node require function values from an increased number of nearest neighbors. The stencil size then becomes a limiting factor when attempting to achieve a given order of accuracy efficiently.

RBF-FD methods using PHSs and polynomials are usually implemented using Cartesian points, hexagonal points, or quasi-uniformly distributed points. A few references that looked into the placement of sampling points for finite-difference methods include [9–13]; however, for RBF-FD methods in 2D, the general strategy has been to generate a set of quasi-uniformly distributed nodes based on repel algorithms in order to achieve a set spacing. The algorithms for computing these scattered nodes can be found in [14–16]. The points in [14,15] are generated using a spatial density function to inform the spacing of the nodes throughout the domain. Similarly, the points in [16] are generated in order to achieve a predetermined average separation between points. In this paper, we consider finding the placement of sampling points for RBF-FD methods using PHSs and polynomials by minimizing a piecewise-defined Lebesgue constant. This will be accomplished by modifying a column-pivoting QR algorithm previously used to find near-optimal sampling points for polynomial interpolation, also known as the approximate-Fekete points.

The modified column-pivoting QR algorithm presented in this work provides a novel sampling method for RBF-FD methods with three major benefits. First, the sampled points mitigate a key computational constraint of RBF-FD methods implemented with PHSs and polynomials. That is, it dramatically reduces the number of nodes per stencil for high-order approximation as compared to other node distributions such as Cartesian or hexagonal points. This reduces the computational requirements of the RBF-FD method while retaining high-order accuracy as it has been shown that the accuracy of these methods depends on the polynomial degree and not the number of nodes in each stencil [6]. The newly sampled points provide sparser differentiation matrices. The second benefit of the modified column-pivoting QR algorithm is the ability to compute sampling points with a simple, robust method. The implementation of the algorithm only requires a set of candidate points and a choice of basis. The basis used in the novel method is chosen to match the basis used for the RBF-FD computations. Thus, once a set of candidate points is chosen and input into the algorithm, a set of sampling points is provided. This provides a simple algorithm for point selection with few variable parameters. Lastly, the algorithm can be used to inform the placement of boundary points for complex 2D regions. These boundary points can then be used in conjunction with scattered repel point algorithms, which provide quasi-uniformly distributed interior points.

We introduce the basics for RBF-FD methods in Section 2. In Section 3, we introduce the novel piecewise-defined Lebesgue constant used to compare sampling point locations for local RBF-FD methods. In this section, we also introduce the modified column-pivoting QR algorithm used to generate sampling point locations for RBF-FD methods. In Section 4, we investigate the behavior of varying sampling point locations in 1D. Section 5 extends the results from 1D into 2D. Lastly, Section 6 implements test cases using the newly generated points.

2. Background
2.1. RBF Setup

A thorough introduction to RBFs methods can be found in [17–20]. The RBF interpolant is a linear combination of translates of a radially-symmetric function denoted by $\phi(\|x - x_j\|)$. In 1D, interpolating through the points (x_j, y_j) gives us the interpolant of the form

$$s(x) = \sum_{j=0}^{n} c_j \phi(\|x - x_j\|), \tag{1}$$

where the coefficients c_k are found by solving the linear interpolation system:

$$\begin{bmatrix} \phi(\|x_0-x_0\|) & \phi(\|x_0-x_1\|) & \cdots & \phi(\|x_0-x_n\|) \\ \phi(\|x_1-x_0\|) & \phi(\|x_1-x_1\|) & \cdots & \phi(\|x_1-x_n\|) \\ \phi(\|x_2-x_0\|) & \phi(\|x_2-x_1\|) & \cdots & \phi(\|x_2-x_n\|) \\ \vdots & \vdots & \ddots & \vdots \\ \phi(\|x_n-x_0\|) & \phi(\|x_n-x_1\|) & \cdots & \phi(\|x_n-x_n\|) \end{bmatrix} \begin{bmatrix} c_0 \\ c_1 \\ c_2 \\ \vdots \\ c_n \end{bmatrix} = \begin{bmatrix} y_0 \\ y_1 \\ y_2 \\ \vdots \\ y_n \end{bmatrix}. \tag{2}$$

Common examples of RBFs are listed in Table 1. Most of these examples, the multiquadric, the inverse multiquadric, and the Gaussian RBFs in particular, contain the presence of the shape parameter, ε. These shape parameters must be tuned in order to balance conditioning and accuracy, and in the case where the number of sampling nodes used becomes large enough, the accuracy stagnates or decreases due to the need to condition the interpolation matrices (alternative options to shape parameter tuning are presented in [20] but are not considered in this study).

Table 1. Example RBFs.

RBF	Basis Function	Parameter
Polyharmonic Spline	$\phi(r) = r^m$	$m \in 2\mathbb{N} - 1$
Multiquadric	$\phi(r) = \sqrt{1 + (\varepsilon r)^2}$	$\varepsilon \in \mathbb{R}$
Inverse Multiquadric	$\phi(r) = \frac{1}{1+(\varepsilon r)^2}$	$\varepsilon \in \mathbb{R}$
Gaussian	$\phi(r) = e^{-(\varepsilon r)^2}$	$\varepsilon \in \mathbb{R}$

Using PHSs and polynomials, we can write the approximation as

$$s(x) = \sum_{j=0}^{n} c_j |x - x_j|^{2m-1} + \sum_{j=0}^{l} c_{j+n+1} x^j. \tag{3}$$

See [20] for more details on RBF approximations with appended polynomials. To implement a 2D finite-difference method with RBFs, the nearest neighbors are to be used in the finite-difference weight calculations. This is accomplished using MATLAB's *KDTree* and *knnsearch* functions. To calculate the RBF-FD weights at a given point, a stencil size is chosen and the nearest neighbors are found. These nearest neighbors are the points used in the RBF-FD calculation for the given point. Figure 1 below illustrates two examples of what these stencils should look like in a complex 2D region, such as the bumped-disk shape. The sampling points are marked by the dots, while the center point is marked by an asterisk with the relevant stencil points being outlined by circles.

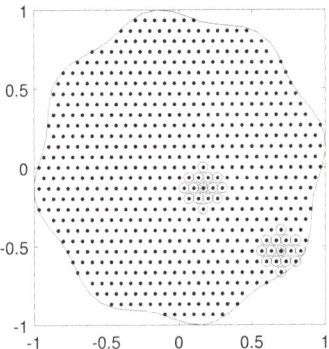

Figure 1. Fifteen-node stencil example for the bumped-disk region.

2.2. Calculating RBF-FD Weights

To find the RBF-FD weights for a given operator $\mathcal{L}$, we first consider the system with strictly RBFs, as shown in Equation (4).

$$\begin{bmatrix} \|\mathbf{x_0}-\mathbf{x_0}\|_2^{2m-1} & \cdots & \|\mathbf{x_0}-\mathbf{x_k}\|_2^{2m-1} \\ \vdots & \ddots & \vdots \\ \|\mathbf{x_n}-\mathbf{x_0}\|_2^{2m-1} & \cdots & \|\mathbf{x_n}-\mathbf{x_k}\|_2^{2m-1} \end{bmatrix} \begin{bmatrix} w_0 \\ \vdots \\ w_k \end{bmatrix} = \begin{bmatrix} L\|\mathbf{x}-\mathbf{x_0}\|_2^{2m-1}|_{\mathbf{x}=\mathbf{x}_c} \\ \vdots \\ L\|\mathbf{x}-\mathbf{x_k}\|_2^{2m-1}|_{\mathbf{x}=\mathbf{x}_c} \end{bmatrix}. \qquad (4)$$

Appending the polynomial terms, as in Equation (3), expands the linear system, as shown in Equations (5) and (6). This calculates the differentiation weights, $w_0, \ldots, w_k$, for the point $\mathbf{x}=\mathbf{x}_c$ using an $k+1$ point stencil with RBFs being appended with the polynomials $1, x, y$. Thus, the top-left sub-matrix is the usual RBF interpolation matrix. We see that a Vandermonde matrix consisting of the same stencil points, but using a monomial basis, is appended to the RBF interpolation matrix. For illustration, the PHSs in this case are combined with polynomials up to the first degree; in most cases, polynomials of higher degree are appended. Here, solving for the weights gives us an RBF-FD approximation for the differentiation operator, L, using PHSs with polynomials.

$$V\mathbf{w} = \mathcal{L}, \qquad (5)$$

$$\begin{bmatrix} \|\mathbf{x_0}-\mathbf{x_0}\|_2^{2m-1} & \cdots & \|\mathbf{x_0}-\mathbf{x_k}\|_2^{2m-1} & 1 & x_0 & y_0 \\ \vdots & \ddots & \vdots & \vdots & \vdots & \vdots \\ \|\mathbf{x_n}-\mathbf{x_0}\|_2^{2m-1} & \cdots & \|\mathbf{x_n}-\mathbf{x_k}\|_2^{2m-1} & 1 & x_k & y_k \\ 1 & \cdots & 1 & 0 & 0 & 0 \\ x_0 & \cdots & x_k & 0 & 0 & 0 \\ y_0 & \cdots & y_k & 0 & 0 & 0 \end{bmatrix} \begin{bmatrix} w_0 \\ \vdots \\ w_k \\ w_{k+1} \\ w_{k+2} \\ w_{k+3} \end{bmatrix} = \begin{bmatrix} L\|\mathbf{x}-\mathbf{x_0}\|_2^{2m-1}|_{\mathbf{x}=\mathbf{x}_c} \\ \vdots \\ L\|\mathbf{x}-\mathbf{x_k}\|_2^{2m-1}|_{\mathbf{x}=\mathbf{x}_c} \\ L1|_{\mathbf{x}=\mathbf{x}_c} \\ Lx|_{\mathbf{x}=\mathbf{x}_c} \\ Ly|_{\mathbf{x}=\mathbf{x}_c} \end{bmatrix}. \qquad (6)$$

As previously mentioned, the system in Equation (5) does not contain any shape parameters, thus eliminating the need to find the optimal value for such a parameter. Instead, the conditioning of the matrix in the left-hand side is achieved by appending the polynomials while using an appropriate stencil size.

2.3. Accuracy Considerations

The convergence rate of RBF-FD methods combining PHSs and polynomials depends on the degree of polynomials used and is independent of both the parameter, m, which defines the PHS, and the stencil size. Thus, approximations converge at the rate of $O(h^p)$, where h is the spacing and p is the degree of polynomials appended. Figure 2 below depicts an example of the convergence rate these RBF-FD methods provide. In this case, a hexagonal nodal set is used on the unit square. A 51 point stencil is used such that there are enough nodes in the stencil to handle the inclusion of polynomials up to degree $p=5$. The PHS used is $\phi(r) = r^3$. The relative error of the approximation of $\frac{d}{dx}(1+\sin(4x)+\cos(3x)+\sin(2y))$ is plotted against the spacing, h, along with the expected convergence rate for each degree of polynomials used in dashed lines.

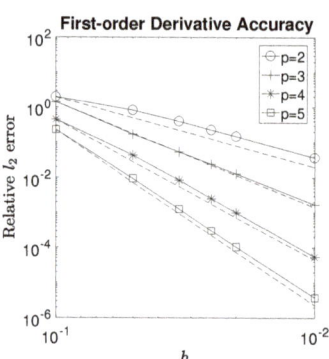

Figure 2. Convergence rates of a first-order derivative approximation using PHS and polynomials.

3. Node Sampling for RBF-FD Methods

To find sampling points for RBF-FD methods, we calculate the set of points with a minimal Lebesgue constant. Lebesgue constants are commonly used to determine the optimality of sets of sampling points for polynomial interpolation. The goal will be to formulate a piecewise-defined Lebegue constant for RBF-FD methods. Previous works [9–12] have looked to find point locations for finite-difference methods but do not formulate piecewise-defined Lebesgue constants and have not focused on RBF-FD methods using PHSs and polynomials. A few works, however, have considered Lebesgue constants for RBF-FD methods for other purposes [21,22].

3.1. The Piecewise-Defined Lebesgue Constant for RBF-FD Methods

To formulate a notion of the Lebesgue constant for RBF-FD methods, we first recall that for polynomial interpolation, given a set of $n+1$ sampling points, $[(x_0,y_0),\ldots,(x_n,y_n)]$, the Lebesgue constant is defined as:

$$\Lambda = \sup_f \frac{\|p_f\|_\infty}{\|f\|_\infty} = \sup_{(x,y)} \sum_{j=0}^{n} |l_j(x,y)|. \qquad (7)$$

Here, p is the approximation of functions $f \in C([-1,1])$ and the l_j's are the cardinal functions which satisfy:

$$l_j(x_k,y_k) = \begin{cases} 1 & k=j \\ 0 & k \neq j. \end{cases} \qquad (8)$$

To define the Lebesgue constant for RBF-FD methods, the cardinal functions must be formulated similarly. Furthermore, the cardinal functions must be considered in a piecewise manner to account for the local nature of RBF-FD methods. This is accomplished by considering the piecewise cardinal functions.

Consider 1D piecewise polynomial interpolation with 4 sampling points, $\left(-1, -\frac{1}{3}, \frac{1}{3}, 1\right)$, using a 3 point stencil. In this case, there are two stencil groupings as outlined in Figure 3 below. For this example, the piecewise cardinal functions are shown in Figure 4. Each piecewise cardinal function contains a discontinuity at $x=0$. For a given x, each piecewise cardinal function, $l_j(x)$, is defined using the 3 closest sampling points.

Thus, for $x \in [-1,0]$, the piecewise cardinal functions are defined using points $\left(-1,-\frac{1}{3},\frac{1}{3}\right)$, while for $x \in (0,1]$, the piecewise cardinal functions are defined using points $\left(-\frac{1}{3},\frac{1}{3},1\right)$.

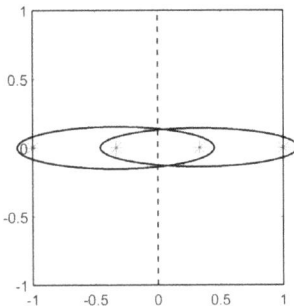

Figure 3. Stencil grouping example.

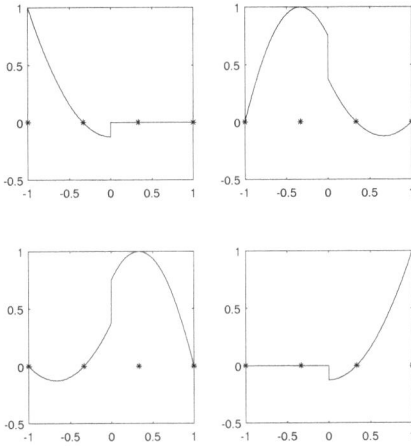

Figure 4. Piecewise cardinal function example. *Clockwise starting from the top left:* $l_1(x)$, $l_2(x)$, $l_4(x)$, $l_3(x)$.

For RBF-FD methods, the piecewise cardinal functions are defined similarly. Consider using a k-point stencil with degree $p=1$ polynomials being appended. The given 2D region is discretized into a fine mesh, Ω, to calculate the piecewise cardinal functions on. Then, the k nearest neighbors from a given set of sampling points, $[(x_0,y_0),\ldots,(x_n,y_n)]$, are found for each point in Ω. The cardinal function coefficients for a given stencil grouping, $[(x'_0,y'_0),\ldots,(x'_k,y'_k)]$, can be calculated by solving the following system in Equation (9).

$$\begin{bmatrix} \|x'_0 - x'_0\|_2^{2m-1} & \cdots & \|x'_0 - x'_k\|_2^{2m-1} & 1 & x'_0 & y'_0 \\ \vdots & \ddots & \vdots & \vdots & \vdots & \vdots \\ \|x'_k - x'_0\|_2^{2m-1} & \cdots & \|x'_k - x'_k\|_2^{2m-1} & 1 & x'_k & y'_k \\ 1 & \cdots & 1 & 0 & 0 & 0 \\ x'_0 & \cdots & x'_k & 0 & 0 & 0 \\ y'_0 & \cdots & y'_k & 0 & 0 & 0 \end{bmatrix} \begin{bmatrix} c_{0,0} & \cdots & c_{0,k} \\ \vdots & \vdots & \vdots \\ c_{k,0} & \cdots & c_{k,k} \\ c_{k+1,0} & \cdots & c_{k+1,k} \\ c_{k+2,0} & \cdots & c_{k+2,k} \\ c_{k+3,0} & \cdots & c_{k+3,k} \end{bmatrix} = \begin{bmatrix} \mathcal{I} \\ 0 \end{bmatrix}. \quad (9)$$

Once the piecewise cardinal function coefficients, $\mathbf{C} = c_{i,j}$ for $i=0,\ldots,k, j=0,\ldots,k$, are obtained, the matrix of piecewise cardinal functions is built according to Equation (10). Here, $(\mathbf{xx_0} \ldots \mathbf{xx_m}) \subseteq \Omega$ denote the points in the fine mesh that have $[(x'_0,y'_0),\ldots,(x'_k,y'_k)]$ as the k-nearest neighbors, and the indices $i_0 \ldots i_k$ are such that $\mathbf{x}_0' = \mathbf{x}_{i_0}' \ldots \mathbf{x}_k' = \mathbf{x}_{i_k}'$. Once this process is repeated for all possible stencil groupings, the full matrix of piecewise cardinal functions will have been tabulated, and the Lebesgue constant for RBF-FD methods is defined as $\Lambda_{RBF-FD} = \sup_{(x,y)} \sum_{j=0}^{n} |l_j(x,y)|$.

$$l = \begin{bmatrix} l_{i_0}(\mathbf{xx_0}) & \ldots & l_{i_k}(\mathbf{xx_0}) \\ \vdots & \vdots & \vdots \\ l_{i_0}(\mathbf{xx_m}) & \ldots & l_{i_k}(\mathbf{xx_m}) \end{bmatrix},$$

$$l = \begin{bmatrix} \|\mathbf{xx_0} - \mathbf{x}_0'\|_2^{2m-1} & \ldots & \|\mathbf{xx_0} - \mathbf{x}_k'\|_2^{2m-1} & 1 & xx_0 & yy_0 \\ \vdots & \ddots & \vdots & \vdots & \vdots & \vdots \\ \|\mathbf{xx_m} - \mathbf{x}_0'\|_2^{2m-1} & \ldots & \|\mathbf{xx_m} - \mathbf{x}_k'\|_2^{2m-1} & 1 & xx_m & yy_m \end{bmatrix} \mathbf{C}. \qquad (10)$$

3.2. Modified Column-Pivoting QR Algorithm (MCpQR Algorithm)

In the previous section, we discussed the metric used to determine the optimality of a set of sampling points for RBF-FD methods using PHSs and polynomials. In this section, we discuss how to sample point locations using this optimality measure. The algorithm proposed is a modification of an algorithm commonly used to find near-optimal sampling points for polynomial interpolation.

Finding optimal and near-optimal sampling points for polynomial interpolation has been studied extensively [23–33]. A robust algorithm for near-optimal polynomial interpolation sampling is modified to be used for RBF-FD methods. This method, which performs column-pivoting QR factorizations, was originally implemented to approximate the Fekete points. These points maximize the denominator of the cardinal function determinant definition shown in Equation (11):

$$l_j(x,y) = \frac{det(V_n[(x_0,y_0),\ldots,(x_{j-1},y_{j-1}),(x,y),(x_{j+1},y_{j+1}),\ldots,(x_n,y_n)])}{det(V_n[\mathbf{x}])}, \qquad (11)$$

where V_n is the Vandermonde matrix defined as:

$$V_n[\mathbf{x}] = V_n[(x_0,y_0),\ldots,(x_n,y_n)]$$

$$= \begin{bmatrix} \phi_0(x_0,y_0) & \phi_1(x_0,y_0) & \phi_2(x_0,y_0) & \ldots & \phi_n(x_0,y_0) \\ \phi_0(x_1,y_1) & \phi_1(x_1,y_1) & \phi_2(x_1,y_1) & \ldots & \phi_n(x_1,y_1) \\ \phi_0(x_2,y_2) & \phi_1(x_2,y_2) & \phi_2(x_2,y_2) & \ldots & \phi_n(x_2,y_2) \\ \vdots & \vdots & \vdots & \ddots & \vdots \\ \phi_0(x_n,y_n) & \phi_1(x_n,y_n) & \phi_2(x_n,y_n) & \ldots & \phi_n(x_n,y_n) \end{bmatrix}.$$

Here, n denotes the number of basis columns in the Vandermonde matrix.

By maximizing this denominator term, the Fekete points provide bounds for the cardinal functions, as well as the Lebesgue constant. These bounds are $\|l_j\|_\infty \leq 1$ and $\Lambda \leq n+1$. To approximate the Fekete points, a greedy algorithm was used in [24,26]. The domain is first discretized into candidate points, $\mathbf{x} = (x_i, y_i)_{i=1}^M \in \Omega$. Then, to select N approximate-Fekete points, the corresponding N-column Vandermonde matrix, $V_{N-1}[\mathbf{x}] \in \mathbb{R}^{M \times N}$, is generated. Finally, the greedy algorithm in Algorithm 1 is applied to $\mathbf{A} = V_{N-1}'[\mathbf{x}]$.

Algorithm 1 Greedy Volume Submatrix Algorithm

- Select ind_1 as the index of the column of $\mathbf{A}$ with maximum length.
- Given indexes $ind_1, \ldots, ind_k$, select ind_{k+1} such that the volume generated by columns $ind_1, \ldots, ind_k, ind_{k+1}$ is maximal.

This greedy algorithm can be easily implemented using a column-pivoting QR factorization. A 1D example is given in Algorithm 2 below. A deeper explanation of approximate-Fekete points can be found in [24,26].

Algorithm 2 Example Column-Pivoting QR Algorithm

$n = 21$; % number of interpolation points
$m = 1000$; % number of candidate points
$xx = linspace(-1,1,m)$;
A = gallery('chebvand',n,xx) % generate Vandermonde matrix with Chebyshev basis
[Q, R, E]=qr(A,'vector')
pts=xx(E(1:n))

A modified Column-pivoting QR Algorithm (MCpQR algorithm) is used to find sampling nodes for RBF-FD methods combining PHSs and polynomials on complex regions in 2D. The proposed method provides a robust algorithm for finding sampling nodes on general complex regions. Furthermore, these nodes display the expected behavior in terms of accuracy and convergence and build upon those results by providing differentiation matrices with increased sparsity through the mitigation of crucial stencil size constraints.

In order to find sampling points using the MCpQR algorithm, a set of candidate points and a basis to populate the matrix upon which we perform the column-pivoted QR factorization is required. Suppose the region is discretized into candidate points $\mathbf{x} = (x_i, y_i)_{i=0}^M \in \Omega$. To find sampling points in the RBF-FD setting, a basis needs to be selected. In the case of RBF-FD methods, the locations of the centers are required in order to determine the RBF basis. Furthermore, changing the location of the sampling nodes also changes the RBF basis. Thus, in order to select a basis to use for the MCpQR algorithm, we first make a starting guess for the sampling node locations. The matrix used in the MCpQR algorithm must also account for this dynamic basis. The obvious choice of basis then becomes the piecewise cardinal function basis. That is, to find $n+1$ sampling points, the matrix calculated using Equation (10), L in Equation (12), is chosen as the matrix to perform the MCpQR algorithm on. Specifically, we perform the algorithm on L' since column selection on L' represents selecting candidate points.

$$L = \begin{bmatrix} l_0(x_0,y_0) & l_1(x_0,y_0) & \ldots & l_n(x_0,y_0) \\ l_0(x_1,y_1) & l_1(x_1,y_1) & \ldots & l_n(x_1,y_1) \\ l_0(x_2,y_2) & l_1(x_2,y_2) & \ldots & l_n(x_2,y_2) \\ \vdots & \vdots & \ddots & \vdots \\ l_0(x_M,y_M) & l_1(x_M,y_M) & \ldots & l_n(x_M,y_M) \end{bmatrix}. \qquad (12)$$

Since the piecewise cardinal function basis is dependent on the sampling node locations, a starting guess is used to populate L. From here, the MCpQR algorithm is iterated. We found that in most cases, 1 iteration is enough to obtain significantly better Lebesgue constants for RBF-FD methods. In some rare cases, up to 5 iterations are required.

It is important to note the computational costs of the MCpQR algorithm. The computational costs of the algorithm may be broken down into two main parts: the calculation of the matrix L and the implementation of the column-pivoting QR factorization. Due to the piecewise nature, the matrix L is sparse. This is depicted in the 1D example shown in Figures 3 and 4. Thus, we can save computational costs by only calculating the non-zero parts of L. Each row in L has the same number of non-zero elements as there are points in the stencil used. Further, candidate points with the same set of nearest neighbors can be grouped together to form a linear system in which the cardinal function coefficients are solved for (Equation (9)). We notice that to solve this system, we must compute the inverse of the RBF-FD matrix of dimensions $(k+1+d) \times (k+1+d)$, where d is the number of polynomials basis functions appended. Thus, for the unique stencil grouping, the cost is $\mathcal{O}(k+1+d)^3$. This approach populates L in a piecewise manner. We note this cost is similar to the computational cost of generating the differentiation matrix for a given set of sampling points, which requires solving the system in Equations (5) and (6).

Once the matrix L is populated, a QR factorization is performed. This factorization costs $\mathcal{O}\left((M+1)(n+1)^2\right)$. This factorization comprises the majority of the computational cost. We see that the algorithm can benefit by limiting the number of candidate points, $M+1$. We discuss in Sections 4 and 5 strategies for reducing this cost.

4. Results in 1D

To study the behavior of node configurations for RBF-FD methods using PHSs and polynomials, we begin with an investigation in 1D. Along with the MCpQR algorithm, we can consider other point locations generated by mappings made possible due to the simpler nature of 1D domains. We compare the points from the MCpQR algorithm with the mapped points in terms of eigenvalue stability and Lebesgue constant optimality.

4.1. Mapped Point Sets

A few references that have looked into the placement of sampling points for finite-difference methods in 1D include [9–12,34]. The strategies used in these works include adding nodes near the boundary to stabilize differentiation operators, moving nodes near the boundary to optimize piecewise polynomial error formulae, and transforming node locations using a mapping to stabilize differentiation operators. It is important to see that these strategies focus on the placement of nodes near the boundaries. We will investigate the effects of similar behavior near the boundary for 1D RBF-FD methods using PHSs and polynomials in this section.

In 1D, we leverage the mapping proposed in [35] to control the placement of nodes near the boundary. This mapping was implemented in [12,34] to generate point sets for 1D finite-difference methods and in [36] for RBF approximations in 1D. The mapping, which we shall refer to as the KTE mapping, is defined as:

$$x_{kte} = \frac{\arcsin(\alpha x_{cheb})}{\arcsin(\alpha)}. \tag{13}$$

x_{cheb} represents a set of Chebyshev points. The outputted x_{kte} approach Chebyshev points as $\alpha \to 0$, while for $\alpha = 1$, x_{kte} are equispaced points. Alternatively, we also consider the inverse of this mapping, which we shall call the IKTE mapping defined by:

$$x_{ikte} = \frac{\sin(\arcsin(\alpha x_{equi}))}{\alpha}. \tag{14}$$

x_{equi} represent a set of equispaced points. The outputted x_{ikte} approach equispaced points as $\alpha \to 0$, while for $\alpha = 1$, x_{kte} are Chebyshev points.

With these two mappings, we have one tunable parameter, α, that controls the spacing of points near the boundary with a set of Chebyshev or equispaced points being the input for the mappings. This allows us to investigate the behavior of point locations near the boundary in terms of Lebesgue constant optimality and eigenvalue stability. In view of the importance that certain eigenvalues have in the analysis of real models formulated by Partial Differential Equations (PDEs), we refer for completeness also to [37,38].

For example, we consider a 37 point stencil, $\phi(r) = r^5$, polynomials up to degree $p = 14$, and 1000 nodes on the interval $[-1,1]$ for RBF-FD calculations. For the KTE and IKTE mapping, we use MATLAB's fmincon to find the α that minimizes the Lebesgue constant. For the KTE mapping, we plot the inputted Chebyshev points and the resulting Dirichlet Laplacian eigenvalues on the left column of Figure 5. The spacing for the points resulting from finding the α that minimizes the Lebesgue constant and the corresponding Dirichlet Laplacian eigenvalues are depicted in the right column of Figure 5. In this case, the Lebesgue constant for the Chebyshev points and the mapped points are $\Lambda_{RBF-FD} = 2.69$ and $\Lambda_{RBF-FD} = 1.82$, respectively. We notice that in this case, the mapped points are equispaced away from the boundary and become clustered close to the boundary. Both sets of points have negative, real eigenvalues; however, the mapped points have eigenvalues of smaller magnitude due to having a larger minimum spacing than the Chebyshev points.

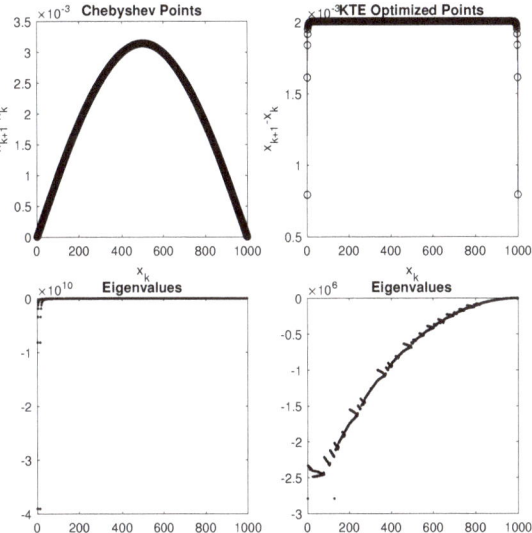

Figure 5. Chebyshev points, KTE optimized points, and their eigenvalues for a 37 point stencil with polynomials up to degree $p = 15$. In this case, both point sets produce purely real eigenvalues.

Alternatively, we plot the results for the IKTE mapping in Figure 6. We notice that for equispaced points, the eigenvalues are imaginary. After using the IKTE mapped points, the eigenvalues return to being purely real. In this case, the Lebesgue constant for the equispaced points and the mapped points are $\Lambda_{RBF-FD} = 4.59$ and $\Lambda_{RBF-FD} = 1.82$, respectively.

One thing we notice is that α that minimizes Λ_{RBF-FD} is very close to 1 from both mappings. That is, the KTE mapping maps the Chebyshev points to points close to equispaced points, and the IKTE mapping maps the equispaced points to points close to the Chebyshev points. This behavior illustrates the fact that the two mappings impact the behavior of clustering near the boundary in different ways, depending on what set of points is being inputted. Figure 7 illustrates the behavior of Λ_{RBF-FD} for different values of α using the IKTE mapping. The subfigure on the right uses a log scale for α to illustrate the behavior of Λ_{RBF-FD} for α close to 1. The optimal α is circled.

The results from the KTE and IKTE mapping in this case lead us to conclude that some clustering near the boundary gives the best results due to the fact that the equispaced points lead to eigenvalues with a non-zero imaginary part, while both mapped sets lead to purely real eigenvalues. Although the KTE and IKTE mappings inform the behavior of the placement of nodes for RBF-FD methods by tuning just one parameter, these mappings cannot be translated to 2D complex regions. As a result, we need a robust algorithm for placing points near the boundary in 2D: the MCpQR algorithm.

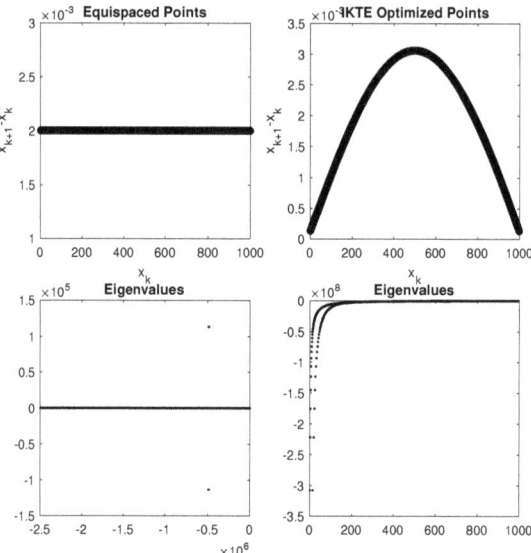

Figure 6. Equispaced points, IKTE optimized points, and their eigenvalues for a 37-point stencil with polynomials up to degree $p = 15$. The equispaced points produce complex eigenvalues, while the IKTE optimized points produce purely real eigenvalues.

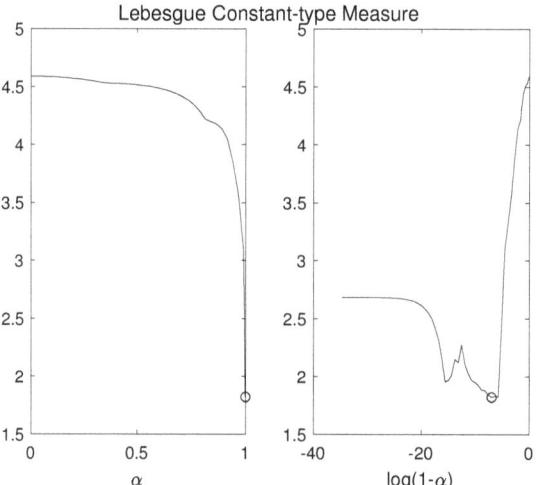

Figure 7. Lebesgue constant for the IKTE mapped points using a 37-point stencil with polynomials up to degree $p = 15$. The optimal α is circled.

4.2. MCpQR Algorithm Point Sets

Based on the results in Section 4.1, we see that a set of points that are equispaced in most of the domain and clustered close to the boundary provide better eigenvalues. As a result, we would like to be able to generate a similar point set using the MCpQR algorithm. This algorithm would then be used to generate point sets for complex 2D regions.

Using the same selections for PHS, polynomial degree, stencil size, and number of points as used in Section 4.1, we implement the MCpQR algorithm to compute point locations for RBF-FD methods. As mentioned previously, we require a starting guess of points to populate the piecewise cardinal function basis. Naturally, from the results

Section 4.1, we use the equispaced points as the starting guess. The spacing for the points computed by the MCpQR algorithm along with the Dirichlet Laplacian eigenvalues are plotted in Figure 8. We notice that the algorithm is again able to compute points with purely real eigenvalues. The eigenvalues closely resemble those from the KTE mapping in Figure 5. In this case, we achieve $\Lambda_{RBF-FD} = 1.85$

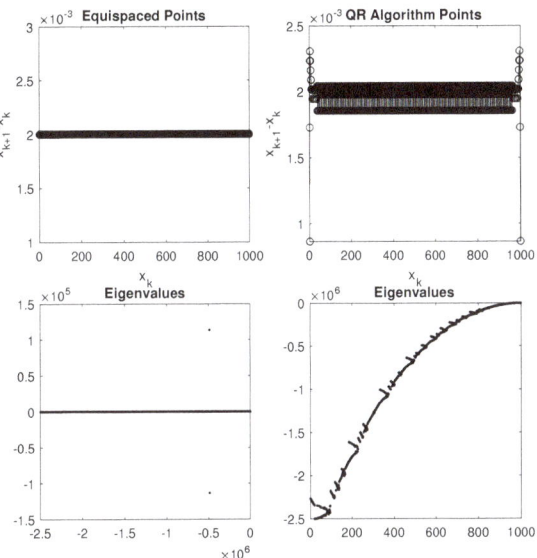

Figure 8. MCpQR algorithm point spacing and Dirichlet Laplacian eigenvalues compared to equispaced points.

We notice that the points computed by the MCpQR algorithm are again nearequispaced for most of the domain and clustered close to the boundary. One way to decrease unnecessary computational costs is to optimize only the points close to the end points of the domain. Thus, we choose a set of equispaced points away from the boundary and keep them fixed. Then, we can choose the spacing of the points near the boundary using our novel algorithm. The candidate points are populated only near the boundary, eliminating the need to incorporate candidate points on the majority of the $[-1, 1]$ interval. This greatly reduces the computational costs outlined for the QR factorization in Section 3.2. Figure 9 illustrates the resulting point set when implementing this boundary-restricted approach. Starting with 1000 equispaced points, we restrict the $1000 - 2k$ interior points and allow the k points closest to -1 and 1 to be moved. The resulting points achieve $\Lambda_{RBF-FD} = 1.85$, the same value that resulted from an unrestricted algorithm. Additionally, we notice that the spacing near the boundary and eigenvalues are similar to the unrestricted algorithm. Thus, we are able to obtain these points for RBF-FD methods by just moving selecting points near the boundary using the MCpQR algorithm.

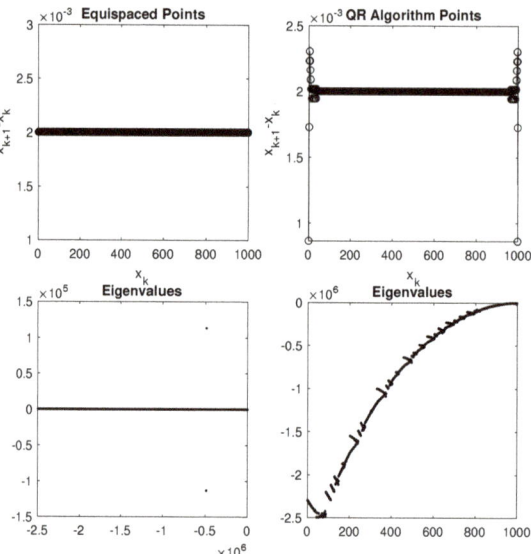

Figure 9. MCpQR algorithm point spacing and Dirichlet Laplacian eigenvalues compared to equispaced points. In this case, the majority of the points are fixed.

5. Results in 2D

Following the results in 1D, we naturally progress to point sets for RBF-FD methods in 2D. The MCpQR algorithm can be used in 2D as long as we have the required basis and candidate points. We begin with rectangular domains and follow with more complex 2D regions. We will demonstrate that the MCpQR algorithm provides a simple, robust algorithm for finding point sets for RBF-FD methods with reduced computational cost.

5.1. Unit Square Results

The first 2D region we consider is the unit square. The unit square allows us to consider the tensor product of resulting 1D point sets. We consider again a 37-point stencil, $\phi(r) = r^5$, polynomials up to degree $p = 4$, and 961 nodes on the interval for RBF-FD calculations. In this case, the 961 nodes are a tensor product of 31 nodes on the $[-1, 1]$ interval. Polynomials up to degree $p = 4$ append 15 polynomial basis functions, the same number appended for polynomials up to degree 14 in 1D. Figures 10–12 plot the resulting QR algorithm points when using tensor product 1D points, hexagonal points, and scattered points as starting guesses. The tensor product 1D points are obtained by taking the tensor product of the points found using the QR algorithm in 1D, as shown in Figure 9.

We notice that for explicit time-stepping, the hexagonal points and the scattered points provide the best eigenvalues. Using these sets for starting guesses, the MCpQR algorithm moves the points near the boundary to decrease the Lebesgue constant while preserving the general behavior of the eigenvalues. This is important, as for complex regions, we can place the hexagonal points within the complex region, draw the boundary of the complex region and move the points near the boundary with the algorithm. This will provide a method similar to the algorithm used to obtain scattered points for complex regions. We note that the tensor product points result in less optimal eigenvalues. The MCpQR algorithm does not move the points near the boundaries for these sets. Thus, these point sets should not be considered.

In Figure 13, we implement the MCpQR algorithm without fixing any nodes. The closely matched results from Figures 11 and 13 show that limiting the algorithm to just moving the points near the boundary produces adequate point sets while eliminating the computational costs required by moving points both close to and away from the boundary.

We notice that the points obtained from the MCpQR algorithm strongly depend on the starting guess. Thus, we can conclude from this that the points from the QR algorithm can only be considered as local minima, not global minima.

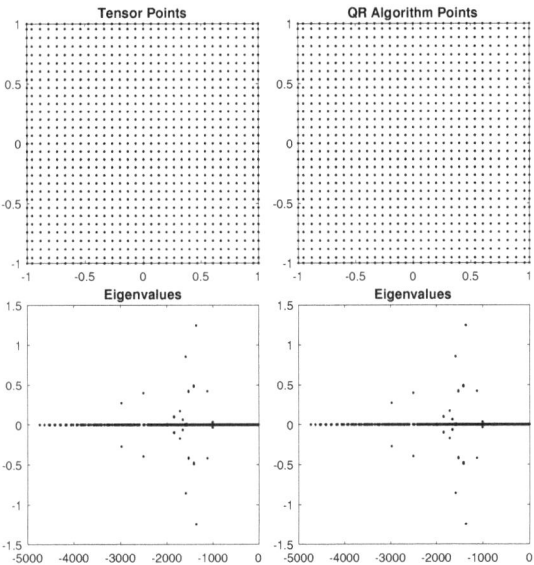

Figure 10. Tensor product points and resulting MCpQR algorithm points with Dirichlet Laplacian eigenvalues.

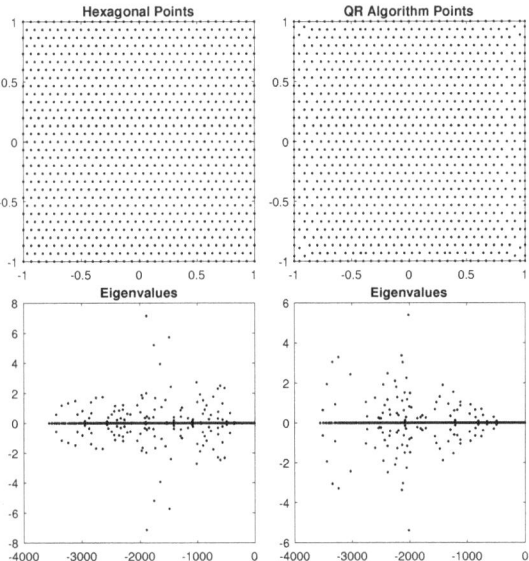

Figure 11. Hexagonal points and resulting MCpQR algorithm points with Dirichlet Laplacian eigenvalues.

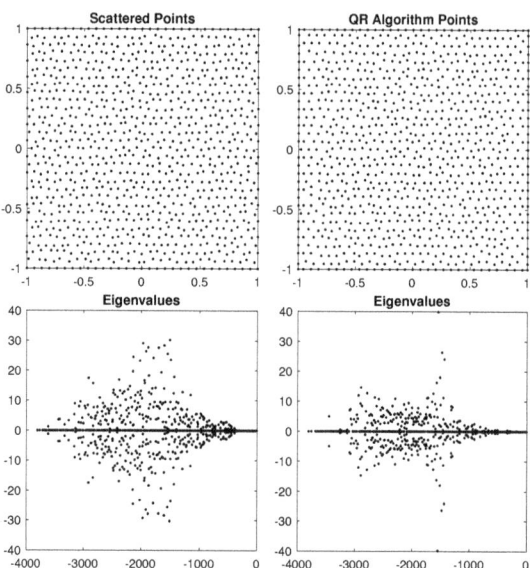

Figure 12. Scattered points and resulting MCpQR algorithm points with Dirichlet Laplacian eigenvalues.

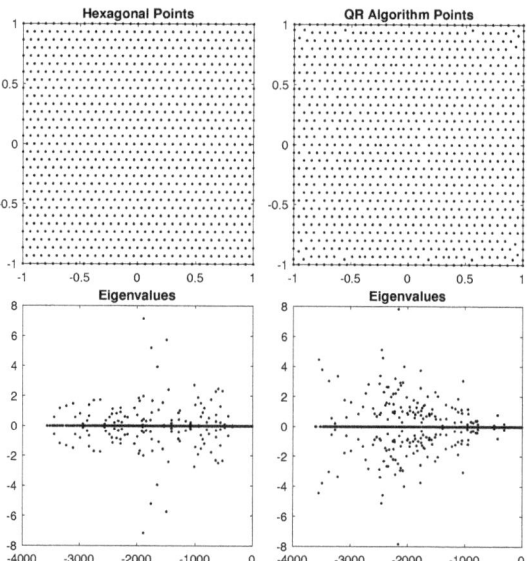

Figure 13. Hexagonal points and resulting MCpQR algorithm points with Dirichlet Laplacian eigenvalues. No interior points are fixed.

Next, we investigate the behavior of point sets for complex 2D regions. For complex regions, we consider the scattered points along with the points resulting from inputting hexagonal points into the QR algorithm since these two sets performed the best on the unit square. We notice three key benefits of using the MCpQR algorithm to generate points for RBF-FD methods in the examples above.

First, the robustness and simplicity of the algorithm allow us to easily generate point sets for any given region. As mentioned previously, the only requirements are a given basis and a set of candidate points. The only tunable parameters in this case are how many candidate points to use since the RBF-FD method already determines the basis used.

Second, in this case, the MCpQR algorithm-generated points produced eigenvalues with smaller imaginary parts. Thus, these points produced eigenvalues closer to the true Dirichlet Laplacian eigenvalues. This implies that for convective PDEs, less hyperviscosity may be needed to be applied in order to handle spurious eigenvalues that arise from the imaginary parts of the Dirichlet Laplacian.

Lastly, the points generated by the MCpQR algorithm allow for a decrease in the stencil size requirements for RBF-FD methods. It has been previously recommended that stencil sizes be at least twice the number of polynomial basis functions appended. Thus, for the example used for the unit square, the stencil size should contain at least 30 points to maintain the conditioning of the system in Equation (5). The use of the points generated by the MCpQR algorithm alleviates the stencil size requirement. For this example, we are able to find points for the RBF-FD method that allow for the use of a 19 point stencil. This is done by first starting with a hexagonal point set using an adequate stencil size (30 in this example), performing the MCpQR algorithm, and using the resulting set as the starting guess to again run the MCpQR algorithm but now with a smaller stencil size. This is then iterated until the conditioning of the system degrades. The resulting point set for a 19 point stencil is shown in Figure 14.

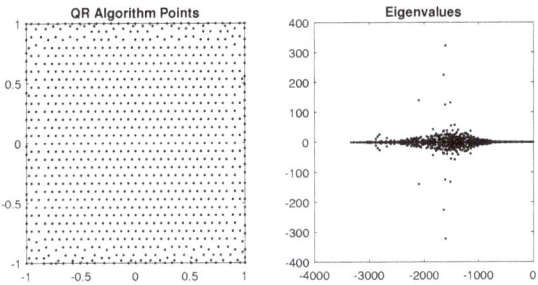

Figure 14. Resulting points for a 19-point stencil with Dirichlet Laplacian eigenvalues. These points were obtained by iteratively applying the MCpQR algorithm for smaller stencils.

5.2. Complex 2D Regions

We adapt the MCpQR algorithm to generate point sets for RBF-FD methods on complex 2D regions. We employ the same strategy: determine a starting point set, fix the interior nodes, and implement the MCpQR algorithm to choose the location of points near the boundary. We use the hexagonal points as the starting guess for the MCpQR algorithm since these points were shown to perform the best in Section 5.1.

For complex regions, we populate the hexagonal points on the unit square, draw the complex region, and keep only the points lying on the interior of the shape. The boundary points of the complex region are then appended to the point set used as the starting guess. In Figure 15, both the starting guess and the resulting MCpQR algorithm sampling nodes for the bumped-disk region are plotted, along with their respective Dirichlet Laplacian eigenvalues. This case considers the bumped-disk region using a 37 point stencil, $\phi(r) = r^3$, and polynomials up to degree $p = 4$. In this example, 734 nodes are used for RBF-FD calculations.

We note that the described method for populating the initial guess produces points that lie too close to each other. This occurs when the boundary points for the shape are located close to the hexagonal grid. As a result, the Dirichlet Laplacian eigenvalues are affected due to the close proximity of certain points. We notice that the MCpQR algorithm is able to remedy the clustering of points near the boundary and improve the behavior of the eigenvalues.

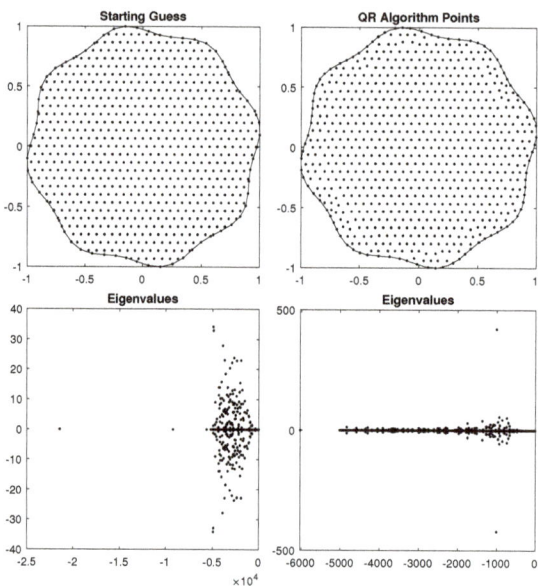

Figure 15. Hexagonal points and resulting MCpQR algorithm points with Dirichlet Laplacian eigenvalues for the bumped-disk region.

Figure 16 displays the results for another complex region: the peanut region. This example also considers a 37-point stencil, $\phi(r) = r^3$, and polynomials up to degree $p = 4$. Here, 830 nodes are used for the RBF-FD calculations. Similar improvements in the spacing of points from the starting guess are observed.

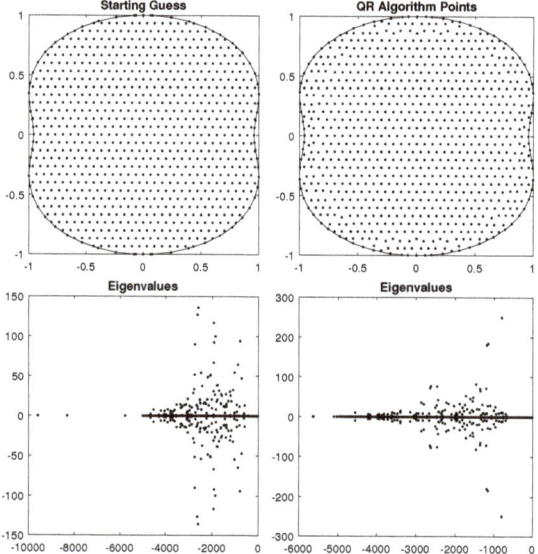

Figure 16. Hexagonal points and resulting MCpQR algorithm points with Dirichlet Laplacian eigenvalues for the peanut region.

We see that the strategy described in this section provides another method for populating point locations for RBF-FD methods on complex regions. The MCpQR algorithm is able to handle complex regions. Furthermore, as mentioned in Section 5.1, the MCpQR algorithm is able to be implemented with a few simple parameter selections (number of candidate points and basis) and allows for decreased computational costs as a result of lower stencil size requirements. Tables 2 and 3 list the stencil size requirement improvement obtained by using the iteratively chosen points for different selections of polynomial degree and PHS degree for both the bumped-disk and peanut regions.

Table 2. Stencil size reduction for different selections of polynomial degree and PHS degree for the bumped-disk region.

Bumped-Disk Region, 734 Nodes			
Polynomial Degree	PHS Degree	Two Times the Number of Polynomial Basis Vectors	Required Stencil Size (Optimized Pts)
deg = 3	r^3	k = 20	15
deg = 3	r^5	k = 20	15
deg = 3	r^7	k = 20	15
deg = 4	r^3	k = 30	19
deg = 4	r^5	k = 30	21
deg = 4	r^7	k = 30	21
deg = 5	r^3	k = 42	31
deg = 5	r^5	k = 42	31
deg = 5	r^7	k = 42	27

Table 3. Stencil size reduction for different selections of polynomial degree and PHS degree for the peanut region.

Peanut Region, 830 Nodes			
Polynomial Degree	PHS Degree	Two Times the Number of Polynomial Basis Vectors	Required Stencil Size (Optimized Pts)
deg = 3	r^3	k = 20	15
deg = 3	r^5	k = 20	15
deg = 3	r^7	k = 20	15
deg = 4	r^3	k = 30	21
deg = 4	r^5	k = 30	25
deg = 4	r^7	k = 30	25
deg = 5	r^3	k = 42	31
deg = 5	r^5	k = 42	31
deg = 5	r^7	k = 42	33

It should be noted that the MCcQR algorithm points and the scattered repel algorithm points perform similarly on complex regions with regards to stencil requirements and eigenvalue stability. Figure 17 plots the repel algorithm points on the bumped-disk region with $\phi(r) = r^3$ and polynomials up to degree $p = 3$. In this case, the repel algorithm points are able to handle a stencil size of 15 as well. Applying the MCpQR algorithm reduces the repel points starting guess measure from $\Lambda_{RBF-FD} = 8.91$ to $\Lambda_{RBF-FD} = 3.56$; however, there is no such improvement with regards to eigenvalue stability.

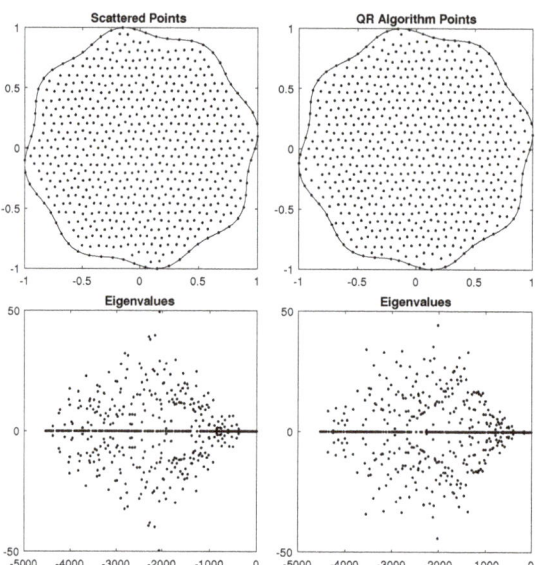

Figure 17. Scattered repel algorithm points and resulting MCpQR points with Dirichlet Laplacian eigenvalues.

The 1D results from Section 4 suggest there should be some point clustering near the boundary of the region. It seems the MCpQR algorithm is not able to recreate the same behavior from 1D in 2D complex regions. After inputting the repel algorithm points (a quasi-uniformly distributed set with no clustering near the boundary) as a starting guess, the MCpQR does not improve the eigenvalues. This may be due to the fact that the algorithm is generating 'local minima' type point sets. As a result, it is concluded that these repel algorithm points perform well on 2D complex regions.

One major benefit the MCpQR algorithm can provide on complex 2D regions is boundary point selection. Currently, the repel algorithm discretizes an equispaced boundary and keeps the boundary points fixed throughout the algorithm [4]. In this case, the algorithm does not inform any selection of boundary points. The MCpQR algorithm can be used in conjunction with the repel algorithm to identify which boundary points to use along with the interior points resulting from the repel algorithm. Consider the bumped-disk region using a 37-point stencil, $\phi(r) = r^3$, and polynomials up to degree $p = 4$. In Figure 18, we see that if we implement the scattered repel algorithm points with too few points on the boundary, the MCqQR algorithm selects additional points to place on the boundary. In this case, the number of boundary points increases from 31 to 63. We notice the improvement in the imaginary part of the eigenvalues. Thus, the MCpQR algorithm can be applied to determine a minimum number of boundary points to use with the scattered repel algorithm points. This again improves eigenvalue stability while decreasing computational cost by keeping the number of boundary points to a minimum. Figure 19 illustrates the same results for the peanut region using a 37 point stencil, $\phi(r) = r^3$, and polynomials up to degree $p = 4$. In this case, the number of boundary points increases from 31 to 68, and the same improvement in the imaginary part of the eigenvalues is observed. We see that the MCpQR algorithm can be used in conjunction with the scattered repel point algorithm to generate a set of boundary points along with a set of quasi-uniformly distributed interior points with reduced computational requirements and improved eigenvalue stability.

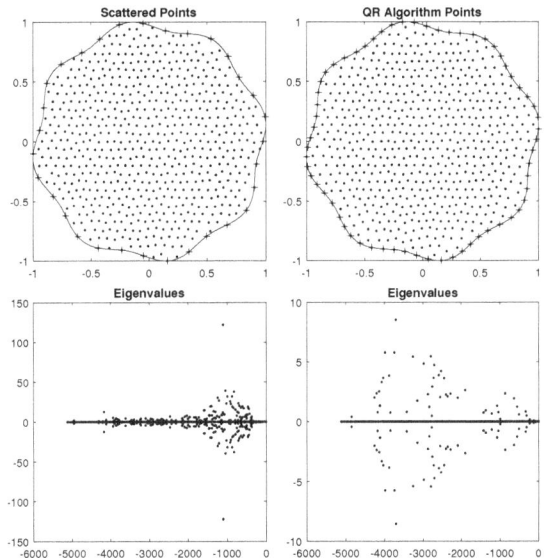

Figure 18. MCpQR boundary selection for scattered repel algorithm points on the bumped-disk region.

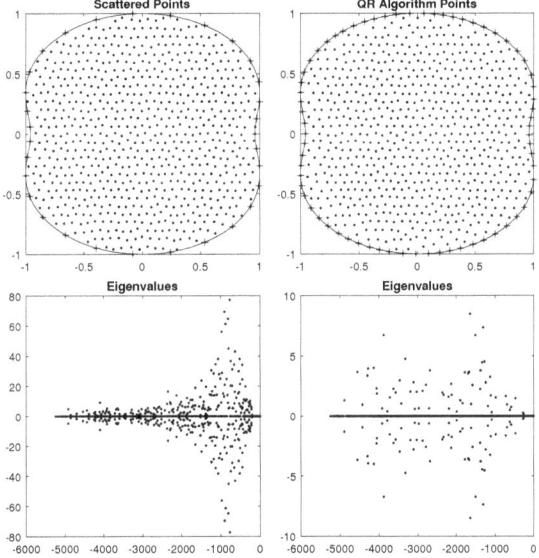

Figure 19. MCpQR boundary selection for scattered repel algorithm points on the peanut region.

6. Test Cases Using MCpQR Algorithm Points

The accuracy of the RBF-FD method implemented with the optimized points is verified by finding the solution to test case PDEs. After implementing the MCpQR algorithm to find sampling points and differentiation matrices for the complex region (peanut and bumped-disk), Ω, we find the solution to each test case listed below. A fourth-order Runge–Kutta method is used for time-stepping.

6.1. Diffusion Equation with Forcing Term

The first test case involves finding the solution, $u(t,x,y)$, at time $t=10$ for the following PDE:

$$u_t = \Delta u + \sin(t), \tag{15}$$
$$u_0 = 0, \tag{16}$$
$$u_{\partial \Omega} = 0. \tag{17}$$

This test case is implemented using a 37-point stencil, $\phi(r) = r^3$, and polynomials of degree $p=4$. The expected rate of convergence is $O(h^4)$ since the rate is dependent on the degree of polynomials used. Running this test case, the same rate of convergence is observed with the optimized points. This is illustrated in Figure 20, which plots the relative error against the average spacing between each sampling point. In this case, a node refinement process is used, and the true solution is taken to be the solution resulting from the case using the finest spacing.

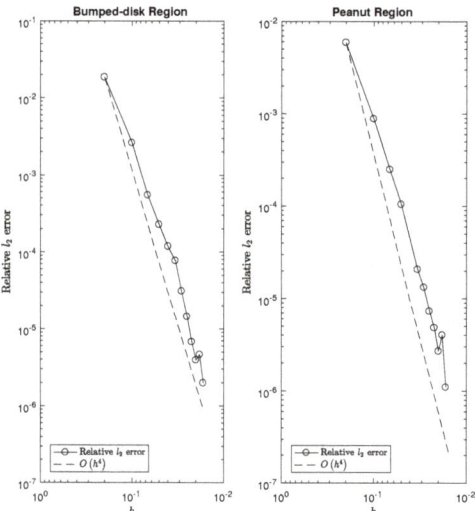

Figure 20. Solution convergence for the diffusion equation with forcing term using optimized points.

6.2. Wave Equation with Hyperviscosity

The second test case requires the implementation of the hyperviscosity methods. This case involves finding the solution, $u(t,x,y)$, at time $t=20$ for the following PDE:

$$u_{tt} = \Delta u, \tag{18}$$
$$u_0 = f(x,y), \tag{19}$$
$$(u_t)_0 = 0, \tag{20}$$
$$u_{\partial \Omega} = 0. \tag{21}$$

Hyperviscosity methods were first introduced in [39] and further studied in [40,41]. These methods allow stable numerical time-stepping for RBF-FD methods. Without hyperviscosity, the differentiation matrices for convective PDEs using RBF-FD methods presented spurious eigenvalues. By damping the spurious eigenvalues while simultaneously preserving the relevant physical properties, the hyperviscosity methods effectively achieve stable numerical time-stepping while still preserving accuracy in the PDE solutions.

The point sets from the MCpQR algorithm are used for hyperviscosity methods. Implementing hyperviscosity methods requires the approximation of high order powers of the Laplacian operator to use as a filter for stable time-stepping. The technique is to then add the high order Laplacian operator to the governing equation of the PDE. As a result, spurious eigenvalues existing in the right (positive, real) half-plane are then shifted into the left (negative, real) half-plane.

Consider the following setup:

$$\begin{bmatrix} u \\ v \end{bmatrix}_t = \mathcal{L} \begin{bmatrix} u \\ v \end{bmatrix}, \qquad (22)$$

where $\mathcal{L}$ is some operator whose differentiation matrix, obtained by implementing RBF-FD methods with PHSs and polynomials, contains spurious eigenvalues. The hyperviscosity method is implemented by redefining the system as:

$$\begin{bmatrix} u \\ v \end{bmatrix}_t = \mathcal{L} \begin{bmatrix} u \\ v \end{bmatrix} + (-1)^{K+1} \gamma h^{2K-1} \begin{bmatrix} \Delta^k u \\ \Delta^k v \end{bmatrix}, \qquad (23)$$

where k denotes the order of the Laplacian used in the hyperviscosity implementation, h represents the average node-spacing, and γ is a parameter that tunes the hyperviscosity filter.

It is important to select a suitable value for the parameter γ. If γ is chosen to be too large, the eigenvalues are forced further out in the left half-plane. Thus, the solution to the PDE will be limited to smaller time-stepping. Furthermore, large values of γ may end up filtering the physically relevant lower modes, thereby, creating accuracy errors. If the hyperviscosity parameter is chosen to be too small, then the possibility of still having eigenvalues existing in the right half-plane, and thus generating an unstable method, remains.

To approximate the higher order Laplacian operators, Gaussian RBFs, $\phi(r) = e^{-(\varepsilon r^2)}$, are used due to the simplicity of higher order Laplacian formulas, which are generalized in [20]. In the case of 2D complex regions, the operators can be approximated by:

$$\Delta^0 \phi(r) = \phi(r), \qquad (24)$$

$$\Delta^1 \phi(r) = \varepsilon^2 \left[4(\varepsilon r)^2 - 4 \right] \phi(r), \qquad (25)$$

$$\Delta^2 \phi(r) = \varepsilon^4 \left[16(\varepsilon r)^4 - 64(\varepsilon r)^2 + 32 \right] \phi(r), \qquad (26)$$

$$\Delta^3 \phi(r) = \varepsilon^6 \left[64(\varepsilon r)^6 - 576(\varepsilon r)^4 + 1152(\varepsilon r)^2 - 384 \right] \phi(r). \qquad (27)$$

The hyperviscosity system for the PDE described in Equations (18)–(21) is then defined as:

$$\begin{bmatrix} u \\ v \end{bmatrix}_t = \begin{bmatrix} (-1)^{K+1} \gamma h^{2K-1} \Delta^k & \mathcal{I} \\ \Delta & (-1)^{K+1} \gamma h^{2K-1} \Delta^k \end{bmatrix} \begin{bmatrix} u \\ v \end{bmatrix},$$

$$= \hat{\mathcal{L}} \begin{bmatrix} u \\ v \end{bmatrix},$$

where $v = u_t$.

This test case is implemented using a 28-point stencil, $\phi(r) = r^9$, polynomials of degree $p = 4$, and Δ^3-type hyperviscosity. Again, the expected rate of convergence is $O(h^4)$ since the rate is dependent on the degree of polynomials used. Running this test case, the $O(h^4)$ rate of convergence is again observed with the optimized points. This is illustrated in Figure 21, where we plot the relative error against the average spacing between each sampling point. For this example, a Bessel function of the first kind on the unit disk is used to provide the initial and boundary conditions. The relative error is then calculated using an exact solution to the PDE.

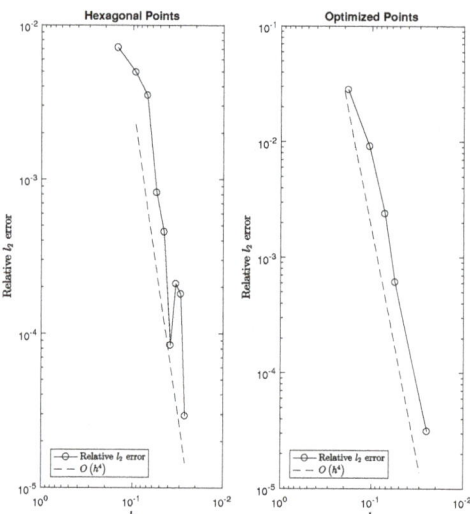

Figure 21. Solution convergence for the wave equation with hyperviscosity for hexagonal and optimized points.

7. Conclusions

A piecewise-defined Lebesgue constant for RBF-FD methods is introduced. Based on the commonly used Lebesgue constant for polynomial interpolation, this measure allows us to sample points for RBF-FD methods combining PHSs and polynomials. We studied the behavior of point sets in 1D, simple 2D regions, and complex 2D regions. Points were generated by modifying a column-pivoting QR algorithm previously used to find optimal points for polynomial interpolation. The resulting points mitigate stencil size restrictions resulting from the use of RBF-FD methods, thus reducing computational cost while preserving accuracy and convergence properties. This method also provides a simple, robust algorithm for point generation with few parameters needing to be tuned. Lastly, we implement the MCpQR algorithm to inform the location of boundary points to be used in conjunction with the scattered repel algorithm points. In the future, 3D regions may be considered as well. One framework for a 3D application is given in [42]. The column-pivoting QR algorithm may be modified to handle RBF-FD methods for 3D by appending corresponding polynomial bases.

Author Contributions: Conceptualization, R.B.P.; Investigation, T.L.; Software, T.L.; Supervision, R.B.P.; Validation, T.L.; Visualization, T.L.; Writing—original draft, T.L.; Writing—review and editing, T.L. All authors have read and agreed to the published version of the manuscript.

Funding: This research received no external funding.

Institutional Review Board Statement: Not applicable.

Informed Consent Statement: Not applicable.

Data Availability Statement: Not applicable.

Acknowledgments: Tony Liu was sponsored by the National Science Foundation (DMS-1502640) for the duration of this work. The authors would like to acknowledge Victor Bayona for providing his code on scattered node generation based on repel algorithms.

Conflicts of Interest: The authors declare no conflict of interest.

References

1. Barnett, G. A Robust RBF-FD Formulation Based on Polyharmonic Splines and Polynomials. Ph.D. Thesis, University of Colorado Boulder, Boulder, CO, USA, 2015.
2. Barnett, G.A.; Flyer, N.; Wicker, L.J. An RBF-FD polynomial method based on polyharmonic splines for the Navier-Stokes equations: Comparisons on different node layouts. *arXiv* **2015**, arXiv:1509.02615.
3. Bayona, V.; Flyer, N.; Fornberg, B. On the role of polynomials in RBF-FD approximations: III. Behavior near domain boundaries. *J. Comput. Phys.* **2019**, *380*, 378–399. [CrossRef]
4. Bayona, V.; Flyer, N.; Fornberg, B.; Barnett, G.A. On the role of polynomials in RBF-FD approximations: II. Numerical solution of elliptic PDEs. *J. Comput. Phys.* **2017**, *332*, 257–273. [CrossRef]
5. Flyer, N.; Barnett, G.A.; Wicker, L.J. Enhancing finite differences with radial basis functions: Experiments on the Navier-Stokes equations. *J. Comput. Phys.* **2016**, *316*, 39–62. [CrossRef]
6. Flyer, N.; Fornberg, B.; Bayona, V.; Barnett, G.A. On the role of polynomials in RBF-FD approximations: I. Interpolation and accuracy. *J. Comput. Phys.* **2016**, *321*, 21–38. [CrossRef]
7. Fasshauer, G.E.; Zhang, J.G. On choosing "optimal" shape parameters for RBF approximation. *Numer. Algorithms* **2007**, *45*, 345–368. [CrossRef]
8. Mongillo, M. Choosing basis functions and shape parameters for radial basis function methods. *SIAM Undergrad. Res. Online* **2011**, *4*, 2–6. [CrossRef]
9. Hagstrom, T.; Hagstrom, G. Grid stabilization of high-order one-sided differencing I: First-order hyperbolic systems. *J. Comput. Phys.* **2007**, *223*, 316–340. [CrossRef]
10. Hagstrom, T.; Hagstrom, G. Grid stabilization of high-order one-sided differencing II: Second-order wave equations. *J. Comput. Phys.* **2012**, *231*, 7907–7931. [CrossRef]
11. Hermanns, M.; Hernández, J.A. Stable high-order finite-difference methods based on non-uniform grid point distributions. *Int. J. Numer. Methods Fluids* **2008**, *56*, 233–255. [CrossRef]
12. Zhong, X.; Tatineni, M. High-order non-uniform grid schemes for numerical simulation of hypersonic boundary-layer stability and transition. *J. Comput. Phys.* **2003**, *190*, 419–458. [CrossRef]
13. Slak, J.; Kosec, G. On generation of node distributions for meshless PDE discretizations. *SIAM J. Sci. Comput.* **2019**, *41*, A3202–A3229. [CrossRef]
14. Fornberg, B.; Flyer, N. Fast generation of 2D node distributions for mesh-free PDE discretizations. *Comput. Math. Appl.* **2015**, *69*, 531–544. [CrossRef]
15. van der Sande, K.; Fornberg, B. Fast variable density 3-D node generation. *arXiv* **2019**, arXiv:1906.00636.
16. Shankar, V; Kirby, R.M.; Fogelson, A.L. Robust node generation for mesh-free discretizations on irregular domains and surfaces. *SIAM J. Sci. Comput.* **2018**, *40*, A2584–A2608. [CrossRef]
17. Wendland, H. *Scattered Data Approximation*; Cambridge University Press: Cambridge, UK, 2004; Volume 17.
18. Fasshauer, G.E. *Meshfree Approximation Methods with Matlab:(With CD-ROM)*; World Scientific Publishing Co Inc.: Hackensack, NJ, USA, 2007; Volume 6.
19. Flyer, N.; Wright, G.B.; Fornberg, B. Radial basis function-generated finite differences: A mesh-free method for computational geosciences. *Handb. GeoMath. Springer Ref.* **2014**, *130*, 1–30.
20. Fornberg, B.; Flyer, N. *A Primer on Radial Basis Functions with Applications to the Geosciences*; SIAM: Philadelphia, PA, USA, 2015.
21. Shankar, V. The overlapped radial basis function-finite difference (RBF-FD) method: A generalization of RBF-FD. *J. Comput. Phys.* **2017**, *342*, 211–228. [CrossRef]
22. Bayona, V. Comparison of moving least squares and RBF+ poly for interpolation and derivative approximation. *J. Sci. Comput.* **2019**, *81*, 486–512. [CrossRef]
23. Bos, L.; Calvi, J.P.; Levenberg, N.; Sommariva, A.; Vianello, M. Geometric weakly admissible meshes, discrete least-squares approximations and approximate Fekete points. *Math. Comput.* **2011**, *80*, 1623–1638. [CrossRef]
24. Bos, L.; De Marchi, S.; Sommariva, A.; Vianello, M. Computing multivariate Fekete and Leja points by numerical linear algebra. *SIAM J. Numer. Anal.* **2010**, *48*, 1984–1999. [CrossRef]
25. Bos, L.; Levenberg, N. On the calculation of approximate Fekete points: The univariate case. *Electron. Trans. Numer. Anal.* **2008**, *30*, 377–397.
26. Briani, M.; Sommariva, A.; Vianello, M. Computing Fekete and Lebesgue points: Simplex, square, disk. *J. Comput. Appl. Math.* **2012**, *236*, 2477–2486. [CrossRef]
27. Caliari, M.; De Marchi, S.; Vianello, M. Bivariate polynomial interpolation on the square at new nodal sets. *Appl. Math. Comput.* **2005**, *165*, 261–274. [CrossRef]
28. Gunzburger, M.; Teckentrup, A.L. Optimal point sets for total degree polynomial interpolation in moderate dimensions. *arXiv* **2014**, arXiv:1407.3291.
29. Guo, L.; Narayan, A.; Yan, L.; Zhou, T. Weighted approximate Fekete points: Sampling for least-squares polynomial approximation. *arXiv* **2017**, arXiv:1708.01296.
30. Narayan, A.; Xiu, D. Stochastic collocation methods on unstructured grids in high dimensions via interpolation. *SIAM J. Sci. Comput.* **2012**, *34*, A1729–A1752. [CrossRef]

31. Narayan, A.; Xiu, D. Constructing Nested Nodal Sets for Multivariate Polynomial Interpolation. *SIAM J. Sci. Comput.* **2013**, *35*, A2293–A2315. [CrossRef]
32. Sommariva, A.; Vianello, M. Computing approximate Fekete points by QR factorizations of Vandermonde matrices. *Comput. Math. Appl.* **2009**, *57*, 1324–1336. [CrossRef]
33. Van Barel, M.; Humet, M.; Sorber, L. Approximating optimal point configurations for multivariate polynomial interpolation. *Electron. Trans. Numer. Anal.* **2014**, *42*, 41–63.
34. Shukla, R.K.; Zhong, X. Derivation of high-order compact finite difference schemes for non-uniform grid using polynomial interpolation. *J. Comput. Phys.* **2005**, *204*, 404–429. [CrossRef]
35. Kosloff, D.; Tal-Ezer, H. A modified Chebyshev pseudospectral method with an O(N-1) time step restriction. *J. Comput. Phys.* **1993**, *104*, 457–469. [CrossRef]
36. Platte, R.B. How fast do radial basis function interpolants of analytic functions converge? *IMA J. Numer. Anal.* **2011**, *31*, 1578–1597. [CrossRef]
37. Li, T.; Pintus, N.; Viglialoro, G. Properties of solutions to porous medium problems with different sources and boundary conditions. *Z. FÜR Angew. Math. Und Phys.* **2019**, *70*, 1–18. [CrossRef]
38. Marras, M.; Piro, S.V.; Viglialoro, G. Lower bounds for blow-up time in a parabolic problem with a gradient term under various boundary conditions. *Kodai Math. J.* **2014**, *37*, 532–543. [CrossRef]
39. Fornberg, B.; Lehto, E. Stabilization of RBF-generated finite difference methods for convective PDEs. *J. Comput. Phys.* **2011**, *230*, 2270–2285. [CrossRef]
40. Larsson, E.; Lehto, E.; Heryudono, A.; Fornberg, B. Stable computation of differentiation matrices and scattered node stencils based on Gaussian radial basis functions. *SIAM J. Sci. Comput.* **2013**, *35*, A2096–A2119. [CrossRef]
41. Shankar, V.; Fogelson, A.L. Hyperviscosity-based stabilization for radial basis function-finite difference (RBF-FD) discretizations of advection–diffusion equations. *J. Comput. Phys.* **2018**, *372*, 616–639. [CrossRef]
42. Gunderman, D.; Flyer, N.; Fornberg, B. Transport schemes in spherical geometries using spline-based RBF-FD with polynomials. *J. Comput. Phys.* **2020**, *408*, 109256. [CrossRef]

Article

Real-Time Data Assimilation in Welding Operations Using Thermal Imaging and Accelerated High-Fidelity Digital Twinning

Pablo Pereira Álvarez [1,2,*], Pierre Kerfriden [2,3,*], David Ryckelynck [2,*] and Vincent Robin [1]

1. Électricité de France Research & Development and Innovation (EDF R&D), 6 Quai Watier, 78400 Chatou, France; vincent.robin@edf.fr
2. Centre des Matériaux (CMAT), MINES ParisTech, PSL University, CNRS UMR 7633, BP 87, 91003 Evry, France
3. School of Engineering, Cardiff University, Queen's Buildings, The Parade, Cardiff CF24 3AA, UK
* Correspondence: pablo.pereira_alvarez@mines-paristech.fr (P.P.Á.); pierre.kerfriden@mines-paristech.fr (P.K.); david.ryckelynck@mines-paristech.fr (D.R.)

Citation: Pereira Álvarez, P.; Kerfriden, P.; Ryckelynck, D.; Robin, V. Real-Time Data Assimilation in Welding Operations Using Thermal Imaging and Accelerated High-Fidelity Digital Twinning. *Mathematics* **2021**, *9*, 2263. https://doi.org/10.3390/math9182263

Academic Editor: Aihua Wood

Received: 20 July 2021
Accepted: 7 September 2021
Published: 15 September 2021

Publisher's Note: MDPI stays neutral with regard to jurisdictional claims in published maps and institutional affiliations.

Copyright: © 2021 by the authors. Licensee MDPI, Basel, Switzerland. This article is an open access article distributed under the terms and conditions of the Creative Commons Attribution (CC BY) license (https://creativecommons.org/licenses/by/4.0/).

Abstract: Welding operations may be subjected to different types of defects when the process is not properly controlled and most defect detection is done a posteriori. The mechanical variables that are at the origin of these imperfections are often not observable in situ. We propose an offline/online data assimilation approach that allows for joint parameter and state estimations based on local probabilistic surrogate models and thermal imaging in real-time. Offline, the surrogate models are built from a high-fidelity thermomechanical Finite Element parametric study of the weld. The online estimations are obtained by conditioning the local models by the observed temperature and known operational parameters, thus fusing high-fidelity simulation data and experimental measurements.

Keywords: data assimilation; model order reduction; finite elements analysis; high dimensional data; welding

1. Introduction

Welding is used extensively in the nuclear industry, for assembly and as a repair technique. It is often used in maintenance operations of different kind that involve various geometries and welding parameter settings. Very high temperatures applied in a localized zone cause expansion and nonuniform thermal contractions, resulting in plastic deformations in the welding and its surrounding areas. Thus, residual stresses and permanent deformations are produced in the welded structure. These could induce a variety of defects such as hot tearing/cracking if the process is not properly controlled. Other defects may appear such as porosity, lack of fusion or lack of penetration that need to be identified after an operation.

The detection of defects in weld beads is usually performed after the welding operation is fully performed [1]. When a defect is identified, the entire operation needs to be done again. Therefore, it would be desirable to obtain real-time estimations of the current mechanical state of the assembly using in situ measurements. With such estimations at hand, welding operations could be stopped and/or controlled whenever predicted or forecasted mechanical states are outside acceptable tolerance regions.

To this end, we propose to develop a digital twinning approach [2] whereby high-fidelity model predictions will be continuously adjusted with respect to thermal images acquired online during the welding operation. The simulation will rely on a state-of-the-art mesoscale transient thermoelastoplastic finite element model. The fusion between FEA and in situ measurements will be done by accounting for well-chosen parametric sources of uncertainties in the simulation model, leaving freedom for the digital twin to react and adapt to the sequence of thermal images. We aim for the data assimilation to be done following a statistical and, if possible, Bayesian framework [3], to enable incorporating engineering knowledge about the uncertain parameters of the model, and allow deriving credible regions for the predictions and forecasts of mechanical states.

Unfortunately, data assimilation problems, i.e., sequential inverse or sequential calibration problems, are notoriously expensive to solve when the numerical models are systems of partial differential equations [4–6]. In the context discussed above, the parameters of spatially detailed nonlinear FEA models would need to be optimized in real-time. This is intractable. We will therefore develop an appropriate piecewise linear meta-modeling technique to achieve real-time efficiency.

Over the last decades, surrogate modeling via model order reduction has been successfully developed for a variety of applications relying on high-fidelity modeling. This is especially true for data-driven approaches based on projecting the high-fidelity model in low-dimensional subspaces [7–13], among which POD-generated subspaces [14–16] can be seen as an extension of the principal component analysis to continuous variables. Projection-based model order reduction methods are known to reduce the computational complexity of high-fidelity models by precomputed candidate solutions corresponding to various points in the parameter domain (snapshot), and reusing the generated information to fasten online solution procedures. However, even by using efficient hyper-reduction schemes [17–20], reduced simulations of welding operations are still computationally too demanding for process monitoring. This is in particular due to (i) the lack of reducibility of moving heat source problems in general [17,21,22] and (ii) the relatively high online cost associated with hyper-reduced models (hyper-reduction generalizes well in large parameter domains owing to the fact that in the online phase, the nonlinear equations of the original simulation model are solved on a reduced integration domain [23,24]).

To circumvent these difficulties, we propose to develop an offline/online metamodeling technique based on a mixture of probabilistic principal component analysis models (PPCA). For any given position of the heat source, the thermomechanical state, augmented by the vector of uncertain parameters, will be postulated to follow a Gaussian distribution with low-rank covariance structure. This model will be identified using the method of snapshots. Offline, we will run the high-fidelity mechanical simulations corresponding to a fine sampling of the parameter space. In a second step, the Gaussian model will be identified using the maximum likelihood approach probabilistic PCA described by Bishop [25]. Online, parameter estimation will reduce to Gaussian conditioning (i.e., the Kalman method), which can be made efficient when the covariance matrix exhibits a low-rank structure. We will show numerically that this strategy allows us to successfully, and for the first time, set up a data assimilation framework for welding operations, blending high-fidelity thermomechanical simulations and thermal imaging in real-time to predict and forecast mechanical states.

In a second stage of developments, we will treat the case where the position of the heat source is to be estimated from thermal imaging. This is of practical interest when the position of the welding torch is not accurately known or tracked during joining operations. To achieve this goal, we build, numerically and offline, a correlation model statistically linking the position of the hottest spots detected on the thermal image to the actual position of the welding torch. Then, the data assimilation method based on the PPCA can be easily adapted, using a probabilistic mixture of PPCA instead of a single Gaussian model. Of course, conditioning the mixture of PPCA remains analytically tractable, and computationally efficient as, as it will be shown, few PPCAs are associated with non-vanishing coefficients at any given time (i.e., the estimation of the position of the torch from thermal images is accurate). Even in this setting, the low-rank structure of the mixture of PPCAs ensures that data assimilation is performed in real-time, without sacrificing the accuracy provided by the high fidelity thermomechanical model.

In all cases, future states may be predicted by conditioning future statistical models to available observations. This is technically done by Gaussian (mixture) conditioning of all future mechanical states to current posterior distributions of unknown model parameters.

The approach proposed in this paper is closely related to other data assimilation methods available in the literature. Filtering methods based on parametrized models (e.g., Kalman or particle filters) typically construct a Markov model for the propagation of parameter distributions in time, progressively assimilating data as they are made available. Our approach can be seen as a degenerate such filter whereby past data are ignored, only the current thermal image influencing the posterior distribution of unknown model parameters. Taking past data into account can be done, for instance, by concatenating mechanical states from current and past assimilation times over a sliding window (i.e., an autoregressive model). For the particular experimental setup considered in this work, taking past data into account brought no significant change in predictive parameter distributions, which justifies our development of a past-agnostic assimilation method. In terms of meta-modeling, we could have used nonlinear meta-modeling techniques such as polynomial chaos or neural network regression [26–28], but the choice of a linear model (PPCA) allows us to solve analytically the conditioning problem, thus bringing robustness to the method, which cannot be expected when using Markov Chain Monte Carlo solvers or sequential importance sampling [5]. Our approach is clearly inspired by the parametrized background data-weak approach [29], which adopts a variational point of view, while our method is Bayesian, thereby delivering credible intervals and not point estimates. Similar to the parametrized background data-weak method, we perform state estimation in large dimensions by using a background covariance matrix generated by parametric variations of a high-fidelity PDE system. The low rank structure of this covariance matrix is used to fasten online Gaussian conditioning, circumventing the usual N^3 complexity issue by making use of standard algebraic techniques [25,30].

Our paper is organized as follows. In Section 2, we present the experimental configuration of our test case. Section 3 aims to present the thermomechanical model and the inverse problem, detailing the known and unknown parameters. In Section 4, we will introduce the construction of the local surrogate models using Probabilistic PCA, and in Section 5, two different use cases are considered: a situation where the heat source position is known and another one where it needs to be estimated. Forecasting is also discussed. Finally, in Section 6 we present results for all the configurations introduced in Section 5 using noisy simulation data when the torch position is known and real experimental data when it is not. Appendices A–C give, respectively, details on the algebraic expressions used to accelerate the computations, some more forecasting results and material parameters of the specimen.

2. Experimental Configuration

To test the proposed approach, we need to find an application that can replicate similar welding conditions (in terms of materials and parameter variety) to those of a real maintenance operation that can be performed in a laboratory environment with proper instrumentation.

The target application is the stress prediction in a 316L stainless steel (The chemical composition and some of the temperature dependent material parameters are given in Appendix C) specimen submitted to Programmierter–Verformungs–Risstest (PVR) hot cracking tests. PVR tests, developed during the 1970s by the Austrian company Boehler, consist of making a fusion line using bead-on-plate TIG-welding with argon shielding while the specimen is uniaxially tensile loaded [31]. It allows the control of the tensile deformation progressively, which means that it can show cracks of different origin, most notably solidification and liquidation cracking. The large surface of the specimen allows easy positioning of thermocouples on both sides of it, as shown in Figure 1, and it is also possible to film the surface of the specimen with an infrared camera. Thus, process parameters and infrared images of the welding operation are the observational data from which 3D stresses and temperatures will be estimated.

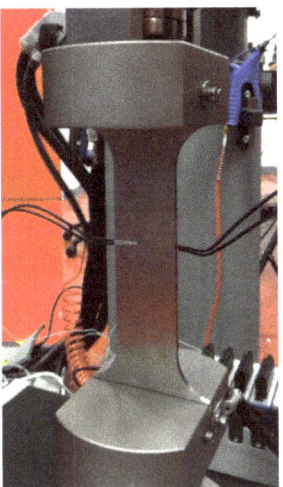

Figure 1. Experimental configuration of a PVR test with two thermocouples.

The experiments are performed with a 6-axes Panasonic robotic arm and equipped with a ValkWelding torch and the tensile loading is applied with a Lloyd Instruments LS100 plus. Two parameters related to the heat source are fixed before the experiment: the speed (v) and the power (Q), which is the product of voltage (U) and current (I). The experiments are instrumented with two thermocouples and filmed with a SC7500 FLIR infrared camera. The infrared video will provide real-time measures on part of the specimen. While both sides of the specimen have thermocouples, only the surface that is not being welded is filmed. This way, reflections from the welding arc and the robot itself are avoided.

3. Digital Twin

3.1. Thermomechanical Model

3.1.1. Thermo-Elasto-Plasticity

Numerical simulation of welding is a very complex problem, as it needs to take into account a great number of parameters to represent multiscale and multiphysics phenomena, using temperature dependent material properties that are not always properly known. Usually, the interactions between the metallurgy, the heat, and the mechanical problems need to be simulated. In the case of 316L stainless steel, the metallurgic interactions are negligible [32], and thus the model is reduced to a weakly coupled nonlinear parametric thermo-elasto-viscoplasticity problem.

We consider a model of unsteady thermo-elasto-viscoplasticity over spatial domain $\Omega \in \mathbb{R}^3$, whose boundary will be denoted by $\partial\Omega$, and time domain $\mathcal{T} \in [0, T]$. For all $(x, t) \in \Omega \times \mathcal{T}$, the unsteady heat equation reads as

$$\rho c_p \frac{\partial T}{\partial t} - \nabla \cdot k \nabla T = q_d \qquad (1)$$

where $T : \Omega \times \mathcal{T} \to \mathbb{R}$ is the temperature field, ρ is the mass density, k is the thermal diffusion coefficient and c_p is the specific heat capacity.

The above equation is complemented by boundary condition

$$T = T_0 \qquad (2)$$

on domain boundary $\partial\Omega_T$. Moreover,

$$-k\nabla T \cdot n = \gamma(T^4 - T_0^4) + \zeta(T - T_0) \qquad (3)$$

on domain boundary $\partial\Omega_q$, where γ is the radiation coefficient and ζ is the thermal convection coefficient. At time $t = 0$, the temperature T is equal to T_0 everywhere in the computational domain.

For all $(x,t) \in \Omega \times \mathcal{T}$, the mechanical equilibrium reads as

$$\nabla \cdot \sigma = 0 \tag{4}$$

This equation is complemented by boundary conditions

$$u = u_d \tag{5}$$

on part of the domain $\partial\Omega_u$, and

$$\sigma \cdot n = 0 \tag{6}$$

on part of the domain $\partial\Omega_t$. The coupled thermomechanical problem is closed by the following constitutive relation:

$$\sigma = C : (\epsilon(u) - \epsilon^p - \alpha(T - T_0)I_d) \tag{7}$$

where $\epsilon(u) := \frac{1}{2}(\nabla u + \nabla u^T)$ is the total strain, ϵ^p is the plastic strain and $\alpha(T - T_0)I_d$ is the thermal strain, with α the coefficient of thermal expansion and I_d the identity tensor. The plastic strain is fully determined by a Von Mises plasticity model with isotropic and kinematic hardening.

$$f(\sigma, R, X) := \sqrt{\frac{3}{2}(\tilde{\sigma} - X) : (\tilde{\sigma} - X)} - R(p) \leq 0 \tag{8}$$

$$\dot{\epsilon}^p = \lambda \frac{df}{d\sigma} \tag{9}$$

$$\dot{p} = \lambda \frac{dR}{dp} \tag{10}$$

$$X = \frac{2}{3}C(p)\alpha \tag{11}$$

$$\dot{\alpha} = \dot{\epsilon}^p - \gamma(p)\alpha\dot{p} \tag{12}$$

$$\lambda \geq 0 \quad \lambda f(\sigma) = 0 \tag{13}$$

where $R(p)$ is the limit of the elastic domain due to isotropic hardening, λ is the plastic multiplier and $\tilde{\sigma} = \sigma - \frac{1}{3}Tr(\sigma)I$ is the deviatoric part of the stress tensor.

In the specimen of a PVR test, the traction boundary conditions are enforced by Dirichlet boundary conditions:

$$u(x,t) = U_d \frac{t}{t_w} \quad \forall x \in \partial\Omega_u \tag{14}$$

The main mechanical quantity of interest is the first principal stress, denoted by σ_I:

$$\sigma_I = \max_{n \in \mathbb{R}^3} \frac{n \cdot \sigma \cdot n}{||n||^2} \tag{15}$$

The first principal stress is the highest principal stress. It has a huge influence on hot cracking during the welding process.

3.1.2. Thermomechanical Load

The choice of a model for the moving heat source is a key point in numerical simulation of welding. The most commonly used one is Goldak's double-ellipsoid model [33], represented in Figure 2. It is important to note that the heat distribution is different in the front and the rear of the heat source, thus the model depends on the current central position of the heat source $P(t) = (x(t), y(t), z(t)) \in \mathbb{R}^3$,

$$P(t) = P_0 + \left(\frac{t}{(P_F - P_0)V} - \epsilon_P(t) \right)(P_F - P_0) \tag{16}$$

In the equation above, $P_0 \in \mathbb{R}^3$ is the position of the heat source at the start of the welding operation, and $P_F \in \mathbb{R}^3$ is the position at the end of the operation. $T = (P_F - P_0)V$ corresponds to the total welding time, which depends linearly on the velocity V of the source. $\epsilon_P(t)$ is a small (normalized) time delay that accounts for potential errors in the control of the welding robot (This is a very simple model, which could easily be modified to include fluctuations of the velocity field.). This parameter will be identified online.

Assuming that the heat source moves in the direction x, the final equation is

$$q(x, y, z, t) = \begin{cases} \frac{12\sqrt{3}\eta Q}{(a_r + a_f) b c \sqrt{\pi^3}} exp\left(\frac{-3(x-X)^2}{a_f^2} + \frac{-3(y-Y)^2}{b^2} + \frac{-3(z-Z)^2}{c^2} \right), & x \geq X. \\ \frac{12\sqrt{3}\eta Q}{(a_r + a_f) b c \sqrt{\pi^3}} exp\left(\frac{-3(x-X)^2}{a_r^2} + \frac{-3(y-Y)^2}{b^2} + \frac{-3(z-Z)^2}{c^2} \right), & x < X. \end{cases} \tag{17}$$

where a_r, a_f, b and c are unknown parameters describing the geometry of the double ellipsoid (In practice, we calibrate a_r, b, c, η and K where K is the ratio between a_r and a_f: $a_r = K a_f$), $Q = UI$ is the power, with U the voltage, I the current and η the efficiency.

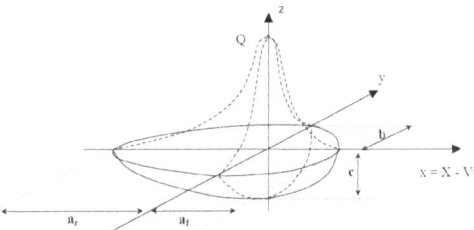

Figure 2. Goldak's double-ellipsoid model.

3.1.3. Space and Time Discretization

The thermomechanical problem is discretized in space using the standard P1 Lagrange finite element method. In the following, $T(t)$ and $\sigma^\square(t)$ will respectively denote the vector of finite element nodal temperature and the vector of components of the stress tensor at the quadrature points.

The thermomechanical problem is discretized in time using the standard backward Euler finite difference scheme.

3.2. "Truth" Online Inverse Problem

We now describe the problem of data assimilation and online forecasting for the welding operation.

3.2.1. Parametrized Probabilistic Setting

The power Q of the heat source and its velocity V are supposed to be controlled with a good degree of accuracy. The mechanical load U_d is also well controlled during the experimental procedure. In order to build the "truth" digital twin, we further assume that the thermal capacity and diffusivity of the material are well characterized. The mechanical behaviour of the structure (thermal expansion, elasticity and plasticity) are also assumed to be well characterized, qualitatively and quantitatively. This is consistent with EDF's decades of experience in modeling, characterizing and simulating such welding processes. Finally, the thermal and mechanical boundary conditions are reasonably well quantified in our experimental setting.

However, several sources of uncertainties negatively affect the predictive capabilities of the simulation model:

- The position P of the center of the heat source is not perfectly well known. We will consider that parameter ϵ_t is random, i.e., is a source of epistemic uncertainty.
- The spatial *distribution* of the heat source is not known with precision. We surmise that that the main contribution to the overall uncertainty of the model is the spatial length-scale of the Goldak model.

Therefore, the parameter space $\mathcal{P} = \mathcal{M} \times \Theta$ space comprises two distinct blocks:

- The set of known parameters $\mu = \{Q, V\} \in \mathcal{M}$, which will vary from one welding operation to the next, but can be controlled with precision.
- The set of unknown parameters $\theta = \{a_r, b, c, \eta, K, \} \in \Theta$, which are given a probability distribution $\theta \sim p_\theta$ that encodes the prior knowledge available about these parameters.

3.2.2. "Truth" Online Bayesian Conditioning

Data will be assimilated at homogeneously distributed times $\bar{\mathcal{T}} = \{t^0 = 0, t^1 = \Delta t, t^2 = 2\Delta t, \ldots t^{N_t} = N_t \Delta t\}$. The assimilation time step Δt is adjusted so that the number of assimilation steps N_t is the same for all simulations, independently of the velocity V of the heat source,

$$\Delta t = \frac{(P_F - P_0)}{N_t} V \tag{18}$$

Online at time $t^k \in \bar{\mathcal{T}}$, we assume that surface temperatures are measured noisily (i.e., with the thermal camera). We will write

$$d_T^k = H_T T^k_{|\theta,\mu} + \epsilon_s^k \tag{19}$$

where H is a fixed Boolean operator acting on the vector of finite element temperature nodal values T^k at time step k. The additive measurement noise is supposed to be zero-mean Gaussian distributed $\epsilon_T \sim \mathcal{N}(0, \Sigma_M = \sigma_M^2 I_d)$. The sources of the temperature measure uncertainty are varied and might include light reflections or lack of knowledge of material parameters like the emissivity at different temperatures [34]. The measure error parameter σ_M^2 will be calibrated empirically comparing the measure from two sensors available in EDF's welding lab.

Assuming that the successive noise vectors $\{\epsilon^k\}_{k \in [0,t] \cap \bar{\mathcal{T}}}$ are independent, the statistical inverse problem to be solved online is the following.

At time $t \in \bar{\mathcal{T}}$, the posterior distribution of unknown parameters given the past measurements is

$$p(\theta | d^k, \mu_k) \propto \Pi_{k' \leq k} \mathcal{L}^{k'}(\theta; d^{k'}) p_\theta(\theta) \tag{20}$$

where

$$\mathcal{L}^k(\theta; d^k) = \mathcal{N}(d^k; H_T T^k_{|\theta,\mu}, \Sigma_T) \tag{21}$$

Unfortunately, evaluating the likelihood function in real-time is unfeasible, even when using standard Markov assumptions.

4. Real-Time Predictions with Gaussian Mixture of Local Surrogate Models

In the previous section, the truth inverse problem was presented, and it was concluded that it is not well suited for real-time applications. To overcome this, in this section the construction of Gaussian local surrogate models from snapshots of a Finite Elements parametric study is discussed.

4.1. Local Multiphysics PCA Model

Our proposal is to make local surrogate models for a linear joint representation of the state, known parameters and unknown parameters for every position of the heat source at the assimilation times. The positions are fixed in a grid between the initial position

P_0 and the final position P_F with a regular separation ΔP. The parametric solutions are clustered by the position $P(t)$ of the heat source and associated to the position $P^k = k\Delta P$ if $|P(t) - P^k| < \frac{\Delta P}{2}$. The local model at position P^k reads as

$$s^k = \begin{pmatrix} T^k \\ \sigma^k \\ \mu^k \\ \theta^k \end{pmatrix} = \bar{s}^k + \Phi^k \alpha^k + \epsilon_s^k \tag{22}$$

with $\alpha^k \sim \mathcal{N}(0, I_d)$ and $\epsilon_s^k \sim \mathcal{N}(0, \sigma^2 I_d)$. Operator Φ^k—a decoder—is fixed for each position and possesses n_ϕ columns. It is obtained by using all parametric solutions of the welding problem corresponding to position P^k. This probabilistic model encodes the dependency between all the state variables and the known and unknown parameters in the form of a multivariate Gaussian:

$$s^k \sim \mathcal{N}(\bar{s}^k, \Phi^k \Phi^{kT} + \sigma_s^2 I_d) \tag{23}$$

The choice of using local models aims at avoiding the use of a global reduced basis, as it is well known that moving heat sources generate irreducible parametric solutions [17]. Notice as well that there is no Markovian assumption in the model. In other words, measurements at times $\{t_l\}_{l<k}$ do not provide any new information about the state at time k.

Using Gaussian assumptions on α^k and ϵ_s follows a certain logic as well. With s^k being Gaussian, we can deduce that d^k, the observed data at time k is also Gaussian distributed. This means that $p(s^k|d^k, \mu)$ will also be Gaussian and can be calculated analytically, allowing us to greatly accelerate the computation times.

4.2. Maximum Likelihood PCA

In order to calibrate the surrogate model, we assume that a snapshot of n_s extended states is available, which we denote by $\mathcal{S} = \{S^1, S^2, \ldots, S^{n_s}\}$. This snapshot should cover the time and parameter domains appropriately.

In 1999, Tipping and Bishop [35] showed that a probabilistic formulation of PCA can be obtained from a Gaussian latent variable model, where the basis vectors are maximum-likelihood estimates. Given Equation (22), there are explicit expressions for the maximum likelihood estimates of the parameters:

$$\Phi = U_q \sqrt{(\Lambda_q - \sigma^2 I_d)} R \tag{24}$$

where U_q contains the left singular vectors associated to the greatest $q < n_s$ singular values of a Singular Value Decomposition (SVD) of the snapshot matrix $\mathcal{S}$. The diagonal matrix Λ_q is also obtained from the SVD, and it contains the greatest q singular values in decreasing order. R is an orthogonal rotation matrix that, in practice, will be chosen to be the identity matrix. Finally, σ^2 is related to the truncation error:

$$\sigma^2 = \frac{1}{n_s - q} \sum_{j=q+1}^{n_s} \lambda_j^2. \tag{25}$$

where λ_j are the smaller singular values.

This means that the generative PCA model can be easily computed from a SVD of the output of a parametric study of the thermomechanical finite elements model.

The snapshots $\mathcal{S} = \{S^1, S^2, \ldots, S^{N_t}\}$ are generated using a straightforward procedure. All the parameters in $\mathcal{P}$ are assumed to be uniformly distributed over a hyper-cube and will be sampled using Latin Hypercube Sampling [36]. The minimum and maximum values for each parameter are given by experts.

The samples are generated as follows. For every (μ, θ) sampled with LHS the finite element simulation is ran over $\mathcal{T}(\mu) = [0, T(\mu)]$, delivering a history of N_t snapshots. Additional postprocessing could be performed depending on the physics that are represented in the state vector s.

5. Online Prediction

In this section, we treat the resolution of the inverse problem. Two cases have to be considered depending on whether the torch position is known or not: The first case, when the heat source position is known at all times, is straightforward and the predictions are obtained using a single well-identified PPCA model. In the second case, the torch position needs to be estimated with some uncertainty. The predictions are now computed from a mixture of PPCAs. Finally, the forecast of future states is discussed.

5.1. Known Position

Let us assume that the heat source position is known at all times. This happens when all measurements are synchronized and knowing how much time has passed from the starting point allows perfect knowledge of the position P^k via the speed or when the robotic arm is equipped with a sensor measuring the advanced distance and this signal is available. This corresponds to a situation where $\epsilon_P(t) = 0$ in Equation (16).

When the position P^k is known, the closest model is also known and it is enough to estimate the current state and unknown parameters. This is done by computing the posterior distribution $p(s^k|d^k, \mu)$, which is Gaussian as seen in Section 4. It can be computed analytically and it is completely determined by its mean $m_{s^k|d^k,\mu}$ and covariance $\Sigma_{s^k|d^k,\mu}$ and it. Introducing the notation $\Sigma_{s^k} = \Phi^k \Phi^{kT} + \sigma_T^2 I_d$, the mean and covariance are

$$\mu_{s^k|d^k,\mu} = \bar{s}^k + \Sigma_{s^k} H^{kT} \left(H^k \Sigma_{s^k} H^{kT} + \sigma_m^2 I_d \right)^{-1} (d^k - \bar{d}^k) \qquad (26)$$

$$\Sigma_{s^k|d^k,\mu} = \Sigma_{s^k} - \Sigma_{s^k} H^{kT} \left(H^k \Sigma_{s^k} H^{kT} + \sigma_m^2 I_d \right)^{-1} H^k \Sigma_{s^k} \qquad (27)$$

where H^k is a Boolean operator acting as the observation function.

The evaluation of this expression is very slow due to the size of the matrices involved. To avoid this cost, we propose to compute the posterior distribution of the reduced coordinates α^k instead. Once again, the posterior distribution $p(\alpha^k|d^k, \mu)$ is Gaussian, so it is determined by its mean $m_{\alpha^k|d^k}$ and covariance $\Sigma_{\alpha^k|d^k}$:

$$m_{\alpha^k|d^k,\mu} = \Phi^{kT} H^{kT} \left(H^k \Sigma_{s^k} H^{kT} + \sigma_m^2 I_d \right)^{-1} (d^k - \bar{d}^k) \qquad (28)$$

$$\Sigma_{\alpha^k|d^k,\mu} = I_d - \Phi^{kT} H^{kT} \left(H^k \Sigma_{\alpha^k} H^{kT} + \sigma_m^2 I_d \right)^{-1} H^k \Phi^k \qquad (29)$$

The mean of the posterior distribution of s^k can be then deduced using the reduced basis Φ^k as follows:

$$\mu_{s^k|d^k,\mu} = \bar{s}^k + \Phi^k m_{\alpha^k|d^k} \qquad (30)$$

As for the covariance matrix, it can be calculated by

$$\Sigma_{s^k|d^k,\mu} = \Phi^k \Sigma_{\alpha^k|d^k,\mu} \Phi^{kT} + \epsilon_s^k \qquad (31)$$

The whole covariance matrix might be impossible to store in RAM if the number of degrees of freedom is large. In this case, only the diagonal of the matrix product $\Phi^k \Sigma_{\alpha^k|d^k,\mu} \Phi^{kT}$ is calculated.

For further details on the acceleration of these computations, see Appendix A.

5.2. Unknown Position

As it is the case in EDF's lab, we may not have access to the exact heat source position for lack of synchronicity between the measure sensors. In this case, $\epsilon_P(t) > 0$ in Equation (16) and the torch position needs to be estimated from the video frame, which adds more uncertainty to the model. The filmed side of the specimen is the opposite side of the weld and there exists a delay between the highest temperature on this side and the torch position. Furthermore, this delay is dependent on material and operational parameters, such as the speed.

Using the finite elements simulations used for the prior generation of the local PPCA surrogate models, we created a Gaussian surrogate model that links the position of the highest temperature measured on the camera side of the specimen x^k and the known parameters μ with the position of the heat source on the welded surface of the specimen P^k. The model was fitted using a Matérn kernel with $\nu = \frac{3}{2}$ as a covariance function. The output of the model is the probability distribution of the estimated heat source position $\hat{P}^k$. We sample this distribution to find possible torch positions.

Not being able to determine the exact position, the data assimilation needs to be adapted using a mixture of PPCAs. The multiphysics state for this video frame is named $s^{\hat{k}}$ to indicate an unknown position. For each sample of the heat source position we assign it the closest local PPCA model and a discrete assignment probability distribution is calculated to find the active coefficients of the mixture. Assuming a total of J local models are active, $p(s^{\hat{k}}|d^k,\mu)$ is computed as a Gaussian mixture of the estimations of each local model weighted by the discrete assignment probability $p_j\ \forall j \in J$:

$$m_{s^{\hat{k}}|d^k,\mu} = \sum_{j \in J} p_j m_{s^j|d^k,\mu} \qquad (32)$$

$$\Sigma_{s^{\hat{k}}|d^k,\mu} = \sum_{j \in J} p_j \Sigma_{s^j|d^k,\mu} \qquad (33)$$

The torch position estimation is sufficiently accurate to ensure that the number of active PPCA models is not too large. The assimilation is still performed in real-time due to the very quick evaluation of each individual PPCA.

5.3. Forecasting of Future States

Despite the lack of Markovian assumption in the model, it can be used to forecast future states without observed experimental data by integrating previously estimated unknown parameter values. Indeed, the unknown parameters are not changed online and, assuming that they were estimated for a position P^k, θ^k could be used to obtain accurate predictions of future states.

The position of the future state is defined by a displacement of $j\Delta P$ in the direction of the weld, where ΔP is the regular separator in the positions grid. This opens two possibilities whether the torch position was known or was estimated from the data frame. If the heat source position was known, the prediction is computed using the PPCA model for position $P^{k+j} = (k+j)\Delta P$, this is $p(\bar{s}^{k+j}|\mu, \theta^k)$.

The other possibility is that the current position was estimated from the experimental data. In this case, the displacement will be added to the position samples so that the uncertainty on the torch position is maintained. Then, the prediction will be calculated using the mixture of PPCAs.

6. Results

In this section, we will show the use of the proposed models. First of all, the experimental configuration of a PVR hot cracking test is explained. Then, details about the parametric finite element study from which the snapshots are generated will be given.

For the numerical tests, two sources of data are considered to show the case where the torch position is well known and the case where and estimation of its position is needed. The first test will use noisy synthetic data obtained from a previously calibrated Finite Element solution as input. The joint multiphysics state and unknown parameters estimation are obtained using a single PPCA model. Then, we will consider a frame of an infrared video of a PVR experiment. The torch position will be deduced from the image and the estimation computed from a mixture of PPCAs.

Last, the forecasting capabilities of the model will be showcased for the prediction of a state situated 12 mm further than the last studied position.

6.1. Experimental Procedure

A PVR hot cracking test was done at EDF R&D welding lab to obtain experimental data. The 316L steel specimen is 200 mm long and 3.5 mm thick. Its width is 80 mm on both ends and 40 mm in the center. The fusion line is 130 mm long. The specimen was placed in a tensile testing machine which was configured to augment the tensile deformation speed progressively from 0 mm/s to a maximum speed of 0.333 mm/s. The welding parameters used in the experiment are shown in Table 1. Thus, the known parameters are $v = 2$ mm/s and $Q_0 = U \times I = 8.4 \times 81 = 680.4$ W. The welded specimen is shown in Figure 3.

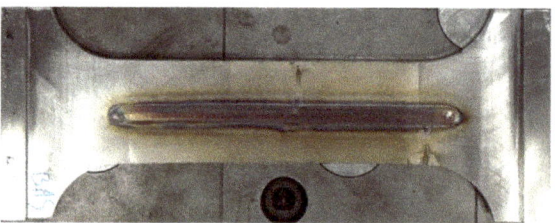

Figure 3. Welded PVR specimen after the experiment.

Table 1. PVR experiment parameter values.

Current	Voltage	Travel Speed	Shielding Gas	Maximal Stroke Rate
81 A	8.4 V	2.00 mm/s	Argon	20 mm/min

The experiment is instrumented with two type K thermocouples and a SC7500 FLIR infrared camera. The thermocouples, one on each side of the specimen (see Figure 1), are placed 110 mm from the bottom and 4 mm to the left from the center. The camera films the surface that is not being welded in order to avoid reflections from the welding arc. Figure 4 shows the projection of a frame of the video on the FE mesh. The resolution of the camera images is 320 × 256 pixels.

Previous uses of this configuration of thermocouples and infrared camera helped with the calibration of the parameter σ_m^2. To estimate the value of σ_m^2, we compare the measures of both sensors on a single point over time. The aim is to obtain an estimation of deviations between measures. Figure 5 shows the comparison between the signals. Here, both measures seem very close but show differences in some areas. Due to the configuration of the camera, only temperatures above 400 °C should be considered. One more thing to note is the peak that appears in the camera measure during the cooling phase, which was probably caused by the reflection of light. Work is being done at the lab to improve the quality of instrumentation and avoid this kind of issue. Considering the thermocouple and camera measures as $\Theta_T(t)$ and $\Theta_C(t)$, respectively, the value of σ_m^2 is estimated as

$$\sigma_m^2 = Var(\Theta_T(t) - \Theta_C(t)) = 323.880246 \qquad (34)$$

which corresponds to a standard deviation of ~18 °C.

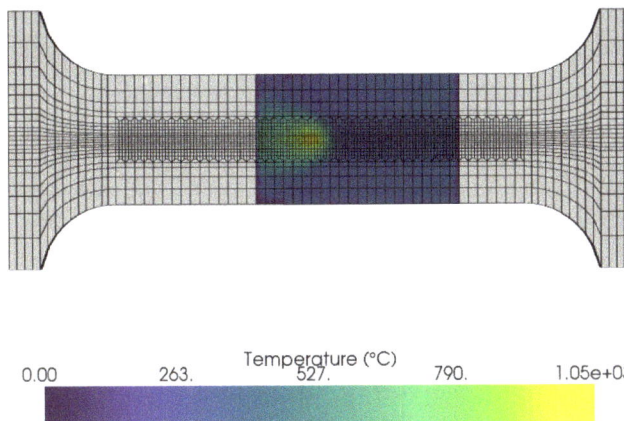

Figure 4. Projection of infrared images on the FE mesh.

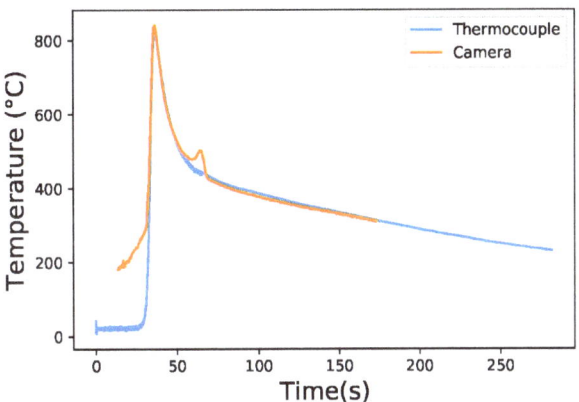

Figure 5. Temperature measures taken with a thermocouple and an infrared camera.

The thermocouple measures, and not the camera data, are also used to calibrate the unknown parameter values for a high fidelity Finite Elements simulation of the experiment that will be used as reference. This is a standard procedure that is done for every simulation of an experiment at EDF and it involves the resolution of an optimization problem. The result of the calibration of θ is shown in Table 2.

Table 2. Deterministic calibration of the unknown parameters.

	a_f	b	c	η	K
Calibrated value	6.657 mm	3 mm	1.5 mm	0.9	1.15

6.2. Prior Generation

In Section 4, we briefly discussed the generation of the prior that is obtained by running finite element simulations for a Latin Hypercube Sampling of the parameter space $\mathcal{P}$, where every parameter is assumed to be uniformly distributed. The minimum and maximum values for each parameter are issued from EDF's decades of experience in both welding and numerical analysis of welding. These values are shown in Table 3 for the known parameters and in Table 4 for the unknown parameters.

Table 3. Minimum and maximum values for the known parameters μ, as determined by the experts.

	Heat Source Speed	Voltage	Current
Min. value	1 mm/s	8 V	70 A
Max. value	3 mm/s	12 V	90 A

Table 4. Minimum and maximum values for the known parameters θ, as determined by the experts.

	a_f	b	c	η	K
Min. value	3 mm	1 mm	0.5 mm	0.75	1.15
Max. value	12 mm	7 mm	2.5 mm	0.95	2.5

A total of 128 simulations of PVR experiments were run with code_aster [37], EDF's open-source thermomechanical simulation software. The selected physical fields for the multiphysics state vector are the temperature (T) and the maximum principal stress σ_I. The maximum principal stress is postprocessed from the stress tensor. It was chosen because it is used in a hot crack criterion developed in a previous PhD [38].

6.3. Numerical Tests

6.3.1. Estimation of the Heat Source Position

Before launching the two tests, the heat source position is identified using the surrogate model presented in Section 5. Figure 6 shows the discrete probability of assignment from 1000 samples of the position distribution.

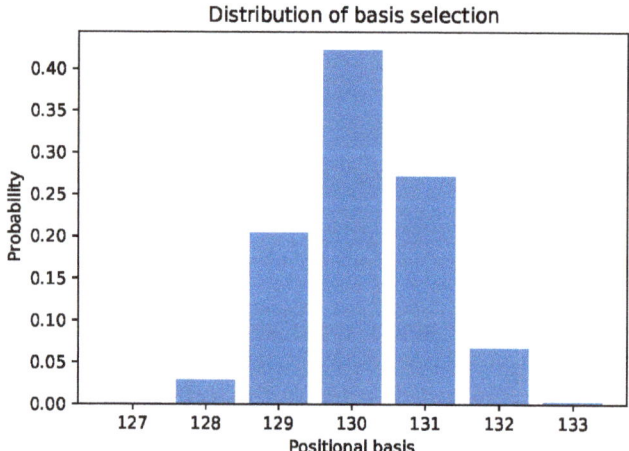

Figure 6. Possible PPCA bases with associated probability.

The most probable basis is 130, which corresponds to 65 mm from the starting position. This position is the one that will be used as known position in the synthetic data test.

6.3.2. Tests with Noisy Synthetic Data

In this first test, the input data come from the calibrated finite elements simulation of the experiment, with $\mu = (2.0, 680.4)$. The snapshot used as input is the one corresponding to the highest probability. White noise of the same amplitude as the measure error estimated for the camera has been added to the data. The observation function H^k restricts the view to a region "seen" by the camera, and it is represented in Figure 7.

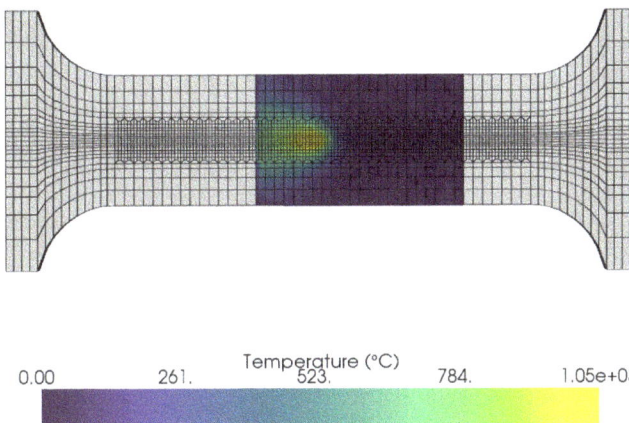

Figure 7. Noisy simulation data on camera area.

The mean of the posterior distribution $m^k_{s_k|d^k}$ contains the estimation of the temperature and maximum principal stress fields, as well as the unknown parameters. We can compare these results to the calibrated finite elements simulation, from which the input data was taken.

The results are compared along a line of 130 mm at the center of both sides of the specimen, which coincides with the the weld line on the torch side. The 95% confidence interval of the estimation is also plotted around the posterior mean in Figures 8 and 9. The reconstruction of the temperature and stress fields is very accurate, with a global relative error of 1.3693% for the temperature and 6.3419% for the principal stress. This relative error is calculated as

$$e^k_T = \frac{||T^k_{|\mu,\theta} - s^k_{|T}||_2}{||T^k_{|\mu,\theta}||_2} \quad (35)$$

$$e^k_{\sigma_I} = \frac{||\sigma^k_{I|\mu,\theta} - s^k_{|\sigma_I}||_2}{||\sigma^k_{I|\mu,\theta}||_2} \quad (36)$$

where $T^k_{|\mu,\theta}$ and $\sigma^k_{I|\mu,\theta}$ are the nodal temperature and principal stress simulation values, respectively, and $s^k_{|T}$ and $s^k_{|\sigma_I}$ are the estimated nodal temperature and principal stress, respectively.

The confidence intervals are thin for the observed temperature data, as it is expected, but it is larger around the peak on the non-observed surface. For the principal stress estimations, although no data was seen, the estimated posterior mean is close to the simulation on both surfaces.

The estimation of the unknown parameters θ is very close to the results of the deterministic calibration, with a relative error inferior to 10%, meaning that the parameter calibration successfully identified the theta values used for the simulation. A comparison of the values obtained by the deterministic calibration and the mean posterior theta estimation is given in Table 5.

Table 5. Posterior estimation of the unknown parameters θ obtained from noisy synthetic data.

	a_f	b	c	η	K
Calibrated value	6.657	3	1.5	0.9	1.15
Estimated value	6.18868114	2.84188617	1.58588521	0.89049624	1.16543712

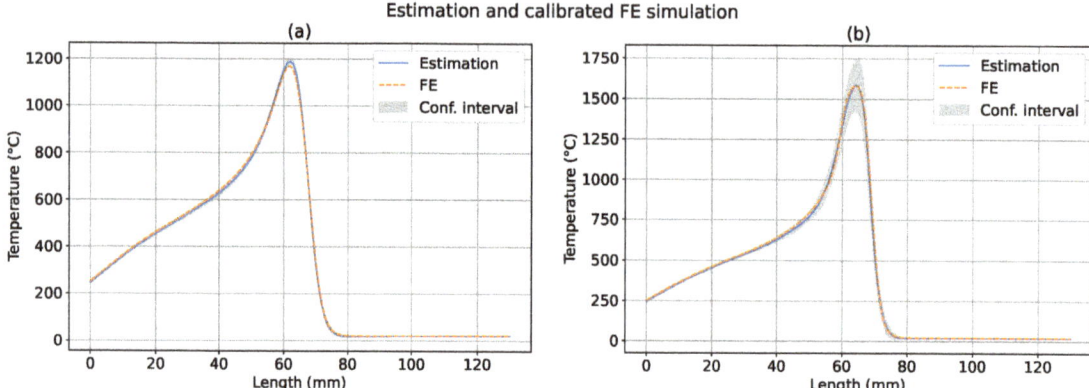

Figure 8. Temperature estimations obtained from noisy synthetic data, FE simulation and confidence interval. (**a**) Camera side. (**b**) Torch side.

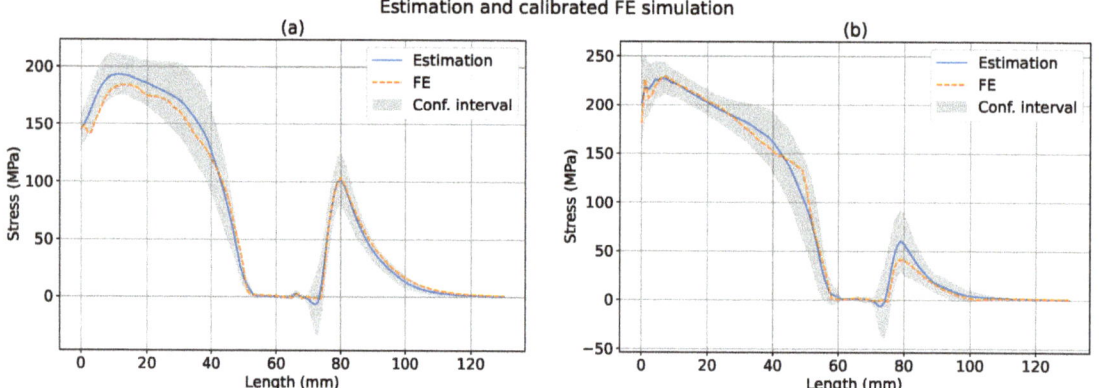

Figure 9. Maximum principal stress estimations obtained from noisy synthetic data, FE simulation and confidence interval. (**a**) Camera side. (**b**) Torch side.

6.3.3. Tests with Real Experimental Data

In this test, we will use the camera frame projected onto the mesh as input data for the model. The camera is configured to capture temperatures above 400 °C, so the observation function H^k is modified to only use data above 400 °C, which is the area represented in Figure 10.

The estimation is calculated as a mixture of the results given by the local PCA models in Figure 6. The amount of active local models is small and each individual conditioning is computed very fast. All the confidence intervals are the 95% confidence intervals.

On the camera side, the estimations can be compared to the measured temperature. In Figure 11, we can see that the model estimates a temperature that follows the experimentally measured one. Interestingly, the estimation deviates from the simulation, as seen in Figure 12, where the temperature is higher for the FE results. We interpret this difference in the estimation and the FE simulation as a model correction given by the partial observation of the temperature. We remind the reader that the FE model was calibrated using only the thermocouple measures and not the infrared camera. This difference between the FE simulation and the measured data in Figure 12 may indicate that the calibration of θ using the thermocouples is not good enough. This is supported by the fact that the estimated efficiency η, shown in Table 6 along with the other unknown parameters, is smaller at ~0.74 while the initial calibration estimated it at 0.9. A smaller efficiency would transfer less temperature to the specimen, explaining the lower estimated temperature.

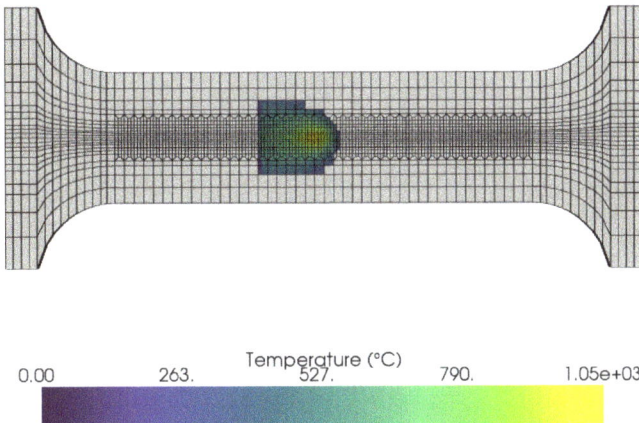

Figure 10. Experimental data—Temperatures above 400 °C.

Table 6. Posterior estimation of the unknown parameters θ obtained from infrared experimental data.

	a_f	b	c	η	K
Calibrated value	6.657	3	1.5	0.9	1.15
Estimated value	6.40815805	3.8549261	1.47004239	0.73615823	1.09763587

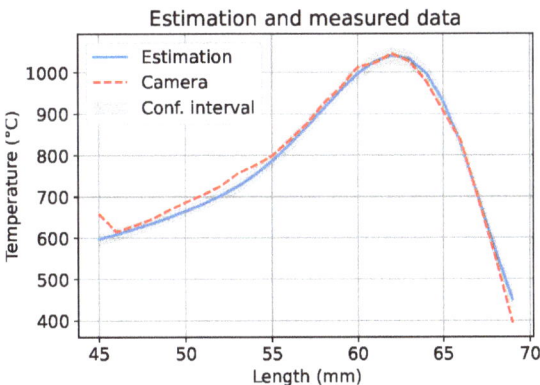

Figure 11. Temperature estimations and camera data.

Note that on the camera side the confidence interval is very small for the temperature estimation, as it is expected from observed data, while on the torch side the confidence interval is larger, especially around the peak position. No mechanical data is observed, and thus the maximum principal stress confidence intervals reflect the uncertainty of the posterior estimation (see Figure 13).

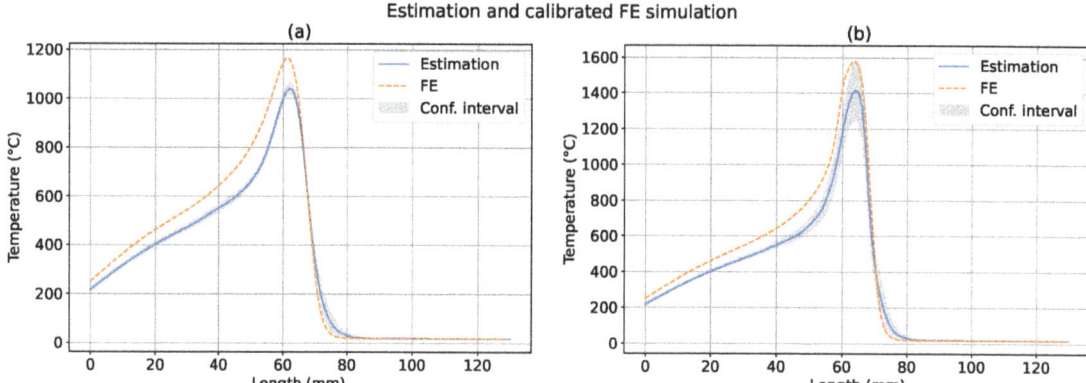

Figure 12. Temperature estimations obtained from infrared experimental data, FE simulation and confidence interval. (**a**) Camera side. (**b**) Torch side.

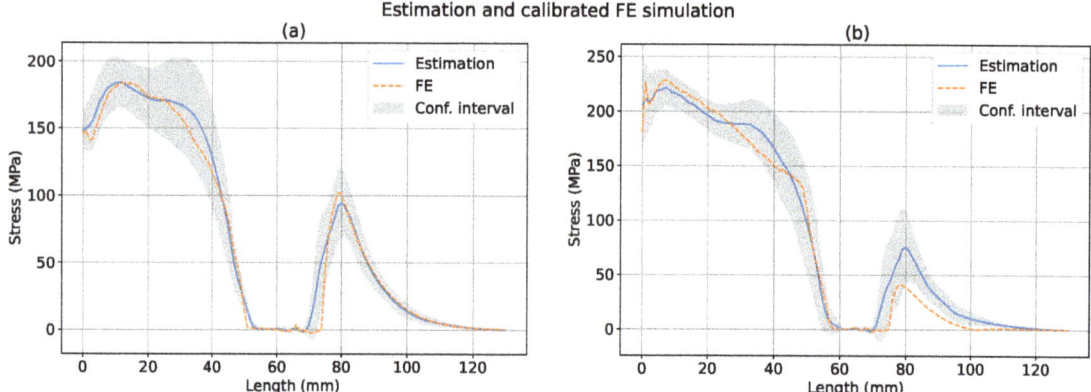

Figure 13. Maximum principal stress estimation obtained from infrared experimental data, FE simulation and confidence interval. (**a**) Camera side. (**b**) Torch side.

6.4. Forecasting

In this last example, the forecasting capabilities of the model will be shown. Let us assume that we want to estimate the temperature and maximum principal stress fields in a future position that has not been observed yet. This position is situated 12 mm further than the one studied for the previous tests. All the information available is the value of the known parameters and the posterior estimation of the unknown parameters with their associated posterior covariance.

Choosing the active coefficient of the mixture of PPCAs for this position $P^j = (k + 12)\Delta P$ is the first problem to solve. When an image is available, the surrogate model for position estimation takes the highest temperature identified in the image and returns the probability distribution of the torch position. In the forecasting problem, we have no video frame to find the highest temperature, so it is assumed to be shifted by the 12 mm and then it is used to obtain the discrete assignment probability. The multiphysics state $\tilde{s}^j$ will be predicted by conditioning its distribution by the known parameters μ and the estimated unknown parameters θ^k with its associated posterior variance.

As with previous tests, the predicted estimations, $p(\tilde{s}^j|\mu, \theta^k)$, are compared to the experimental data (when it is possible) and to a Finite Elements simulation over a line on both sides of the specimen. Figure 14 shows a frame of the video where the highest temperature position is situated 12 mm to the right of the highest temperature position found in Figure 10. The measures corresponding to that frame are compared to the predictions in Figure 15. The estimation is very close to the camera data but the 95% confidence interval is very large compared to the one in Figure 11, a case where the data was observed, reducing the uncertainty.

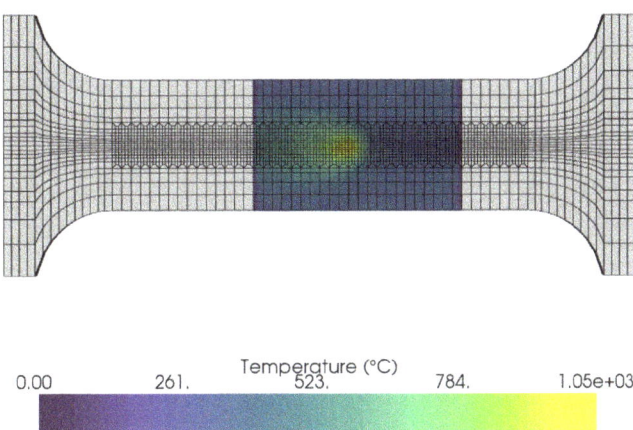

Figure 14. Camera data of the future state.

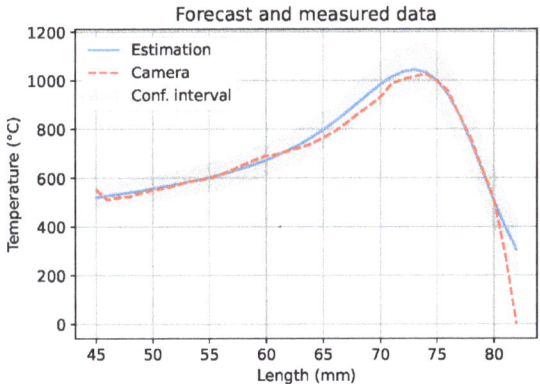

Figure 15. Forecasted temperature and camera data.

The Finite Elements simulation was used as reference in Figures 16 and 17, where we observe generally large confidence intervals on both sides of the specimen. Overall, knowledge of the known and unknown parameters is enough to obtain predictions that represent the behaviour of the temperature and maximum principal stress fields but with a high degree of uncertainty, which can be greatly reduced by observation of the thermal images. This is not the case when only the known parameters are observed. Additional figures are shown in the Appendix B to support this claim.

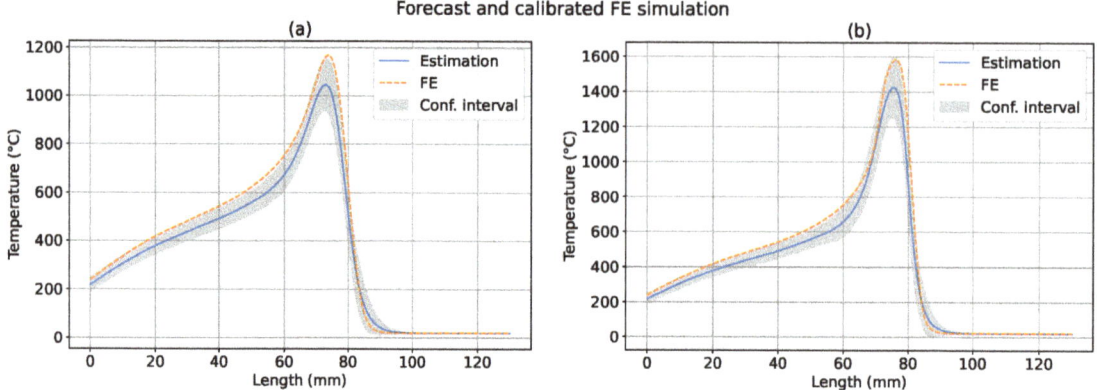

Figure 16. Forecasted temperature, FE simulation and confidence interval. (**a**) Camera side. (**b**) Torch side.

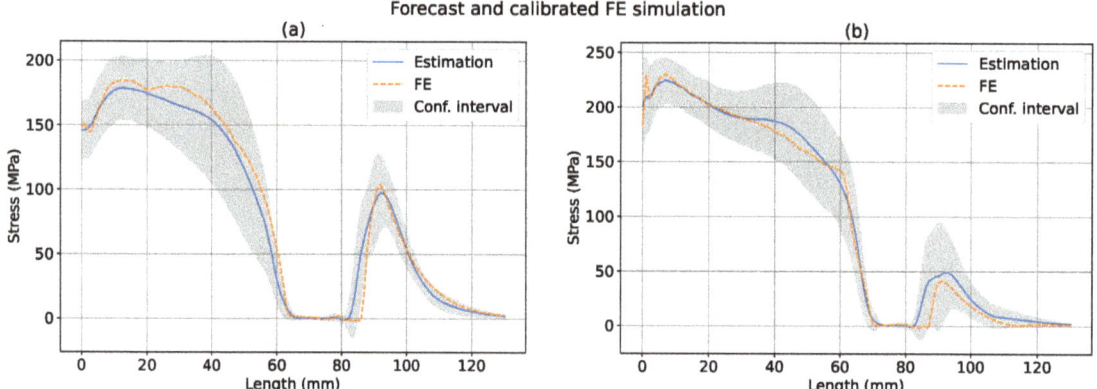

Figure 17. Forecasted maximum principal stress, FE simulation and confidence interval. (**a**) Camera side. (**b**) Torch side.

7. Conclusions

We have proposed a novel data assimilation methodology for automatized welding operations monitored using thermal imaging. The approach is based on digital twinning, a physically detailed thermomechanical finite element model assimilating online data and predicting unseen mechanical quantities of interest such as stress fields. Sources of variability are clearly identified and modeled using appropriate random parameters, the distribution of which are designed based on expert knowledge. Data assimilation then consists in finding the posterior distribution of the uncertain parameters given the sequence of thermal images available at current process time.

The data assimilation framework is made in real-time by deploying and offline–online meta-modeling technology. Sparse linear models are postulated for every position of the welding torch, linking observations and hidden mechanical quantities to the parameters of the welding process and to the random parameters. The coefficients of the linear models are identified using the method of snapshots, using hundreds of prior high-fidelity computations. Online, predicting unseen mechanical states operation reduces to simple Gaussian conditioning with a background covariance matrix exhibiting a low-rank structure. This is made computationally efficient using standard algebraic manipulations.

Thanks to this model/data fusion technology, we have shown that we were, for the first time, able to predict mechanical stress fields during welding in real-time, the predictions being continuously adjusted based on thermal imaging. We have shown that our digital twin may produce predictions that differ significantly from those produced by a high-fidelity model that is precalibrated using standard thermal sensing via a set of thermocouples. We interpret this finding as an online model correction, the data acquired online being richer, corresponding to thermal sensing closer to the regions of intense thermal gradients.

Finally, we were able to forecast stress predictions into the future of welding operations, by simply using current posterior distributions of unknown parameters and perform uncertainty propagation. We have shown that such predictions where reasonably sharp and could be used to stop welding operations in advance, when forecasting inadmissible levels of stresses.

The developments presented in this paper are expected to be the foundations for further theoretical and applied research. Future experiments may include additional instrumentation to obtain more data such as measures of the strain field by digital image correlation. In terms of algorithmic efficiency, the proposed meta-modeling methodology is piecewise linear, homogeneously along the trajectory of the welding torch. We are now investigating a data-driven clustering approach of the welding parameter domain, which we expect will help us minimize the memory required to store the various meta-models. In terms of application, the present developments focus on the prediction of stresses during the operations, but could easily be extended to the prediction of residual stresses. The exploration of parameter spaces describing geometrical variations of joining operations is also of high practical interest. Finally, the proposed approach may constitute the computational core of a model-based control technology aimed at adjusting the process parameters online in order to ensure that the joining operations produce assemblies that are of acceptable mechanical qualities.

Author Contributions: Conceptualization, P.K. and D.R.; Data curation, P.P.Á.; Investigation, P.P.Á., P.K. and D.R.; Methodology, V.R.; Resources, V.R.; Software, P.P.Á.; Supervision, P.K., D.R. and V.R.; Visualization, V.R.; Writing—original draft, P.P.Á.; Writing—review & editing, P.P.Á., P.K., D.R. and V.R. All authors have read and agreed to the published version of the manuscript.

Funding: This work has been partially funded by ANRT and EDF through an EDF-CIFRE contract N° 2018/1457.

Institutional Review Board Statement: Not applicable.

Informed Consent Statement: Not applicable.

Data Availability Statement: Not applicable.

Acknowledgments: The authors would like to thank Josselin Delmas and Charles Demay for their helpful remarks and support.

Conflicts of Interest: The authors declare no conflict of interest.

Appendix A. Fast Computation of Mean and Covariance

The mean and covariance of the posterior distribution have the following explicit expression:

$$m_{\alpha^k|d^k,\mu} = \Phi^{kT} H^{kT} \left(H^k \Phi^k \Phi^{kT} H^{kT} + (\sigma_M^2 + \sigma_s^2) Id \right)^{-1} (d^k - \bar{d}^k) \tag{A1}$$

$$\Sigma_{\alpha^k|d^k,\mu} = Id - \Phi^{kT} H^{kT} \left(H^k \Phi^k \Phi^{kT} H^{kT} + (\sigma_M^2 + \sigma_s^2) Id \right)^{-1} H^k \Phi^k \tag{A2}$$

where $\Phi^k \in \mathbb{R}^{\mathcal{N} \times N_M}$ is the PPCA basis, $H^k \in \mathbb{R}^{\mathcal{N} \times N_C}$ is the Boolean observation function, $d^k \in \mathbb{R}^{N_C}$ is the observed data, and σ_M and σ_s are the parameters guiding the measure and truncation errors. These matrices are of very high dimension, so evaluation is potentially slow.

Let us introduce the notation $X := H^k \Phi^k$ et $Z := (\sigma_M^2 + \sigma_s^2) Id$. Equations (A1) and (A2) become

$$m_{\alpha^k|d^k,\mu} = X^T \left(XX^T + Z \right)^{-1} (d^k - \bar{d}^k) \tag{A3}$$

$$\Sigma_{\alpha^k|d^k,\mu} = Id - X^T \left(XX^T + Z \right)^{-1} X \tag{A4}$$

Appendix A.1. Mean of the Posterior Gaussian Distribution

In order to accelerate the evaluation, the following algebraic identity [25] is used:

$$\left(P^{-1} + B^T R^{-1} B \right)^{-1} B^T R^{-1} = PB^T \left(BPB^T + R \right)^{-1} \tag{A5}$$

where $P \in \mathbb{R}^{M \times M}$, $R \in \mathbb{R}^{N \times N}$ and $B \in \mathbb{R}^{N \times M}$. This expression can be verified by multiplying both sides by $(BPB^T + R)$. The left side of the equation is quicker to evaluate when $M \ll N$. If $N \ll M$, the right side is quicker to evaluate.

It is easy to see that $X^T (XX^T + Z)^{-1}$ in Equation (A3) corresponds to the right side of Equation (A5) when $P = Id$, $B = X$ and $R = Z$. In our case, the left side of Equation (A5) is faster because N_M, the number of PPCA modes, is smaller than N_C, the number of mesh nodes where the infrared camera is observed ($N_M \ll N_C$). The expression that should be evaluated is

$$m_{\alpha^k|d^k,\mu} = \left(X^T Z^{-1} X + Id \right)^{-1} X^T Z^{-1} (d^k - \bar{d}^k) \tag{A6}$$

$$\Sigma_{\alpha^k|d^k,\mu} = I - \left(X^T Z^{-1} X + I \right)^{-1} X^T Z^{-1} X \tag{A7}$$

Notice that Z is a diagonal matrix, which means that computing Z^{-1} is trivial and $X^T Z^{-1}$ is very fast.

Appendix A.2. Covariance of the Posterior Gaussian Distribution

In the case of the covariance matrix, it is convenient to use the Woodbury identity [30]:

$$(A + UBV)^{-1} = A^{-1} - A^{-1} U \left(B^{-1} + VA^{-1} U \right)^{-1} VA^{-1} \tag{A8}$$

where $A \in \mathbb{R}^{M \times M}$, $U \in \mathbb{R}^{M \times N}$, $B \in \mathbb{R}^{N \times N}$ and $V \in \mathbb{R}^{N \times M}$.

If $A = Id$, $U = X^T$, $B = Z^{-1}$ and $V = X$, then we recognize that the right side of Equation (A8) is the expression in Equation (A4). It is quicker to evaluate the left side of Equation (A8) due to the dimensions of the problem. Thus, the covariance matrix is computed as

$$\Sigma_{\alpha^k|d^k,\mu} = \left(X^T Z^{-1} X + I \right)^{-1} \tag{A9}$$

Notice that Equation (A7) is part of Equation (A6), meaning that it only needs to be computed once.

Appendix A.3. Example

We will show the results of two examples indicating the values for the different dimensions. The first example uses input on all nodes on the whole surface of the specimen ($N_C = 8093$) and 20 PPCA modes ($N_M = 20$). The second example uses input on the area seen by the infrared camera (the surface shown in Figure 10, with $N_C = 3418$) and 20 PPCA modes ($N_M = 20$).

The results shown in Table A1 are the average of 7000 runs for both examples using each of the expressions presented previously. The computation time is greatly reduced using the expression in Equations (A6) and (A7).

Table A1. Computation time with both expressions with two different observation functions.

	Old Expression	New Expression
Whole surface	1.55 s	0.00377 s
Camera area	0.337 s	0.00212 s

Appendix B. Forecasting Results

In this appendix, we show some complementary forecasting results. In particular, we want to compare a forecast performed with and without observing the posterior theta values, as well as a comparison between the posterior covariance of a forecast and an estimation obtained observing an infrared image.

We have shown that using the previously estimated unknown parameters θ, forecasting a future state is possible. Indeed, parametric uncertainty is the main source of uncertainty in our model. The first result we want to show is a comparison between a forecast obtained using only the known parameters μ and another one observing the mean θ posterior. Figure A1 shows that without observing the mean θ posterior, the estimation is not good and forecasting is not possible. This indicates that the prior distribution of the unknown parameters might not be well chosen or that it reflects a spectrum of values that is too large for this case.

Additionally, we want to compare the posterior covariance of a forecasted estimation and an estimation obtained after observing an infrared image. It is clear that observing the temperature field, the uncertainty is greatly reduced, as seen in Figure A2.

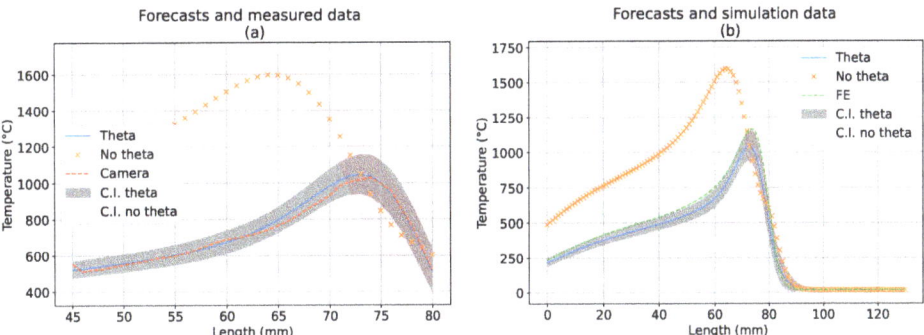

Figure A1. Forecasted temperature conditioning by the known parameters (μ) and by the known and unknown parameters (μ, θ) with their correspondent confidence interval. (**a**) Compared to experimental camera data. (**b**) Compared to FE results.

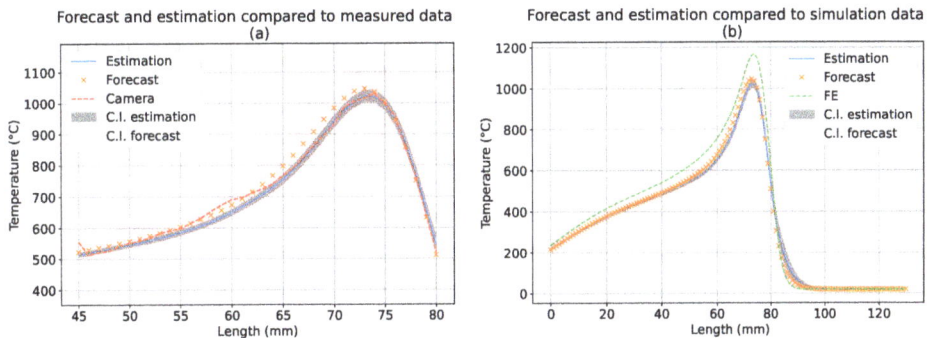

Figure A2. Forecasted temperature conditioning by the known and unknown parameters (μ, θ) and posterior estimation using the infrared camera data with their correspondent confidence interval. (**a**) Compared to experimental camera data. (**b**) Compared to FE results.

Appendix C. Relevant Material Parameters

In this final appendix, more details on the 316 stainless steel specimen is given. Most of the thermal and mechanical material properties of steel are temperature dependent and, thus, it is of great importance to take the variations into account. In Table A2, the temperature dependent values of four material properties used in the simulations are shown. These were obtained in previous research works by EDF R&D in collaboration with CEA Saclay and Université de Bretagne Sud.

The chemical composition of the material is directly related to the occurrence of cracks. In Table A3, the detailed chemical composition of the used specimen is given. The data was directly obtained from the manufacturer.

Table A2. Temperature dependent material parameters: thermal conductivity, volumetric heat capacity, thermal expansion coefficient and Young's modulus.

Temperature (°C)	$\lambda \left(\frac{W}{mm\,K}\right)$	$\rho c_p \left(\frac{J}{mm^3\,K}\right)$	$\alpha \left(\frac{1}{K}\right)$	E (MPa)
20	14×10^{-3}	3.784×10^{-3}	15.5×10^{-6}	190×10^3
100	15.2×10^{-3}	-	16×10^{-6}	-
200	16.6×10^{-3}	4.036×10^{-3}	16.6×10^{-6}	-
300	17.9×10^{-3}	-	17.1×10^{-6}	-
400	19×10^{-3}	4.302×10^{-3}	17.5×10^{-6}	-
500	20.6×10^{-3}	-	18×10^{-6}	-
600	21.8×10^{-3}	4.557×10^{-3}	18.4×10^{-6}	140×10^3
700	23.1×10^{-3}	-	18.7×10^{-6}	-
775	-	-	-	56.3×10^3
800	24.3×10^{-3}	4.823×10^{-3}	19×10^{-6}	-
850	-	-	-	56.3×10^3
900	26×10^{-3}	-	19.2×10^{-6}	-
1000	27.3×10^{-3}	5.072×10^{-3}	19.4×10^{-6}	-
1150	-	-	-	37.3×10^3
1200	29.9×10^{-3}	5.327×10^{-3}	-	-
1250	-	-	-	20.3×10^3
1384	-	5.572×10^{-3}	-	-
1390	-	7.823×10^{-3}	-	-
1394	-	11.175×10^{-3}	-	-
1400	32.5×10^{-3}	-	19.6×10^{-6}	-
1404	-	22.35×10^{-3}	-	-
1420	-	44.70×10^{-3}	-	-
1425	-	52.15×10^{-3}	-	-
1450	-	5.7×10^{-3}	-	-
1600	-	5.7×10^{-3}	19.7×10^{-6}	-
2400	32.5×10^{-2}	-	-	-

Table A3. Chemical composition of the 316L stainless steel specimen (in percentages except for boron).

B (ppm)	C	Mn	Si	P	S	Cr	Ni	Mo	Co	Cu	Al	N
39 ± 4	0.016	1.59	0.540	0.027	0.0011	17.25	10.03	2.05	0.080	0.106	0.043	0.052

References

1. Rampaul, H. *Pipe Welding Procedures*, 2nd ed.; Industrial Press: New York, NY, USA, 2003.
2. Glaessgen, E.H.; Stargel, D. The digital twin paradigm for future nasa and us air force vehicles. In Proceedings of the 53rd Structural Dynamics, and Materials Conference: Special Session on Digital Twin, Honolulu, HI, USA, 16 April 2012; pp. 1–14.
3. Dashti, M.; Stuart, A.M. The Bayesian Approach to Inverse Problems. In *Handbook of Uncertainty Quantification*; Springer International Publishing: Cham, Switzerland, 2017; pp. 311–428. [CrossRef]

4. Kalman, R.E. A new approach to linear filtering and prediction problems. *J. Basic Eng.* **1960**, *82*, 35–45. [CrossRef]
5. Liu, J.S.; Chen, R. Sequential Monte Carlo methods for dynamic systems. *J. Am. Stat. Assoc.* **1998**, *93*, 1032–1044. [CrossRef]
6. Navon, I.M. Data Assimilation for Numerical Weather Prediction: A Review. In *Data Assimilation for Atmospheric, Oceanic and Hydrologic Applications*; Springer: Berlin/Heidelberg, Germany, 2009; pp. 21–65. [CrossRef]
7. Prud'Homme, C.; Rovas, D.V.; Veroy, K.; Machiels, L.; Maday, Y.; Patera, A.T.; Turinici, G. Reliable Real-Time Solution of Parametrized Partial Differential Equations: Reduced-Basis Output Bound Methods. *J. Fluids Eng.* **2001**, *124*, 70–80. [CrossRef]
8. Amsallem, D.; Farhat, C. Interpolation method for adapting reduced-order models and application to aeroelasticity. *AIAA J.* **2008**, *46*, 1803–1813. [CrossRef]
9. Amsallem, D.; Cortial, J.; Carlberg, K.; Farhat, C. A method for interpolating on manifolds structural dynamics reduced-order models. *Int. J. Numer. Methods Eng.* **2009**, *80*, 1241–1258. [CrossRef]
10. Rozza, G.; Huynh, D.B.P.; Patera, A.T. Reduced basis approximation and a posteriori error estimation for affinely parametrized elliptic coercive partial differential equations. *Arch. Comput. Methods Eng.* **2007**, *15*, 1. [CrossRef]
11. Ryckelynck, D.; Chinesta, F.; Cueto, E.; Ammar, A. On the a priori model reduction: Overview and recent developments. *Arch. Comput. Methods Eng.* **2006**, *13*, 91–128. [CrossRef]
12. Kerfriden, P.; Gosselet, P.; Adhikari, S.; Bordas, S.P.A. Bridging proper orthogonal decomposition methods and augmented Newton–Krylov algorithms: An adaptive model order reduction for highly nonlinear mechanical problems. *Comput. Methods Appl. Mech. Eng.* **2011**, *200*, 850–866. [CrossRef] [PubMed]
13. Chinesta, F.; Ladeveze, P.; Cueto, E. A short review on model order reduction based on proper generalized decomposition. *Arch. Comput. Methods Eng.* **2011**, *18*, 395. [CrossRef]
14. Aubry, N.; Holmes, P.; Lumley, J.L.; Stone, E. The Dynamics of Coherent Structures in the Wall Region of A Turbulent Boundary-Layer. *J. Fluid Mech.* **1988**, *192*, 115–173. [CrossRef]
15. Volkwein, S. *Model Reduction Using Proper Orthogonal Decomposition*; Lecture Notes; Institute of Mathematics and Scientific Computing, University of Graz: Graz, Austria, 2011. Available online: http://www.uni-graz.at/imawww/volkwein/POD.pdf (accessed on 4 March 2021).
16. Willcox, K.; Peraire, J. Balanced model reduction via the proper orthogonal decomposition. *AIAA J.* **2002**, *40*, 2323–2330. [CrossRef]
17. Cosimo, A.; Cardona, A.; Idelsohn, S. Improving the k-compressibility of hyper reduced order models with moving sources: Applications to welding and phase change problems. *Comput. Methods Appl. Mech. Eng.* **2014**, *274*, 237–263. [CrossRef]
18. Zhang, Y.; Combescure, A.; Gravouil, A. Efficient hyper reduced-order model (HROM) for parametric studies of the 3D thermo-elasto-plastic calculation. *Finite Elem. Anal. Des.* **2015**, *102*, 37–51. [CrossRef]
19. Cosimo, A.; Cardona, A.; Idelsohn, S. Global-local HROM for non-linear thermal problems with irreversible changes of material states. *C. R. Mécanique* **2018**, *346*, 539–555. [CrossRef]
20. Lu, Y.; Jones, K.K.; Gan, Z.; Liu, W.K. Adaptive hyper reduction for additive manufacturing thermal fluid analysis. *Comput. Methods Appl. Mech. Eng.* **2020**, *372*, 113312. [CrossRef]
21. Favoretto, B.; de Hillerin, C.; Bettinotti, O.; Oancea, V.; Barbarulo, A. Reduced order modeling via PGD for highly transient thermal evolutions in additive manufacturing. *Comput. Methods Appl. Mech. Eng.* **2019**, *349*, 405–430. [CrossRef]
22. Kerfriden, P.; Schmidt, K.M.; Rabczuk, T.; Bordas, S.P.A. Statistical extraction of process zones and representative subspaces in fracture of random composites. *Int. J. Multiscale Comput. Eng.* **2013**, *11*,1940-4352. [CrossRef] [PubMed]
23. Dinh Trong, T. Modèles Hyper-réduits Pour la Simulation Simplifiée du Soudage en Sustitut de Calcul Hors D'atteinte. Ph.D. Thesis, Mines ParisTech, Paris, France, 2018.
24. Rocha, I.; Kerfriden, P.; van der Meer, F. Micromechanics-based surrogate models for the response of composites: A critical comparison between a classical mesoscale constitutive model, hyper-reduction and neural networks. *Eur. J. Mech.-A/Solids* **2020**, *82*, 103995. [CrossRef]
25. Bishop, C.M. *Pattern Recognition and Machine Learning*; Springer: Berlin/Heidelberg, Germany, 2006.
26. Ghanem, R.G.; Spanos, P.D. *Stochastic Finite Elements: A Spectral Approach*; Courier Corporation: North Chelmsford, MA, USA, 2003.
27. Rasmussen, C.; Williams, C. *Gaussian Processes for Machine Learning*; Adaptive Computation and Machine Learning; MIT Press: Cambridge, MA, USA, 2006; p. 248.
28. Peherstorfer, B.; Willcox, K.; Gunzburger, M. Survey of multifidelity methods in uncertainty propagation, inference, and optimization. *Siam Rev.* **2018**, *60*, 550–591. [CrossRef]
29. Maday, Y.; Patera, A.T.; Penn, J.D.; Yano, M. A parameterized-background data-weak approach to variational data assimilation: Formulation, analysis, and application to acoustics. *Int. J. Numer. Methods Eng.* **2015**, *102*, 933–965. [CrossRef]
30. Henderson, H.V.; Searle, S.R. On deriving the inverse of a sum of matrices. *Siam Rev.* **1981**, *23*, 53–60. [CrossRef]
31. Kannengiesser, T.; Boellinghaus, T. Hot cracking tests: An overview of present technologies and applications. *Weld. World* **2014**, *58*, 397–421. [CrossRef]
32. Depradeux, L. Simulation Numérique du Soudage-Acier 316L: Validation sur Cas Tests de Complexité Croissante. Ph.D. Thesis, Institut National de Sciences Appliquées, Villeurbanne, France, 2004.
33. Goldak, J.; Chakravarti, A.; Bibby, M. A new finite element model for welding heat sources. *Metall. Trans. B* **1984**, *15*, 299–305. [CrossRef]

34. Hunnewell, T.S.; Walton, K.L.; Sharma, S.; Ghosh, T.K.; Tompson, R.V.; Viswanath, D.S.; Loyalka, S.K. Total hemispherical emissivity of SS 316L with simulated very high temperature reactor surface conditions. *Nucl. Technol.* **2017**, *198*, 293–305. [CrossRef]
35. Tipping, M.E.; Bishop, C.M. Probabilistic principal component analysis. *J. R. Stat. Soc. Ser. B (Stat. Methodol.)* **1999**, *61*, 611–622. [CrossRef]
36. Fang, K.T.; Li, R.; Sudjianto, A. *Design and Modeling for Computer Experiments*; CRC Press: Boca Raton, FL, USA, 2005.
37. Code_aster. Available online: https://code-aster.org/spip.php?rubrique1 (accessed on 12 April 2021).
38. Tran Van, G. Determination of a Liquation Hot Cracking Criterion as a Function of Boron Content and Its Location for 316L Austenitic Stainless Steel. Ph.D. Thesis, Université de Bretagne Sud, Lorient, France, 2018.

MDPI
St. Alban-Anlage 66
4052 Basel
Switzerland
Tel. +41 61 683 77 34
Fax +41 61 302 89 18
www.mdpi.com

Mathematics Editorial Office
E-mail: mathematics@mdpi.com
www.mdpi.com/journal/mathematics

www.ingramcontent.com/pod-product-compliance
Lightning Source LLC
LaVergne TN
LVHW070506100526
83820ZLV00014B/1802